Time Matters

SUNY Series in Jewish Philosophy
Kenneth Seeskin, editor

Time Matters

Time, Creation, and Cosmology in Medieval Jewish Philosophy

T. M. Rudavsky

State University of New York Press

Published by
State University of New York Press, Albany

Printed in the United States of America

For information, address State University of New York Press
State University Press, Albany, New York 12246

Production by Dana Foote
Marketing by Anne Valentine

Library of Congress Cataloging-in-Publication Data

Rudavsky, Tamar, 1951–
Time matters : time, creation, and cosmology in medieval Jewish philosophy /
T. M. Rudavsky.
p. cm. — (SUNY series in Jewish philosophy)
Includes bibliographical references and index.
ISBN 0–7914–4453–8 (hc. : alk. paper) — ISBN 0–7914–4454–6 (pbk. : alk. paper)
1. Philosophy, Jewish. 2. Philosophy, Medieval. 3. Time. 4. Creation—History of doctrines. 5. Jewish cosmology. I. Title. II. Series.
B755 .R83 2000
113'.089'924—dc21 99–088797

10 9 8 7 6 5 4 3 2 1

Cover photo: "The Sundial". This sundial is found atop the Immigrant Shelter building (*Hachnasat Orchim*). The building was founded by S. Levi in 5668 in Jerusalem and is still maintained by the author's family.
Photo by Nathaniel Rudavsky-Brody.

Contents

Illustrations

Acknowledgments

The focus of this book, the interrelations between theories of creation, time, and infinity in medieval Jewish thought, is part of an ongoing project devoted to exploring medieval attempts to reconcile the tensions between philosophy, medieval science, and theology. This topic has interested me since my graduate school days and has been expressed in a series of publications pertaining to how Jewish philosophers understand God's interactions with the world. In 1982 I attended an NEH Seminar directed by Professor Norman Kretzmann; it was at this seminar that I first studied extensively the works of scholastic thinkers related to issues of divine omniscience, infinite divisibility, and *insolubilia* literature. This interest led to my first edited volume, *Divine Omniscience and Omnipotence in Medieval Jewish, Christian and Islamic Philosophy* (Reidel, 1984). In subsequent articles I have extended this interest further, with an eye to exploring issues of creation and cosmology. It is this cluster of questions that forms the basis of this study.

Portions of several chapters of this book have appeared in preliminary form as separate articles. Chapter 1 contains material from "Philosophical Cosmology in Medieval Judaism," *Early Science and Medicine,* (1997) pp. 149–84. Chapter 2 draws upon material published in "Time in Maimonides and Crescas," *Maimonidean Studies,* (1990) and "Creation, Time and Infinity in Gersonides," *The Journal of the History of Philosophy* (1988), pp. 25–45. Chapter 4 draws upon material appearing in "Divine Omniscience and Future Contingents in Gersonides," *The Journal of the History of Philosophy* (1983), pp. 513–36 and "Maimonides and Averroes on God's Knowledge of Possibles," *Da'at* (1984), pp. 27–44. In addition, I have adapted and reconceptualized material from two articles: "Divine Omniscience, Contingency and Prophecy in Gersonides," in my volume *Divine Omniscience and Omnipotence in Medieval Philosophy* (Reidel, 1984), and "Creation and Time in Maimonides and Gersonides," in *God and Creation,* ed. B. McGinn and D. Burrell (Notre Dame, 1990). I thank the editors and publishers of these publications for their assistance.

I also thank the many individuals who have offered suggestions and commented upon portions of this work over the years: in particular I owe thanks to Robert Turnbull, Alfred Ivry, Seymour Feldman, Idit Dobbs-Weinstein, Tzvi Langermann, George Pappas, Ivan Boh, Josef Stern, and

Calvin Normore, who have offered comments at various stages. Thanks go to Kenneth Seeskin, editor of the series in which this volume appears, and Mr. James Peltz and Ms. Dana Foote of SUNY Press who have been consistently patient and helpful throughout the process of preparing the manuscript for publication. A special thank-you to OSU graduate students who have pored over texts with me in graduate seminars, and especially to Mr. Kevin Osterloh and Ms. Sarah Pessin who have labored with me over notes and bibliographical minutae. I thank as well the Melton Center for Jewish Studies, The Center for Medieval and Renaissance Studies, and the College of Humanities at The Ohio State University for generous research and publication support.

Finally, I am extraordinarily grateful to my husband, Richard Brody, as well as to my two children, Miriam and Nathaniel, without whose ongoing support this book would not exist. Both their assurances and their belief that this project will come to fruition have been invaluable.

Introduction

In this study I trace the development of the concepts of time, cosmology, and creation in medieval Jewish philosophy. I first examine the formative sources found in classical Greek thought and then turn to the transmission of these sources in late Hellenistic texts as well as in representative rabbinic sources. The heart of the book, an analysis of representative figures in medieval Jewish philosophy, focuses upon the following set of issues: Is time finite or infinite? Is time composed of indivisible units? Did God create the universe in time, or did God create time? Can there be a first instant to creation? Is God's knowledge of temporal events simultaneous with these events? These issues comprise the intellectual agenda of medieval Jewish philosophers, as well as their Christian and Muslim peers, who must construct a philosophical theology that is sensitive to religious constraints while at the same time incorporating compelling elements of (the current) science and philosophy. As such, this volume contributes to recent studies which focus on the tensions and accommodations accruing between religion, science, and philosophy.

Have Jewish thinkers contributed to the world a particularly "Jewish" perspective on time? Although many scholars have argued that there is a Jewish conception of time, they have disagreed over what it is. Steven Jay Gould, for example, has identified time's arrow as "the most important and distinctive contribution of Jewish thought, for most other systems, both before and after, have favored the immanence of time's cycle over the chain of linear history."[1] Others, like Néher, claim that time for the early Hebrews is cyclic. Néher starts out his analysis by pointing out that the Jewish way of measuring time is *"me-'erev 'ad 'erev,"* (from eve to eve) which is immutably rooted in the account of creation. Time is hence measured in cycles, with its origin in creation.[2] Others have pointed to the importance of time in Jewish law (*halakhah*). That many of the commandments are subject to rigorous time conditions which affect the moment and nature of their performance has resulted in a well-ordered structure in the flux of time, a comprehensible pattern in the cycle of the seasons.

My primary concern in this work is not to argue that there exists a unique theory of time that can be associated primarily with Jewish thinkers in contradistinction to those theories associated with Christian or Greek philosophical writers. Following Ravitsky, I agree that approaching these Jewish philosophical texts with the presumption that their authors were completely isolated from their non-Jewish intellectual environment is to misunderstand the underlying tension inherent in the very texts.[3] On the contrary, Jewish philosophers, both medieval and modern, have had to wrestle with the project of reconciling these "outside" sources and influences with their understanding of scriptural dicta. Sometimes this wrestling results in a mode of discourse which obfuscates the full force of the compromise. Nowhere is this seen more clearly than in the case of Maimonides. I shall argue in chapter 2 that Maimonides was constrained to accommodate traditional Jewish beliefs with what he took to be the strongest points in Aristotelian cosmology and cosmogony. This accommodation, I shall argue, precludes his tacit approval of an unwavering faith in the scriptural account of creation. The force of my study is to assess the coherence of emerging theories of temporality of Jewish philosophers such as Maimonides in light of the concomitant theological, scientific, and philosophical constraints placed upon them.

If there is a subsidiary theme to this work, it is that the medieval Jewish attempts to understand time and cosmology are relevant to contemporary thinking about these issues. We too live under the aegis of time, and we too are forced to think of its passage. Our attempts to construct theories of temporality and cosmology that are sensitive to the constraints of philosophy, modern science, and theology are analogous to the attempts of our philosophical predecessors. We too are attempting to reach an accommodation between the domain of modern science and that of theology. Living in the post-Newtonian age, we have become accustomed to thinking in terms of a relativized space-time continuum in which time is seen as relative to the observer. As described by Hawking, observers must have their own measure of time as recorded by a clock carried within them, and that time depends upon their location in the space-time grid.[4] Further, not only do space and time affect events, but they are affected by everything that happens in the universe.[5] Time and events in time have become ontologically interdependent. We have grown more comfortable with paradoxes resulting from modern physics; the realization that there exists no absolute time, while at times perplexing, is nevertheless intellectually acceptable. Even talk of time-travel can be placed into an appropriate context. And yet, how these scientific theories of time affect and are affected by theological concerns remains an issue in modern times.

In the first chapter I explore the basic paradigms of time found in classical Jewish sources as well as in Greek philosophical texts. These paradigms have been interpreted by modern scholars in a variety of ways, reflecting a basic dichotomy between two conceptions of time. On the one pole of the dichotomy, time's arrow is the linear view of time according to which "history is an irreversible sequence of unrepeatable events."[6] On the other pole, time's cycle, events are seen as immanent in time, never changing, parts of repeating cycles; time has no direction.[7] Both poles, Gould reminds us, are present in the Bible: "Judeo-Christian traditions have struggled to embrace the necessary parts of both contradictory poles."[8] Each of these poles can be further elucidated. For example, the cyclic view of time has incorporated not only the notion that reality is unchanging, but also the idea of recurring cycles of separable events precisely repeated.[9] So too, time's arrow can be both bidirectional as well as unidirectional. Gould states that the strength of this dichotomy is that Western thinkers must grapple with both conceptions, for "time's arrow is the intelligibility of distinct and irreversible events, while time's cycle is the intelligibility of timeless order and law-like structure. We must have both."[10] Physicists speaking of "time's arrow" find themselves obliged to explain why it is that time moves in one direction rather than another. In other words, why is it that we remember the past and not the future; why is it that time is future-directed rather than past-directed if there is no ontological difference between past, present, and future.[11]

The recognition of this dichotomy between linear and cyclical time finds expression in numerous discussions, such as Eliade's seminal work *The Sacred and the Profane.* In this work, best known for its portrayal of sacred and profane time, Eliade argues that only in nineteenth and twentieth centuries has the notion of time's arrow taken hold; according to Eliade, most people throughout history have held fast to time's cycle.[12] It is the cyclic pattern of time that is associated with the sacred, whereas the linear historicity of time's passage is associated with the ordinary flow of events. In chapter 1 I shall explore the theological and metaphysical ramifications of this distinction between linear and cyclic time.

The contrast between the flow of events, associated with the human domain, and a timeless domain associated with divinity is a dominant motif in the medieval period and provides an additional thrust to the present work. More specifically, we say of time itself that it passes, but of things in time that they endure, or have duration: time as flowing does not endure, whereas time as experienced, or lived, is duration. But duration without succession is nothing but an unchanging present.[13] The move from future, to present, to past, represents, therefore, a move of succession that precludes duration. The successiveness of time can be viewed as relational in the sense that the passage of time is relative to things in time. We are aware of the passage of time by virtue of changing events and objects. Time can be construed as slower or faster depending upon our awareness of this passage. But the very awareness of the relativity of time's passage carries with it the suggestion that there exists an absolute time independent of our perception, an independent, objective time in terms of which we measure the relativeness of our perception of time's passage.[14]

These questions underlying the ontology of time provide a focal point for chapters 2, 3, and 4. Many of the discussions found in medieval texts concerning the ontology of time have to do with the discontinuity between past and future. Clearly the past appears to be fixed in a way that the future is not. More bluntly, the past is actual whereas the future is possible. But from the divine perspective there is no ontological difference between past and future: all events exist in an "eternal now" for God, and so what is possible from the human perspective is actual from God's eternal gaze. All events, known from God's perspective, happen simultaneously. But what does it mean for two things to happen "at the same time," and what is the effect of God's knowledge of simultaneously known events?[15] In my first edited volume I focused on how God's knowledge affects the objects of his knowledge, in particular, acts of human free will. In an important way the present work, in particular chapter 4, expands upon these implications of God's eternal knowledge by examining the corollary to God's eternity, namely temporality.

Another way to understand the relation between eternity and temporality is to examine the notion of creation. Traditionally, God the creator is said to be eternal, or outside of time, whereas creatures are construed as being in time, or subject to the flow of time. By understanding the notion of creation and how an eternal, timeless creator created a temporal universe, we may begin to understand how the notions of eternity and time function. What are the metaphysical implications of these issues? "If moments have no distinction, then they have no interest."[16] The pole of time's cycle reiterates that nothing is distinctive because every event is recurrent. No event can tell us where we are, for nothing anchors us to a particular point in time, but only to a particular stage in the cycle of time. Borges' story "The Book of Sand" exemplifies the dilemma of the infinity of time's cycle: for if every event is recurrent and must happen again, history is destroyed. If time is infinite, we may be at any place in time.[17] On the other hand, time's arrow requires that time "move inexorably forward, and that one truly cannot step twice into the same river."[18] In the irreversibility of time's arrow, Gould stresses the "unrepeatable uniqueness of each step in a sequence of events linked through time in physical connection."[19] And so these issues all refer back to the pivotal notions of eternity, infinity, and temporality.

The concept of eternity, first intimated by Parmenides, was developed by Plato in the *Timaeus* and then systematized by Aristotle in both the *Physics* and the *Metaphysics*. Wolfson distinguishes the Platonic view that eternity represents the antithesis of time from the Aristotelian view that eternity refers to endless time.[20] But of course many additional strands must be considered when examining medieval discussions. I consider these strands in the first and second chapters. Mediaeval philosophers writing on the creation of the world distinguish eternity from time by emphasizing Aristotle's characterization of time as a series of indivisible units or instants. They were, as well, immensely affected by Scripture and in particular the creation account found in Genesis 1–2. When trying to prove that the world was created by God in time, medieval philosophers who wanted to support a biblical theory of creation *ex nihilo* had to reject Aristotle's position that time is infinite. For if, as Aristotle claimed, time is the measure of motion, and motion is of material stuff, then to postulate infinite time implies that the universe has no finite beginning—is endless. But inasmuch as adopting the eternity of the universe is tantamount to rejecting God's efficacy and involvement in human affairs, medieval Jewish philosophers were constrained to reject the Aristotelian characterization of time while at the same time continue to accept his overall philosophical authority.[21] In an article on Ockham's theory of creation, Kretzmann sketches three major explanations of the world's existence as the big bang theory, the standard theological account found in Genesis , and Aristotle's "beginningless theory" according to which the world has existed from eternity.[22] Kretzmann then suggests that Ockham attempts to reconcile the first and third views, claiming that the world is "both divinely created and beginninglessly existent."[23] In chapter 2 I explore Jewish attempts to reconcile these positions, and I suggest that Crescas' theory of creation comes close to reflecting Ockham's formulation.

An ancillary issue has to do with infinity and infinite divisibility, which forms the focus of chapter 3. The problem of infinite divisibility has vexed philosophers since the time of Zeno. It centers around whether both time and space are infinitely divisible. If only one is divisible, we are left with a discontinuity between space and time; and yet if both space and time are infinitely divisible, then numerous paradoxes result, such as Zeno's celebrated "Achilles and the Tortoise." Jewish philosophers did not, as a rule, concern themselves with Zeno's paradoxes of time and space. However, a number of philosophers did address the problem in the context of other related discussions, most notably that of creation. Both Saadia Gaon and Gersonides, for example, offered their solutions to Zeno in the context of developing a general theory of creation.

Most scholars agree that medieval philosophers inherited from Aristotle two distinct concepts of infinity: the quantitative and the substantive notions. That the quantitative notion of infinity is to be found in Aristotle is fairly uncontroversial. We need only turn to Aristotle's discussions in *Physics* 3.4.204a where Aristotle distinguishes four types of quantitative infinite, and in *Metaphysics* 11.10. 1066a35–1066b, where quantitative infinity is characterized further. Wolfson, for example, notes the "generally accepted and well-known medieval concepts" of infinity as having two meanings that go back to Aristotle: an accident either of magnitude or number, or an essence. In the first sense, infinity is contrasted with finitude and refers to "an unlimited distance or length, something that has no end or boundary."[24] As an accident of number, infinity refers to something that is endlessly add-ible or divisible. In the second sense, infinity refers to a self-existent substance, one whose essence is so unique and so incomparable that it cannot suffer any form of limitation.[25]

Jewish conceptions of infinity can be correlated to both these Aristotelian characterizations. In classical Hebrew texts, different expressions were used to express the idea of infinity. Not until the writings of Ibn Gabirol in the eleventh century, however, is the term *'ein-sof* (without-end) used. In the thirteenth century we find a split in terminology: Qabbalists appropriate the term *'ein-sof,* whereas philosophers primarily employ the expression *bilti ba'al takhlit.*[26] By the sixteenth century the term *'ein-sof* was practically synonymous with infinity. Why this split in terminology between *'ein-sof* and *bilti ba'al takhlit,* and how did the term *'ein-sof* come into use? These are questions that Scholem has addressed briefly in the context of his history of Qabbalah. Clearly in both the *Sefer Yetzirah* and *Sefer Bahir,* infinity is the principal property of thought, but neither text mentions *'ein-sof* as a term designating God.[27] It is Isaac the Blind, writing in the 1300s, who introduces three stages in the mystery of the deity: *'ein-sof,* thought, and speech. For Isaac the Blind, *'ein-sof* designates "a domain of the divine that is above all reflective contemplation, indeed above the divine Thought itself, a domain called by Isaac the Blind 'the cause of Thought' and designated by a new term: *'ein-sof.*"[28] Scholem hypothesizes that possibly the term comes from a Saadianic influence, although he admits that "no major influence on Isaac the Blind can be ascribed to Saadianic ideas."[29] My own hypothesis, to which I shall return in chapter 3, is that it might come from the influence of Ibn-Gabirol, whose eclectic form of Neoplatonism forced him literally to invent a new terminology to deal with the concept of infinity.

In chapter 5 I explore the influence of Jewish philosophy upon the early modern period, concentrating upon the work of Spinoza as representing the synthesis of motifs and arguments covered in previous chapters. Spinoza's interactions with Descartes, and his foreshadowing of Newton's theory of time, represent the bridge between late medieval and early modern philosophy. In my analysis I emphasize arguments and considerations that reflect Spinoza's medieval heritage, as well as modernist tendencies. My discussion of theories of immortality in Gersonides and Spinoza reinforces the pivotal role played by Spinoza in this study.

Finally my interest in these topics has been strengthened by contemporary thinking about the relation between religion, philosophy, and science. My concerns in chapters 2, 3, and 4 are twofold: first, to understand exactly how medieval Jewish thinkers expound the notions in question, and just as importantly, to examine how they articulate the very project of reconciling scriptural accounts with the current scientific model of their day. In other words, my interests are as much methodological as metaphysical. That issues of creation and cosmology are of contemporary as well as medieval interest should go without saying. The importance of assigning temporal causes to events in general, and to the universe as a whole, has been a preoccupation of both medieval and modern philosophy. Although the causal theory of time is associated primarily with Kant, who argues that we discover time order by examining causal order as distinct from perceptual order, nevertheless the importance of causal sequence was recognized centuries earlier by Kalām philosophers. Within the paradigm of causality we must articulate such questions as: Did time begin? Can time exist independently of the universe? Does God exist in time? And does God's knowledge of temporal events, events in time, carry with it causal efficacy? Modern theories of creation (ranging from the big bang theory to the steady state theory) have presupposed a particular position with respect to causal theories and the interconnectedness of matter and time. The paradoxes that result from notions of infinity still haunt today's mathematicians and philosophers of science. The enduring questions remain.

Despite the importance of time and cosmology to Western thought, surprisingly little attention has been paid to them in histories of Jewish philosophy. One early work, Efros' *The Concept of Space in Jewish Medieval Philosophy*[30] devotes a chapter to theories of infinity in Jewish philosophers. In the scholastic tradition, Kretzmann's volume *Infinity and Continuity in Ancient and Medieval Philosophy*[31] explores an important collection of Christian philosophical texts dealing with this issue. And recently Davidson has published a comprehensive study on the themes of eternity and creation in Jewish and Islamic philosophy.[32] Apart from these works, however, there has been no sustained examination of the cluster of issues surrounding time and cosmology in Jewish philosophical writings. Hence my study represents a contribution to a broader understanding of the notions of time, cosmology and temporality in Jewish philosophy, within the context of Western thought. This study should thus further an appreciation of the complex interactions between medieval science, philosophy, and Judaism.

A final note to the reader concerns the style of this work. The book itself is addressed to three sets of readers: historians of philosophy who are trained in analytic methods and are interested in learning more about Jewish contributions to these important issues; scholars of Jewish philosophy who are familiar with the texts I discuss but may not be familiar with issues of time and cosmology; and general enlightened readers who have no technical training in philosophy, but are interested in the issues and the texts. Much of the book represents a close reading of philosophical arguments. At the same time, however, in those cases in which I feel that analytic argument and critique are essential to understanding the philosophical text in question, I have introduced a brief nontechnical summary of each argument I present. These summaries are set apart from the text and enable the general reader to benefit from the analytical exposition without getting bogged down in technical detail. In this way I have tried to provide the critical analysis that is so important, without losing more nontechnical readers. In so doing, I reiterate my view, mentioned above, that I regard Jewish philosophers as engaged in a dialogue with perennial philosophical issues, issues that have emerged in every age. It is to these issues that we now turn.

Chapter One

Time and Cosmology in Athens and Jerusalem

Introduction

My purpose in this chapter is to trace the development of the concept of time in early Jewish and Greek writings with an ultimate eye to the impact of these texts upon the medieval period. I shall first examine biblical and rabbinic discussions relevant to our theme, and then turn to conceptions of time in Greek philosophy. In short, I shall concentrate upon those canonical texts within Jewish and Greek tradition which reflect an understanding and awareness of time and temporality. My concern in this chapter is not so much to present a comprehensive study of time in the Rabbis and in Greek philosophy, but rather to emphasize those elements that influence and reappear in medieval Jewish writings.

Biblical Conceptions of Time

For reasons having to do as much with contemporary theological concerns as with pedagogical research, modern biblical scholars have devoted much time trying to uncover a "theory of time" in the Hebrew scriptures. The enterprise has been fraught with frustration, however, and has not reached a scholarly consensus. That linguistic analysis of biblical temporal terms has yielded little fruit is obvious, especially in light of the ostensible lack of significant discussion in Scripture having to do with time per se. The word *'et* is the most important word in biblical Hebrew for time and tends to mean the moment or point of time at which something happens.[1] The point of time can change over into a longer period of time. Other words used for time indications include *mo'ed, zeman, 'olam,* and *yom.* The term *mo'ed* means "place of meeting," and with reference to time, comes to mean "appointed time." It emphasizes—more than *'et*—conscious designation and arrangement. The term *zeman* occurs only three times in the latest period (Nehemiah 2:6; Esther 9:27, 31) and means, like *mo'ed,* appointed day. By Mishnaic times the term *zeman* takes over and it is the most commonly used term in medieval texts.[2] The term *'olam,* frequently translated as "world" in rabbinic texts, refers as well to time or duration in the Bible.[3] The term *yom* can be used to indicate the era of an important event or the day of someone's birth, death, and so on. Additional temporal words that are occasionally used include *rega',* *qetz,* *aḥarit,* *qedem,* and *dor.*

Despite the overwhelming use of such temporal terms, there is very little discussion, however, concerning the underlying ontology of time. Within biblical scholarship we can discern three interpretative schools that reiterate the distinction drawn in the introduction between cyclical and linear time. On the first reading, primitive societies, represented by the Jews and biblical culture, are presumed to reflect a cyclical view of time rooted in the cyclicity of nature and the repeatedness of natural events. Modern society, on the other hand, is associated with an abstract linear conception of time rooted in history and temporal flow. Some scholars place the Greeks and Jews both into the cyclical camp, others place the Jews into the linear camp, and yet others place the Greeks alone into the linear camp as forerunners of an abstract theory of time.[4] Thus Gurevich, for example, maintains that whereas for the Greeks reality is static and unchanging, it is "with the Jews that time as of ontological import is introduced."[5] On this conception, it is the Jews, through Scripture, who introduced the ontology of time and temporality to Western thought. A more moderate version of this thesis is that while the early Hebrews and Greek philosophers were both interested in temporality, they developed theories of time which differed conceptually in important ways. Malina follows Gurevich and others in contrasting a modern, technological, linear oriented temporality, on the one hand, to a premodern, agrarian, cyclical temporality, on the other. Malina suggests that whereas modern society is future-oriented in that "people live achievement-directed lives focused on relatively distant goals [in which] the present always serves as a means to some more distant end,"[6] premodern peasant societies, in contrast, value the present over the future: they "have the present as first-order temporal preference. Secondary preference is past. The future comes in as third choice."[7] Based on this distinction, Malina then contrasts the peasant view with our modern, abstract view of time, which he characterizes as "linear, a unidirectional, past, present and future, that is separable into various discrete compartments."[8] According to Malina, the Mediterranean cultures, best represented by biblical society, knew nothing of this abstract time; clock time came into existence during the Industrial Revolution.[9]

Barr, however, rejects these typologies of time altogether. Barr points out that, in contradistinction to the commonly held view, the Greeks did not always hold a cyclic view of time, the Hebrews sometimes did hold a cyclic view of time, and what the Greeks viewed as cyclic (if they did) was not the same thing as what the Hebrews regarded as a straight line (if they did).[10] Further, Barr argues that we can no more deduce a Hebrew view of time from the verbal tense system of Hebrew than we could deduce the Platonic and Aristotelian philosophies of time from the tense system of Greek.[11] It is "the very serious shortage within the Bible of the kind of *actual statements* about 'time' or 'eternity' which could form a sufficient basis for a Christian philosophical-theological view of time,"[12] which has forced Biblical exegetes into trying to get a view of time out of the *words* themselves. But such an approach, according to Barr, has led to disastrous claims, for example, that because the Hebrew verb does not express the temporal differences between past, present, and future, there is no actual temporal difference in Hebrew.[13]

It is important to keep in mind Barr's caveats when looking for a metaphysics of time in Biblical and Rabbinic texts. For the Rabbis were not philosophers and were not interested in elucidating a philosophical theory of time per se. With the exception, perhaps, of Ecclesiastes, there is very little speculation of a specifically metaphysical nature in biblical texts, little awareness of time as a metaphysical construct. Nevertheless, let me suggest that the historicity of the Bible associated with linear temporality can be underscored through three defining moments. The first is represented by the moment of creation, the very instant in which God gave momentum to temporality, thrusting it forward. Secondly, the people Israel is marked by a covenant with God; this covenant defines the ongoing, linear relationship between a Deity and its people. The eschatological tenor of the prophets reinforces those moments in which the covenant is in jeopardy. Religious eschatology culminates in the kingdom of heaven, which represents the final consummation of past and present into the future.[14] The third defining moment occurs during the revelation at Sinai, the moment at which the Israelites receive the Tablets of the Law and commence the journey as the "chosen" people of God.

The linear thrust of history has its metaphysical counterpoint in Ecclesiastes, a work devoted, among other things, to expressing the futility of temporal flow. Chapter 3 of Ecclesiastes can be read in several ways. Most obvious is the prevalence of God's predetermination of all human events: that "everything has its appointed time and there is a season for every event under the sky" (Eccl. 3.1) points to the futility of human striving in light of God's predetermining of all events in their appointed time. This predetermination is reinforced in (Eccl. 3.11), "Everything He has made proper in its due time." And yet the predetermination of temporal events brings with it the comfort of cyclicity as well as the recognition that a providential deity controls human affairs. In recognizing that "there is a time to be born, a time to die, a time to plant and a time to uproot," (Eccl. 3.2) our sage underscores the comforting reality that events do not happen randomly, out of sequence. Rather events have both an inner and outer sequence that is repeated on a cyclic basis. Planting and uprooting, living and dying, mourning and dancing, these all occur and recur with constant regularity, reinforcing the motif of time as recurrence. Death too is a constant motif, underscoring the futility of human endeavors. "Again, I saw that beneath the sun the race is not to the swift, nor the battle to the brave, nor is bread won by the wise, nor wealth by the clever, nor favor by the learned, for time (*'et*) and accident overtake them all," (*Eccl.* 9:11) and human beings are "trapped in a time of misfortune, when it befalls upon them suddenly" (Eccl. 9:12). Time, then, comes to represent not only the predetermined order into which human beings are thrust, but the cruel means by which they are yanked out of this order into nonexistence, notwithstanding all efforts to the contrary. Ecclesiastes emphasizes the futility of marking time in light of its repetitiveness: "What has been, already exists, and what is still to be, has already been, and God always seeks to repeat the past" (Eccl. 3.15).

Rabbinical Models of Time and Creation

What we have then is a model of time that transcends simple binary dualism, that recognizes that temporality can be construed as both linear and cyclic in one and the same textual tradition. Both the linearity of historical events and the cyclicity of natural cycles must be recognized in this model.[15] Given the pre-eminence of ritualized events in Judaism, the marking of time assumes overwhelming importance in the rabbinic period. Inasmuch as the Rabbis are equally concerned with the daily rituals and events that are performed at specific times, so the exact determination of temporal demarcations, for example, "day," "twilight," "cycle," becomes of paramount importance in rabbinic literature. The following passage attests to the extent to which the Rabbis attempt to define a temporal unit:

> Until what time do they plow an orchard [of fruitbearing trees] during the sixth year of the Sabbatical cycle]? The House of Shammai say, "so long as [the plowing continues] to benefit the produce [of the sixth year. That is, after the crop of the sixth year has ripened and been harvested, the farmer no longer may plow in his orchard]." But the House of Hillel say, "[One may continue to plow] until Pentecost [of the sixth year]." And the opinion of the one is close to the opinion of the other.[16]

In this passage the Rabbis are concerned with determining when the sabbatical year (the seventh year in which fields must be allowed to remain fallow) technically begins: How do we demarcate the end of one year and the beginning of the next? That is, do we need a fixed time, or can utilitarian considerations be used to define the beginning of the seventh year? Although the two camps disagree over what ought to constitute the end of the sixth harvest cycle, it is interesting to note that neither group presupposes an absolute, fixed criterion of temporality. Both Shammai and Hillel opt for pragmatic considerations having to do with the actual harvest. Such rabbinic discussions anticipate the later scholastic *incipit-desinit* literature in which the "Oxford calculators" work at determining the exact starting and ending points of an infinitely divisible continuum.[17] In both cases, obviously, what is at stake is the precise instant at which a unit (temporal or quantitative) is said to begin. It is this problem that occupies generations of Jewish philosophers as well, in the context of the issue of creation of the universe.

Not surprisingly, early rabbinical texts evince earnest grappling with the scriptural account of creation. The act of creation represents the positing of conceptual order upon a disordered reality. This initial ordering action occurred at the first instant of creation, when God brought order into a disordered state. Rabbinical sayings find their way into the philosophical corpus early on. Although the Rabbis were not technically philosophers, they nevertheless raised many questions which were incorporated into philosophical discussions.[18] Scholars have pointed to three cosmological traditions that emerged within rabbinic texts: creation from an eternal matter, creation *ex nihilo,* and emanation theory. Each of these traditions has had far-reaching implications with respect to medieval philosophical texts. The first tradition, creation of the world out of a pre-existing matter, is rooted in Gnostic writings as well as in Plato's *Timaeus*; in response to these authorities, the Rabbis were concerned to determine whether the world was created out of a primordial matter or out of nothing.[19] That the first statement in Genesis could be read to support a theory of pre-existent formless matter was recognized early on by the Rabbis. According to one rabbinical dictum the phrase *be-re'ishit* (in the beginning) refers to the fact that before the actual creation there pre-existed a number of things. Numerous rabbinical texts suggest that the presently existing world came into being after a series of worlds that had been created and destroyed:[20]

- Seven things were created before the world, viz. the Torah, repentance, the Garden of Eden, Gehenna, the Throne of Glory, the Temple, and the name of the Messiah.[21]
- Six things came before the creation of the world, some created, some at least considered as candidates for creation . . .[22]

- He has come to receive the Torah, answered He to them. Said they to Him. "That secret treasure, which has been hidden by Thee for nine hundred and seventy-four generations before the world was created."[23]
- It is taught: R. Simeon the Pious said: These are the nine hundred and seventy four generations who pressed themselves forward to be created before the world was created, but were not created.[24]

The Rabbis clearly had no religious compunctions against suggesting that our world did not represent the first creative effort on the part of God; rather, they emphasized that entire worlds or generations pre-existed the creation of the universe. This interpretation would explain the emptiness and void (*tohu va-vohu*) that appeared to exist already when God initiated his original creative act. Rashi, for example, along with many other Rabbis, interpreted the first two sentences of Genesis as meaning that "when God created the heavens and earth, the earth was (already) empty and void (*tohu va-vohu*), and darkness (*ḥoshekh*) was upon the face of the deep."[25] The explicit implication of this reading is that God created the universe out of a pre-existing *tohu, vohu,* and *ḥoshekh.* This pre-existent stuff was the result of at least one prior world. Ibn Ezra, however, is a bit more circumspect, suggesting that, in contradistinction to the commonly accepted notion of creation *ex nihilo (yesh me-'ayin),* the meaning of the term *bara'* is "to cut (*ligzor*) or set a boundary (*vele-shum gvul nigzar*). The intelligent person will understand what I am alluding to."[26] By this Ibn Ezra intimates that a pre-existent matter is cut, or limited by form.

The second and third doctrines presumably found their adherents as well. Several references in scripture, for instance, 2 Maccabees 7:28 to creation "out of things non-existent," have been utilized by scholars to support a rabbinic doctrine of creation *ex nihilo.* But as Altmann and Goldstein emphasize, this tradition emerged late in Jewish sources as a response to Gnosticism and was not incorporated fully into Jewish thought until the medieval period.[27] In a similar vein, David Winston offers compelling argument supporting his claim that "not only was an unambiguous doctrine of creation *ex nihilo* missing in Jewish-Hellenistic literature, but even in Rabbinic literature such a doctrine appeared at best only in a polemical context."[28] Winston argues further, drawing upon the work of Wolfson, that the first explicit formulation of the notion of creation *ex nihilo* appeared in second century Christian literature and was amplified by both Saadiah Gaon and Maimonides in their reference to creation "not from a thing" (*lā min shay'*) as opposed to creation from "no-thing" (*min lā shay'*).[29] As we shall see in chapter 2, Crescas' formulation of creation *lo' mi-davar* (not from a thing) reflects this former expression.

Yet a third motif is the doctrine of emanation developed in some midrashic sources. According to this theory, the original light in Genesis I.3 was created as an emanation from God's glory, or from God's garments. On this theory, light is actually created first, as an archetype, from which the world unfolds. For example, in the following text, "Rab Judah further said that Rab said: 'Ten things were created the first day, and they are as follows: heaven and earth, *Tohu, Vohu,* light and darkness, wind and water, the measure of day and the measure of night,' " light is clearly introduced as one of the original ten ontological fundaments out of which everything else is created.[30] As Altmann has argued, such passages served as a prototype for later Neoplatonic writings, culminating in the mystical doctrines of the *Sefer Bahir, Sefer Yetzirah,* and Zoharic texts.[31]

With the postulation of pre-existent materials of creation in both the first and third creation theories, these texts therefore raise the second difficult question of whether time itself pre-existed creation. Does the phrase *be-rei'shit* already imply temporality, that is, does creation occur in time, or is time created along with the creation of the universe? In the following passage, it is suggested that time could have existed before the existence of the universe: "Said R. Tanḥuma, 'The world was created at the proper time. The world was not ready to be created prior to this time.' "[32] One way to understand the phrase "prior to this time" is to posit the eternity of time. That is, introducing a temporal indicator to talk about the time when the world was created itself suggests that time antedated creation. But other Rabbis claim that time was created. In the text mentioned earlier, namely "Rab Judah further said that Rab said: 'Ten things were created the first day, and they are as follows: heaven and earth, *Tohu, Vohu,* light and darkness, wind and water, the measure of day and the measure of night,' "[33] inasmuch as light and darkness, and the measure of day and night represent temporal markers, they come to represent the creation of time. This point is reiterated when Naḥmanides, in commenting on the phrase "And God called the light day," claims that "He states here that time was created and He fixed the span of day and span of night."[34] Naḥmanides reiterates the creation of time in commenting on the phrase "and God divided the light from the darkness":

> It is also possible for us to explain that when the heavens and earth emerged from nothingness into the something which is mentioned in the first verse, time came into being. For although our time consists of minutes and hours which belong to light and darkness, yet from the moment when the "something" came into existence "time" was attached to it. On this assumption heaven and earth were created and remained as they were for the space of a night without light; and He said "let there be light" and there was light, and

> He decreed that it should remain for the same space of time as the preceding (night) and later should be removed from the elements and so "there was evening and there was morning."[35]

In this passage Naḥmanides incorporates several motifs: the notion that the original created time differs from our own "clock-measured" time, that created substance and created time must exist simultaneously, and that the original evening was measured by the absence of light.

A final question concerns how to understand the word "day" (*yom*), on day one, if those astronomical markers most associated with measuring day and night have not been created until day four. There are several distinct problems involved here. The first concerns how to reconcile the method of reckoning a day (i.e., a day and a night) from sunrise to sunset with the accepted Israelite practice of connecting the day-time with the preceding night, that is, the custom of regarding sunset as the starting point of the day.[36] A second issue has to do with the ontological status of days one to three with respect to the other days of creation. If the temporal markers were not created until the fourth day, it makes no sense to designate measured temporality prior to the fourth day. Rashi, for example, emphasizes the uniqueness of day one on the grounds that on the first day, the entire universe becomes existent. Naḥmanides' exoteric interpretation of the text states that "the 'days' which are mentioned in the account of the creation were at the creation of heaven and earth real days, composed of hours and minutes, and there were six like the six days of work, as is indicated by the simple sense of the verse."[37] This reading is reinforced by the astronomical reminder that "day one" can refer to the "circuit of the heavenly sphere round the whole earth in twenty-four hours. For every minute of them is morning at different places and evening in the places opposite to them."[38] But Naḥmanides also offers an esoteric interpretation, suggesting that the term "day" can refer to the *Sefirot*, on the grounds that "every change (*ma'amar*) producing an existence is called a day."[39] This latter interpretation does not, however, account for any difference between the days.

In his commentary upon Genesis, Sforno addresses the problem of measuring the first days of creation a bit differently. Sforno interprets the phrase *be-rei'shit* as postulating "a first indivisible moment (*rega' ri'shon bilti mitḥaleq*) before which there is no time."[40] The term *bara'* (created) suggests creation from nothing, and hence a period "in which no time occurs at all."[41] Turning then to the issue of light in Genesis 1.5, Sforno argues that even though there was no actual time of light and dark, God adopted our terminology in describing the difference between day and night: "Even though he separated the light and the dark so that they would be used for different time periods without the help of rotating spheres, he differentiated them gradually in (such) a way that there was between them a time of evening when night came and a time of morning when day came."[42] Finally Sforno turns to the comprehensive process of creation. Commenting upon the significance of the seventh day, he states: "(God completed all creative activity) at the beginning of the seventh day, which is an indivisible moment (*ha-rega' bilti mitḥaleq*) which [marks] the beginning of future time but is not a part of it. As our Sages said, may they be remembered for a blessing, 'one enters into it by a hair's breadth.'"[43] As Samuelson has suggested, Sforno refers to a time instant so brief that it has no duration at all. The first six days of creation have no duration and so *measurable* time does not begin until the seventh day.[44] If by measurable Sforno means passage from past to future, the seventh day represents the first *measurable* day of the universe. The time span from the first to the seventh day is an indivisible atom of time, a *rega' mitḥaleq* which is identical to the first day or temporal instant.

Finally let me turn to the architectonic of time according to which time assumes a mythos of its own. This architectonic is captured by what Higgins has called "liturgical *curcus*" which is most applicable to religious ritual and practice. In medieval Christian practice, for example, liturgical time refers to the specific times for reciting prayer, the ringing of the ecclesiastical bells at appropriate times, the setting of the ecclesiastical calendar, and so on.[45] A similar point can be made with respect to Jewish attitudes toward liturgical time. Yerushalmi has suggested that while in the Bible there is a sense of chronology, a genuine sense of the flow of historical time, the Rabbis in contrast "seem to play with Time as though with an accordion, expanding and collapsing it at will."[46] Even a casual glance at rabbinic texts supports Yerushalmi's claim, for in rabbinic texts the ordinary chronological barriers of time have truly been ignored, and all ages interact with one another. Witness, for example, the importance of ritualized, recursive, readings of Scripture, during which temporality becomes atemporal.[47] Each reading of the weekly Torah portion hearkens back not only to other weekly portions read sequentially during the year, but to previous years' readings as well, thus elevating the event to an atemporal plane. The cyclicity of temporally individuated events is emphasized in their yearly, monthly, or weekly repetition.

The religious calendar orients celebrants in time through the use of regularly repeated rituals; this temporal system structures the life of the community.[48] Such time is not just chronological time but is connected with repetitions: temporal repetition is one essential attribute of ritual. In short, myth and ritual are connected through temporality. Calendars are not restricted to purely practical functions, that is, to refer to points in time and to time durations. They also represent a process of human cognition in which the experience of time is conceptualized, structured and comprehended. "Calendars make sense of the dimension of time by imposing a rational, human structure upon it. Calendars represent at once a way of describing time and of establishing conceptual order amidst a seemingly disordered world."[49]

Time, Order, and Creation in the Greek Philosophical Tradition

Obviously we cannot, in this work, present a complete account of time in the Greek philosophical tradition.[50] Nevertheless, I should like to emphasize several motifs and arguments that are extremely influential upon subsequent medieval Jewish discussions. The earliest sustained philosophical discussions of time occur in the fragments of the Presocratic philosophers Heraclitus (540 BCE) and Parmenides (515 BCE) in the context of trying to account for change in reality. Heraclitus attempts to support his contention that flux and becoming are alone real, and that permanence and constancy are merely apparent. Every physical object is subject to temporal change, hence Heraclitus' emphasis upon the eternal flux of reality, and his insistence that all is in flux.[51] Parmenides, however, disagrees with Heraclitus and argues that only the permanent and enduring are real, and all time, flux, motion and change are unreal. In contradistinction to Heraclitus who emphasized the ontological priority of change, Parmenides denies that change can occur.[52]

It is Plato who, against the backdrop of his Presocratic predecessors, tries to resolve the paradoxes of change and permanence. The influence of Parmenides can be felt most fully when we turn to Plato's most explicitly cosmological dialogue, the *Timaeus;* it is this dialogue that has exerted an enormous influence upon Jewish cosmogonic and cosmological writing. In the *Timaeus,* Plato distinguishes between eternity (*aionios*) and everlastingness (*aidios*): evelastingness is "the nearest approach to eternity of which sensible things are capable."[53] On Plato's account, the Demiurge created the cosmos out of his goodness, not jealousy.[54] Desiring all things to be like him, the Demiurge instills order into inherent chaos. More specifically, the Demiurge creates the cosmos according to an eternal model that is independent of him. There can be only one cosmos since uniqueness is itself a perfection.[55] The Demiurge is not omnipotent, however: he works with eternal forms that he did not create, and he inherits the domain of chaos, which is independent of his creative powers. The Demiurge is ultimately only concerned with the world of becoming. That the Demiurge is not identical with the Form of the Good is evidenced by the fact that the Demiurge uses the idea of good in order to impose order, but the Demiurge is not good in himself.

We now turn to the heart of our discussion, namely the relation of time to creation. First, Plato distinguishes between the sphere of eternity and that of time. Because of the importance of this passage to subsequent Neoplatonist thought, I shall present it in its entirety:

> So as that pattern is the living Being that is everlasting (*aidios*), he sought to make this universe also like it, so far as might be, in that respect. Now the nature of that Living Being was eternal (*aionios*), and this character it was impossible to confer in full completeness on the generated thing. But he took thought to make, as it were, a moving likeness of eternity (*aion*); and at the same time that he ordered the Heaven, he made, of eternity that abides in unity, a likeness moving according to number which is eternal (*aionios*)—that to which we have given the name Time.[56]

In this passage it is clear that the Living Being, identified with the domain of pattern or forms, is eternal and hence not subject to time. The world-soul and the world-body, which characterize the world of becoming, cannot be eternal because they comprise "motion"; hence the Demiurge makes them a "moving likeness of eternity." Time is then defined as a likeness of eternity which is the measure of the world-soul and world-body, or more specifically an "everlasting likeness" moving according to number. The contrast between eternity and time is characterized more fully as Plato turns to the creation of days and nights. Plato states that the domains of "was" and 'will be" belong to becoming and hence to time, whereas the domain of 'is' "alone really belongs to it [everlasting being (*aidios*)] and describes it truly."[57] Unlike the domain of "was" and "will be," the domain of "is" denotes "that which is always (*aei*) in the same state immovably . . . nothing belongs to it of all that Becoming attaches to the moving things of sense."[58] Eternity (*aion*) is distinguished from time in that the domain of eternity is outside time, and change through time, altogether. What is not clear from these passages is whether Plato's depiction of eternity comprises duration.[59] We shall return to the importance of this issue shortly, in the context of Plotinus' theory of time.

When time actually comes into being, however, is a problematic issue and subject to much interpretative ink. In *Timaeus* 38b Plato suggests that inasmuch as time came into being with the heavens, it would appear that before creation there is no time: "Be that as it may, Time came into being together with the Heaven, in order that, as they were brought into being together, so they may be dissolved together."[60] Other early passages in the *Timaeus* suggest that time itself was created along with the cosmos as a whole.[61] But later passages in the dialogue support the interpretation that time may pre-exist the creation of the heavens.[62] In contradistinction to those passages adduced earlier that suggested *ex nihilo* creation, other passages suggest that time existed, in some ontological measure, before creation.[63] The very fact that temporal words are used in (38b) points to the existence of time. Sorabji and Vlastos both offer the suggestion that while orderly time began with the orderly universe, before that there existed a domain of "disorderly time," along with motion and matter.[64] If we take seriously, however, Plato's definition of time as "a likeness moving according to number," which is aligned with the creation of the heavenly spheres, and if we emphasize the importance of these celestial spheres with respect to the measure of time, then it is not possible that time pre-exist the creation of the heavens. Sambursky rightly emphasizes the lack of symmetry between time and space in that whereas space is intermediate between Being and Becoming, time is at the level of things becoming.[65] As we shall see below, this asymmetry is eliminated by Plotinus, who introduces the domain of intelligible space, which functions as the analogue of intelligible time.[66] It is here that the discontinuity between time and space appears most explicitly in the *Timaeus.* For unlike time, which is the measure of motion, space is the necessary precondition for the coming into being of motion. And so while it is perfectly possible for *chora* to pre-exist the creation of the heavens, it is not possible for time to so pre-exist.

Like that of Plato, Aristotle's theory of time and motion is profoundly affected by cosmological considerations. For Aristotle, the prime example of time and motion is the relation between time and the circular motion of the heavens. Time is therefore a circle, measured by the circular motion of the heavens.[67] When Aristotle turns to issues of creation and time, he does so against the fabric of Plato's *Timaeus.* In *De Caelo* 1.10 he summarizes the positions of his predecessors, in particular that of Plato, who claimed that the cosmos had a beginning but is everlasting. Aristotle presents a number of arguments against Plato, claiming, for example on the basis of observation, that generated things are *seen* to be always destroyed. A second argument is that if the cosmos came into being, it must have a cause of change; but if that cause is present already, it could function again and allow for the dissolution of the cosmos. Aristotle then argues that a universe that is ungenerated and indestructible at all times is not capable of nonexistence and rules out the possibility in which things exist for an infinite time and then cease to exist for an infinite time.[68] Aristotle then applies these and other considerations to Plato's arguments in the *Timaeus,* arguing that it is impossible for something to be at once generated and indestructible.[69]

The eternity of the cosmos is integrally related to Aristotle's conception of time. In answer to the question whether time was generated, Aristotle develops Plato's notion of the instant or "now" (*to nûn*) as a basic feature of time. The instant is defined as the middle point between the beginning and end of time. Since it is a boundary or limit, it has no size and hence cannot be considered to exist: it is a durationless instant. Since instants do not in and of themselves exist, it might be argued that time itself does not exist. That is, the past and future do not now exist, and the present "now" is not a part of time since, as we have already noted, it is sizeless. Because the extremity, or limit, of time resides in the instant, Aristotle claims that time must exist on both sides of it: "Since the now is both a beginning and an end, there must always be time on both sides of it."[70] And in *Metaphysics* 12.6 Aristotle claims that there can be no "before" or "after" if time does not exist, for both terms imply the existence of relative time. "For there could not be a before and an after if time did not exist."[71] For these reasons time must be uncreated. Aristotle's basic argument, centered on his definition of the "now" as the midpoint between "before" and "after," is that in order to account for the coming into existence of any present instant, there must exist a prior actual instant; but in the case of the first instant, there could be no prior instant, actual or potential.[72]

Aristotle's emphasis upon an ungenerated cosmos is reinforced in his analysis of the relation between time and motion. It is this relationship that reappears throughout the history of medieval and early modern philosophy. Aristotle claims in a number of texts that time is defined in terms of motion and so there can be no time without motion. For example, in *De Caelo* Aristotle argues that time is an integral part of the cosmos. He had already postulated that there can be no body or matter outside of the heavens, since all that exists is contained within the heavens. Since, however, time is defined as the number of movement, and there can be no movement without body, it follows that there can be no time outside of the heavens. "It is obvious then that there is neither place nor void nor time outside the heaven, since it has been demonstrated that there neither is nor can be body there."[73] Whatever does exist beyond the heavens must be "changeless and impassive."[74] A similar point is propounded in *Physics* 4.12.

Aristotle develops this characterization further in his classic discussion on time in *Physics* 4.10–14.[75] Having asked of time whether "it belongs to the class of things that exist or that of things that do not exist,"[76] he rejects various considerations which might lead one to think that time does not exist. Time, he claims, is connected with movement, noting that

> when the state of our own minds does not change at all, or we have not noticed its changing, we do not realize that time has elapsed, any more than those who are fabled to sleep among the heroes in Sardinia do when they are awakened; for they connect the earlier 'now' with the later and make them one, cutting out the interval because of their failure to notice it. So just as, if the 'now' were not different but one and the same, there would not have been time, so too when its difference escapes our notice the interval does not seem to be time.[77]

The interesting question raised by this example, one to which medieval commentators will return,[78] is not, as Bostock has argued, whether we can notice if time has passed when we notice that things have not changed, but rather whether changeless times can be in fact perceived at all. Bostock claims that Aristotle's premise is false because "we notice that time has passed not only when we notice that things have changed but also when we notice that they have not: that is, to be aware of rest, no less than to be aware of movement, is equally to be aware of the passing of time." He gives as an example noticing a clock's ticking, and noticing the "nothing" that happens between each tick.[79] But Bostock's counterexample to Aristotle misses the point, for it is precisely the "noticing the nothing between each tick" that, in Aristotle's paradigm, would constitute motion or change. Aristotle's point is that we notice time *only* when we notice change, but the noticing itself can constitute a change.

From the epistemological point that "we perceive movement and time together," Aristotle draws an ontological conclusion, namely that "hence time is either movement or something that belongs to movement."[80] It is not just that we cannot perceive changeless time, but that time itself does not exist when there is no change. This leads to a definition of time in terms of the movement of the "now:" "When we do perceive a 'before' and an 'after,' then we say that there is time. For time is just this—number of motion in respect of 'before' and 'after' . . . time is only movement in so far as it admits of enumeration. . . . Time then is a kind of number."[81] What we see in this important and influential passage is that time is defined as the "number of motion;" without something to measure, and without a unit of measure, there can be no time. But what does it mean to say that time is a kind of number? Aristotle distinguishes two meanings to the term number: what is counted, or the countable, and that by which we count, and then associates time with the first kind of number, claiming that time is that which is counted, and not the measure used to count.[82] Thus, to know how long a process takes is a matter of being able to count or measure its duration. In other texts, however, Aristotle seems to conflate these two functions.[83]

Aristotle then stipulates two important qualifications to his characterization of time in terms of movement. First he points out that "not only do we measure the movement by the time, but also the time by the movement, because they define each other."[84] Further, he argues that time is the measure not only of motion but of rest as well. "For all rest is in time. For it does not follow that what is in time is moved, though what is in motion is necessarily moved. For time is not motion, but 'number of motion:' and what is at rest also can be in the number of motion."[85] We shall return to the importance of this passage in chapter 3 when we examine Crescas' emphasis upon the measure of rest. Finally, Aristotle raises an important query concerning the relationship between time and the rational perceiver:

> Whether if soul did not exist time would exist or not, is a question that may fairly be asked; for if there cannot be some one to count there cannot be anything that can be counted, so that evidently there cannot be number; for number is either what has been, or what can be counted. But if nothing but

soul, or in soul reason, is qualified to count, there would not be time unless there were soul, but only that of which it is an attribute . . .[86]

This passage raises the important question of whether time exists if there is "no soul" to perceive it. Aristotle himself does not provide an answer to this query, but his suggestion is fairly straightforward: inasmuch as time is a kind of number and its function lies in counting, there can be no time if there is "no soul" doing the counting.[87] Later commentators, however, latched upon the issue and it became the basis for subsequent idealist descriptions of time.[88] For Aristotle, then, time falls into the category of accident which exists in motion. What this means is that we have a perception of time only when we perceive motion.

Ancient Greek Astronomy and Cosmology

These Platonic and Aristotelian conceptions of time reappear in the context of ancient astronomy and cosmology. That cosmology and astronomy comprised separate disciplines was already explicitly indicated in early textual traditions. In his commentary upon Aristotle's *Physics,* for example, Simplicius reflected the long-standing tradition of distinguishing between the two, stating that:

> It is the business of physical inquiry to consider the substance of the heaven and the stars, their force and quality, their coming into being and their destruction, nay, it is in a position even to prove the facts about their size, shape, and arrangement; astronomy, on the other hand, does not attempt to speak of anything of this kind, but proves the arrangement of the heavenly bodies by considerations based on the view that the heaven is a real kosmos, and further it tells us of the shapes and sizes and distances of the Earth, Sun and Moon, and of eclipses and conjunctions of the stars, as well as of the quality and extent of their movements. . . . The things, then, of which alone astronomy claims to give an account it is able to establish by means of arithmetic and geometry.[89]

In the medieval world, two rival cosmologies, that of Aristotle and that of Ptolemy, competed for acceptance. Influenced by Aristotle's physical and natural works, many cosmologists followed Aristotle rather than Ptolemy in their quest to provide a comprehensive theory of the universe.[90] The formative classical texts included Aristotle's *De Caelo,* supplemented by relevant passages from the *Metaphysics, Physics,* and *De Generatione et Corruptione.* Plato's *Timaeus* and commentaries upon Genesis presented an additional dimension to this corpus.

In the Aristotelian cosmology, the universe is a finite sphere whose center is at the earth and bounded by the sphere of the fixed stars. Nine primary concentric spheres (in turn divided into subsidiary spheres) rotate around the earth; these spheres form a compact whole, much like the skins of an onion, with no vacuum. First came the other three terrestrial elements, water, air, and fire. Surrounding the sphere of fire were the crystalline spheres in which were placed the seven planets: the Moon, Mercury, Venus, the Sun, Mars, Jupiter, and Saturn. Beyond the last planet came the fixed stars, identified with the sphere of the "prime mover," and then nothing.[91] The sphere of the moon separated the universe into the sublunar, or terrestrial, region, and the superlunar or celestial region. The superlunar heavens differed in composition from the sublunar bodies in that the former were composed of a single incorruptible element, *aether*, while the earth was comprised of the four elements. One of the purposes of *aether* was to account for the movements of the celestial bodies that Aristotle argued could not move in the same way as did the terrestrial elements. For according to Aristotle, elements in the sublunar realm were subject to the four kinds of change, while elements comprised of aether only underwent one kind of change, eternally uniform motion in a circle. Another reason was that the four terrestrial elements could not account for the vast distance between the earth and the outermost sphere; only an element not subject to contraries, Aristotle argued, could exist sufficiently long to fill this space.[92]

The ultimate source of motion in this Aristotelian system is God, or the unmoved first mover.[93] But did God move the first moving sphere as an active, efficient cause, or as a passive, final cause? Aristotle had attributed to all the celestial spheres a mover, the ultimate source of motion being God. Medieval thinkers, however, introduced immobile created intelligences to explain celestial motion. These separate intelligences move the orbs with both intellect and will.[94] Each sphere has a soul or internal moving source; Maimonides identifies these spheres with angels.[95]

And yet, although cosmology and astronomy represented separate disciplines, as it were, nevertheless they intersected in the area of theory formation. As Pederson has argued, tensions centered around the metaphysical status of mathematical theories in science.[96] For on the one hand the universe, as described by Aristotle in *De Caelo*, was a material entity based on the laws of physics. On the other hand, mathematical astronomy made use of geometrical devices that violated these very physical laws. More specifically, both Aristotle and Ptolemy agreed that there must be a plurality of spheres to account for the motion of each planet. These spheres, as we have seen, were nested contiguously. On Aristotle's model there was a series of concentric orbs, each moving in a natural, uniform, circular motion, all sharing the earth as a common center. Ptolemy, however, recognized that

Aristotle could not account for variations in the observed distances of the planets. This recognition led to the postulating of an alternative cosmological scheme.

In his two astronomical works *Almagest* and *Hypothesis of the Planets,* Ptolemy argued that the planets were carried about by a system of eccentric and epicyclical spheres.[97] In the *Almagest* Ptolemy had proposed that his astronomical theory was merely a method the purpose of which was to "save the appearances," or account for the observed phenomena.[98] In his *Hypothesis of the Planets* he provided the mechanical explanation for his system; this work, although not available in a Latin translation, reached Western Europe most likely through Arabic translations.[99] On Ptolemy's model, each concentric planetary orb contained at least three partial eccentric and epicyclical spheres. That this system of eccentric and epicyclical spheres contravened the concentric spheres of Aristotle was not lost upon Ptolemy or his followers. In particular, Ptolemy's insistence that partial eccentric orbs had centers other than the earth violated Aristotle's dictum that all celestial spheres move around the earth with uniform motion.[100] And yet most medieval astronomers found that Ptolemy's system did a better job of "saving the appearances" of astronomical data. As Grant has argued, "the medieval conflict between the Aristotelian and Ptolemaic systems centered on efforts to demonstrate that eccentric and epicyclical orbs did not imply consequences that were subversive and destructive of Aristotelian cosmology and physics."[101] In other words, medieval philosophers were faced with a dilemma: they could either reject the earth's centrality and abandon a vital part of Aristotelian physics in the name of astronomical and mathematical purity, or they could accept a cosmology that was untenable from the perspective of the astronomers.[102]

Plotinus and the Neoplatonist Tradition

The last important Greek philosophical influence upon Jewish philosophy is Neoplatonism, which was largely based on the writings of Plotinus and Proclus. Jewish Neoplatonism dates from the ninth century and provided the philosophical context for the thought of many cultivated Jews of the eleventh and twelfth centuries; during the Arabic period it was complemented by elements stemming from Islamic religious traditions. For serious Jewish thinkers, the speculations of certain Neoplatonist philosophies provided epistemological and metaphysical notions that were quite compatible with their own attempts to characterize the nature of God and his nature and relation to humans. Although not all Jewish thinkers supported Neoplatonism, it was extremely influential on the formation of Jewish thought during the late Hellenistic, Roman, and medieval periods.[103]

The Islamic school of Neoplatonism most clearly influenced medieval Jewish writers. The work of Plotinus was transmitted in a variety of ways, most notably through the *Theology of Aristotle* (a paraphrase of books 4, 5, and 6 of the *Enneads*), and through doxographies, collections of sayings of Plotinus that were circulated among religious communities. The *Theology of Aristotle* exists in two versions. The shorter (vulgate) version, belonging to a later period and found in many manuscripts, was the version first published by F. Dieterici. The second, longer version exists in three fragmentary manuscripts in Hebrew script, discovered by Borisov in Leningrad.[104] Two other influential works are worthy of note as well. Proclus' *Elements of Theology* was transmitted to Jewish thinkers in the period between the early ninth and late tenth centuries through an Arabic translation, *Kalām fī māḥd al-khair.* Known to Latin thinkers as the *Liber de causis,* this work was translated in the twelfth century from Arabic into Latin most likely by Gerard of Cremona and was generally attributed by medieval philosophers to Aristotle.[105] And finally, the *Book of Five Substances* attributed to Empedocles was originally written in the ninth century in Arabic and translated into Hebrew in the fourteenth–fifteenth centuries. Published by David Kaufmann in 1899, this pseudo-Empedoclean work represents a variant of Ibn Ḥasday's Neoplatonism and was highly influential upon the work of Ibn Gabirol, especially in its placement of "spiritual matter" as the first of the five substances.[106] Because of the importance and influence of this tradition upon subsequent Jewish philosophy, a brief examination of Plotinus' theory of time is thus in order.

Plotinus' distinction between time and eternity is carried out against the background of Plato's characterization of time in the *Timaeus* as the "moving image of eternity," as well as Aristotle's description of time in *Physics* 4 as the measure of motion. Plotinus attacks Aristotle's theory of time, which he sees as the only main alternative to a Platonic theory. The main distinction in Plotinus is between that which is outside of time altogether, and that to which temporal predicates apply.[107] In contrast to the Aristotelian view we have just elaborated, Plotinus and his followers develop a theory of time according to which time does not depend upon external objects and their motion for its existence. On this view, the essence of time is not motion but rather duration. In *Enneads* 3.7 Plotinus rejects the view that makes time dependent upon physical motion. Rather, he connects it with the "the Life of the Soul in a motion of change from one stage of life to another."[108]

Plotinus begins with a critique of his predecessors' conception of eternity. He first states and rejects the view of Plato that eternity is the "intelligible substance itself," identified with the whole of the cosmos, on the grounds that it rests on a faulty view of predication.[109] Turning then to Aristotle's view that "eternity exists in virtue of Rest," Plotinus rejects this view as well for a number of reasons: it does not allow for motion to be eternal, it does not explain how rest contains within itself the concept of 'always,' and it fails to note the critical difference between time, which is sometimes extended, and eternity which is never extended.[110]

Plotinus is now ready to offer his own conception of eternity. Plotinus defines eternity as

> A life which remains always in the same state, always having the whole present to it—not one thing now and then another, but everything at once, and not different things now, and afterward different things, but a partless completion, as if all things existed together in a single point, and never flowed forth, but remained there in the same state, and did not change, but were always in the present, because none of it has gone by, nor shall it come to be, but it is just what it is.[111]

In this important and influential definition, we note that eternity always has the whole present to it; it is changeless, always present. By this Plotinus means that lacking in nothing, all future [unactualized states] are irrelevant to the life of intellect. Hence "was" and "will be" are inapplicable to it. The domain of eternity does not undergo affection or change, nor does it expend any of its energy. Eternity is "all at once, and is everywhere full yet unextended."[112] Plotinus is ostensibly aware of the dilemma of speaking about eternity using temporal predicates. Because eternity always is, it never "is not" and cannot be other than it is. What Plotinus seems to want to say is that "always" (*aei*) implies completeness and not temporality. Because it is partless and does not derive its being from time, the terms "before" and "after" do not apply to it. Only temporal beings have need of a future. Eternity is totally devoid of temporality: it "does not get its being from any extent of time, but is prior to any extent of time."[113]

In contrast to eternity, then, time represents the domain of incompleteness. Temporality is the image of the eternal. Plotinus examines and rejects the three conventionally accepted notions of time that tie time to motion and measure: against the Stoics he argues that time cannot be identified with movement; against Aristotle he argues that time is not the number of movement; and against Epicurus he argues that time is not an attribute of movement. In all three cases, these philosophers have failed to distinguish time itself from that which is in time or measured in time.[114] For Plotinus, time is a function of the movement of the life of the soul: it is "the product of the spreading out (*diastesis*) of life."[115] Time is dependent upon soul, and so upon the return of the soul to the One, time itself will disappear. But the origin of time, and the soul, is unclear in Plotinus. Originally time "was not yet time, but it too was at rest in Eternity."[116] Due, however, to the "officious nature" of world-soul, the world-soul moved away [and down] from eternity, and time moved with it; in this move away from Eternity the world-soul "produced time as the image of eternity" when it produced the sensible world in imitation of the intelligible world.[117] What is interesting about this myth of generation is that in the beginning, time was both in eternity and yet distinguishable from it. Plotinus does not explain what accounts for the initial discontent of the soul, nor does he explain why time moves along with the soul away from eternity.[118]

Both Plotinus and Plato agree that time came into existence with the universe, but for Plotinus the reason has to do with the co-generation of the soul. They also agree that day and night were created as markers of time. Time itself, however, is not a measure independent of the soul. Just as eternity exists in the intelligible domain, so too time exists in soul and with soul. Inasmuch as this spreading out or duration of soul is unmeasured and undetermined, it is ultimately incomprehensible.[119] One implication of the fact that time is not tied to the external world is that time acquires a subjective existence in the mind of its cognizers. Augustine has captured this notion succinctly in his famous dictum that "it seems to me that time is nothing other than an extension (*distentio*) but of what it is an extension I do not know. It would be surprising if it were not an extension of the mind (*animus*) itself."[120]

Conclusion

Let me briefly summarize the main features of the various systems we have examined. I have depicted in this chapter those elements within biblical and rabbinic thought, as well as within Greek philosophy, which pertain to the themes of time and creation. Both the Jewish and Greek philosophical traditions have contributed important ingredients. From Scripture comes the unambiguous statement of a "Beginning." From the Rabbis comes an understanding of the nuances inherent in interpreting the first instant of creation. Plato's theory of temporality must be read against the backdrop of the distinction between the domain of being and that of becoming. Reflecting the influence of Parmenides, Plato emphasizes the timelessness of the domain of being, in contradistinction to the flux and temporal change of the domain of becoming. Time, associated with motion, is created along with the heavens. There is room, however, for the suggestion that a "pre-existent" time exists in conjunction with the "pre-existent chaos" out of which the Demiurge creates the cosmos. In contradistinction to Plato, we can articulate the following features of time in Aristotle's work. First, time is inconceivable without motion, implying as it does the existence of a corporeal object in motion. Eternal incorporeal beings cannot have time associated with them. Because of this close association to matter and motion, time cannot exist prior to their existence; neither time nor the heavens is generated. And finally, it is clear, despite the indication of dependence upon human consciousness, that time does have a kind of reality outside the mind.

The similarity of certain passages to Plato's *Timaeus* is striking and did not go unnoticed by later Jewish thinkers. In fact, as we shall see, later medieval Jewish philosophers were able to capitalize upon these similarities in order to emphasize the harmonization of Scripture and Greek philosophy. Like the Rabbis, Plato too worried whether the existence of the universe at a time implied the creation of time itself. Within the panoply of ancient Greek cosmologists, Plato, as we have seen, is the first to identify time with the movement of the heavens. We have seen that Plato defines time as "an eternal likeness moving according to number—that to which we have given the name Time."[121] This everlasting likeness is of the Living Being which is itself eternal. With respect to the thorny question of whether time was created along with the heavens or whether it pre-existed creation, we have seen that Plato offers multiple responses that have led scholars to postulate multiple interpretations of the text.[122] What is clear, though, is that created time measures the circular motion of the heavenly spheres. The sun, moon, and planets were "made to define and preserve the numbers of Time."[123] The periods of time—the day, month, and year—are the basic units of measurement which cannot exist without the motions of the celestial bodies, the heavenly clock as it were. Neither time nor the celestial bodies can exist without the other.

What we have then is a striking confluence of themes in both the *Timaeus* and in Genesis. Both texts postulate the existence of a creator. Both impute to this creator the urge to create, the willful choosing to bring the universe into existence. Both recognize the importance of temporality in this creation process: the scriptural author(s) by focusing on the importance of the term "day" (*yom*) in the creation account, and Plato by introducing time as the ontological divide between the superlunar and sublunar spheres. And finally, both accounts allow for the *possibility* of creation occurring out of a "pre-existent matter," a chaotic, formless stuff upon which order is imposed.[124] These similarities will play a crucial role in subsequent Jewish discussions of creation, for they allow thinkers such as Albo, and possibly even Maimonides, to reconcile a scriptural account of creation with a philosophically minded account without undermining their religious presuppositions.

Chapter Two

Time, Creation, and Cosmology

Introduction

Of the many philosophical perplexities facing medieval Jewish thinkers, perhaps none has been as challenging or as divisive as determining whether the universe is created or eternal. No Jewish philosopher denied the centrality of the doctrine of creation to Jewish belief. Jews were enormously affected by Scripture and in particular by the creation account found in Genesis 1–2. But like their Christian and Moslem counterparts, Jewish thinkers did not always agree upon what qualifies as an acceptable model of creation. In the context of this topic, perhaps the most important phrase of Scripture is *be-rei'shit,* "in the beginning." The very term *be-rei'shit* designates the fact that there was a beginning, that is, temporality has been introduced if only in the weakest sense that this creative act occupies a period of time.[1] But medieval Jewish philosophers thinking about creation were influenced as well by Aristotle's model of an eternally existing world. When trying to prove that the world was created by God in time, philosophers who wanted to support a biblical theory of creation in time had to reject Aristotle's position that time is infinite.

We have seen in chapter one that Aristotle posits an eternal universe in which time is potentially, if not actually infinite.[2] That is, Aristotle argues that since there can be no "before" to time, neither time nor the universe was created. Jewish philosophers, however, almost without exception are committed to the belief that God created the universe. At the same time they want to accept certain aspects of Aristotle's theory of time and the universe. For if, as Aristotle claimed, time is the measure of motion, and motion is of material stuff, then infinite time implies the eternity of the universe. But inasmuch as accepting the eternity of the universe qualifies the role God plays in determining the act and materials of creation, Jewish thinkers were motivated to reject the Aristotelian characterization of time while at the same time continuing to accept his overall philosophical authority.

In this and the following chapter I shall focus upon Jewish philosophical attempts to clarify what is entailed by postulating a theory of creation. I shall begin in this chapter with those medieval Jewish thinkers who, influenced by those rabbinical texts adduced in the first chapter, respond to the Aristotelian challenge posed with respect to the issue of creation. In chapter 3, I shall turn specifically to the philosophical perplexities surrounding the notion of a "first instant" of creation. Both Kalām and Neoplatonic schools will be examined. In general I shall argue that our thinkers are committed to reconciling traditional Jewish beliefs with what they feel are the strongest points in Aristotle in an attempt to explain the existence of the universe in time. Further, in contradistinction to recent scholars who suggest that the question of time is not the primary concern of Maimonides in this controversy, I shall argue that issues of temporality are critical to understanding the theory of creation both in Maimonides, as well as in subsequent Jewish philosophers.[3]

Astronomy and Cosmology: The True Perplexity Revealed

It is clear that by the twelfth century the influence of Aristotle, mediated through Neoplatonist eyes, has taken hold among Jewish philosophers. These Jewish philosophers attempt to reconcile a philosophical theory of time with biblical and rabbinic dicta. Nowhere is this attempted reconciliation felt more keenly than in the topic of creation of the universe. Jewish philosophers must respond to an Aristotelian framework according to which reality is a continuous plenum in which time and matter are infinitely divisible. In this eternal plenum, time is potentially, if not actually, infinite; neither time nor the universe was created. Jewish philosophers, however, must interpret Aristotle in light of the first verses of Genesis, and they must therefore determine whether this universe was created simultaneous with or subsequent to the creation of time. Furthermore, they must analyze the significance of the term *be-rei'shit* in the context of an Aristotelian theory of time. The writings of Maimonides and his successors, to whom we now turn, attest to this tension.

Maimonides' major philosophical work *The Guide of the Perplexed* gives eloquent testimony to the tension apparent in trying to reconcile Aristotelian cosmology and physics with traditional Jewish belief. In the introduction to the *Guide* Maimonides distinguished two levels of interpretation, exoteric and esoteric, and suggested that it is sometimes incumbent upon a philosopher to conceal his own esoteric position behind the veil of exoteric doctrine. He further swore his own student not to divulge his secret teachings to others. Maimonides therefore introduced an ambiguity into the very reading and understanding of his texts, an ambiguity that has persisted to this day.[4] The tensions surrounding how to read the *Guide*, and how to reconcile the *Guide* with more explicitly religiously focused works such as the *Mishneh Torah*, come to the forefront in his discussion of issues pertinent to philosophical cosmology. For one of the topics explicitly mentioned by Maimonides as being deliberately obfuscated in the *Guide* is the "Account of Creation" which is identified by Maimonides with Aristotelian physics. Philosophical cosmology comprises the relation between astronomy and cosmology, between astrology and astronomy, as well as his well-known discussion of creation.

Maimonides' analysis of astronomy and its relation to cosmology occurs in two works: in *Mishneh Torah* 3–4 and *Guide* 1.72 and 2.19–24. These two sets of texts present conflicting accounts; unraveling the source of conflict will enable us to determine more fully Maimonides' attitude toward astronomy. In the *Mishneh Torah* Maimonides offers the following description of the universe. The finite universe, comprising the four heavens (*Shamayim, Raqi'a, Zevulun*, and *'Aravot*), contains nine concentric spheres that circle the earth: the Moon, Mercury, Venus, the Sun, Mars, Jupiter, Saturn, the stars, and the sphere that moves from east to west. This ninth and largest sphere "includes and encircles all things."[5] Each of the eight other spheres is divided into subspheres "like the several layers of onions."[6] These spheres are "clear and transparent" and are contiguous, where "no vacuum intervenes." There are eighteen such spheres, all of which revolve around the earth. In addition there are eight "small spheres" that do not revolve around the earth, but that are fixed in the larger spheres that do so revolve. These small spheres correspond to the epicycles introduced by Ptolemy. Ascertaining the science of these eighteen spheres, including their number, lines of movement, and courses in the heavens is what Maimonides terms the "science of mathematical astronomy (*ḥokhmat ḥeshbon tequfot u-mazalot*), on which the Greeks composed many treatises."[7] The ninth sphere, divided into the twelve constellations, has no division or stars but reflects the stars in the eighth sphere. Each sphere and star has a soul that is "endowed with knowledge and intelligence."[8] Situated half-way between humans and angels, their knowledge is "less than that of angels and greater than that of human beings."[9] The sublunar realm contains the four elements, fire, air, water, and earth, again in contiguity to one another with no intervening vacuum. These four bodies have no soul, no knowledge, and hence are lifeless. They have a governing unalterable principle (*minhag*) and they form the basic elements of all created things.[10] The substances formed by combinations of these four elements are called "individuals."

On the face of it, this extended description in *Mishneh Torah* captures the current Ptolemaic cosmology: the spheres are organized in an orderly fashion, with no vacuum obtaining between spheres; epicycles are introduced to account for variation in revolution; and spheres are endowed with intelligible souls responsible for their orderly motion. And yet, what complicates matters is that Maimonides offers a decidedly anti-Ptolemaic account in the *Guide* that appears incompatible with the *Mishneh Torah* account. Let us turn, then, to those passages in the *Guide* that contravene the account given in the *Mishneh Torah.*

In *Guide* 1.72 Maimonides presents a cosmological scheme that shares some affinities with the picture found in the *Mishneh Torah.* The sphere of the outermost heaven is comprised of the heavens, the four elements, and their composites. In this sphere there is no vacuum: it contains many spheres, "one contained within the other, with no hollows between them and no vacuum in any way whatever."[11] These spheres are spherical and all move in a "circular uniform motion," with some spheres moving more rapidly than others. The heaven itself (which encompasses the universe) moves all the other heavens "simultaneously with itself."[12] Maimonides goes on to point out that the heavens have different centers: "The center of some of them is identical with the center of the world, while the center of others is eccentric to the center of the world."[13] It is here that he clearly acknowledges the existence of eccentric orbits. The total number of spheres is at least eighteen. It is a matter of speculation, Maimonides avers, "whether there are epicycles, that is, spheres that do not encompass the world."[14]

This cosmological picture is then amplified in *Guide* 2.19–24. Maimonides argues, following Aristotle, that both the matter and the form of the spheres differ from that of the four elements, as reflected in the different types of motion exhibited by them. But he then points to several problems with Aristotle's attempts to explain why the sphere moves from the East and not from the West, and why some spheres move faster than others. Maimonides rejects Aristotle's explanations on the grounds that "the science of astronomy was not in his [Aristotle's] time what it is today."[15]

Having rejected Aristotle's analysis, Maimonides presents his own version in *Guide* 2.24. His main thesis is that the underlying premise of Ptolemy's *Almagest,* namely that "everything depends on two principles; either that of the epicycles or that of the eccentric spheres or on both of them,"[16] is untenable. Maimonides' own contention is that these two principles are "entirely outside the bounds of reasoning and opposed to all that has been made clear in natural science."[17] In other words, Maimonides rejects Ptolemaic astronomy on the grounds that it conflicts with Aristotelian physics. The first principle is rejected on the grounds that the existence of epicycles implies that that the "epicycle rolls and changes its place completely," hence undermining the Aristotelian dictum that things in the heavens are immovable.[18] He then offers other considerations, in the name of Abu Bakr, against accepting the doctrine of epicycles.[19]

Following this analysis, Maimonides presents the following theoretical perplexity:

> If what Aristotle has stated with regard to natural science is true, there are no epicycles or eccentric circles and everything revolves round the center of the earth. But in that case how can the various motions of the stars come about? Is it in any way possible that motion should be on the one hand circular, uniform, and perfect, and that on the other hand the things that are observable should be observed in consequence of it, unless this be accounted for by making use of one of the two principles, or of both of them? This consideration is all the stronger because of the fact that if one accepts everything stated by Ptolemy concerning the epicycle of the moon and its deviation toward a point outside the center of the world and also outside the center of the eccentric circle, it will be found that what is calculated on the hypothesis of the two principles is not at fault by even a minute. . . . This is the true perplexity.[20]

That Maimonides characterizes an astronomical conundrum as the "true perplexity" in his work devoted to defusing perplexities has not escaped scholars.[21] Maimonides responds to this perplexity by defining the function of the astronomer fairly precisely. This characterization forms an extended argument that I have reformulated as follows:[22]

1.1 The purpose of the astronomer is not to "tell us in which way the spheres truly are, but to posit an astronomical system in which it would be possible for the motions to be circular and uniform and to correspond to what is apprehended through sight, regardless of whether or not things are thus in fact."[23]

1.2 Aristotle himself never mentioned the eccentricity of the sun because he did not know about it. Had he known the thesis to be true, he himself "would have become most perplexed about all his assumptions on the subject."

1.3 Maimonides agrees that what Aristotle has to say about sublunar existence "is in accordance with reasoning."

1.4 However, regarding knowledge of superlunar existence, "man grasps nothing but a small measure of what is mathematical; and you know what is in it."[24]

1.5 Maimonides supports this epistemological insight concerning mathematical rigor with "poetical preciousness," that is, with a quotation from Scripture.

1.6 Maimonides then asserts that only God knows the true reality of the heavens while humans do have knowledge of sublunar existence.
1.7 The heavens are "too far away from us and too high in place and in rank" for humans to know.
1.8 Even God's existence cannot be concluded and known on the basis of superlunar events and knowledge.
1.9 To fatigue the mind with matters that "cannot be grasped by them" is a defect in ones inborn disposition.
1.10 Let us therefore leave such matters that "cannot be grasped by reasoning" to one who has been "reached by the divine overflow."
1.11 It is possible that somebody else may "find a demonstration by means of which the true reality of what is obscure for me will become clear to him."
1.12 At this point Maimonides avers that he has not heard "a demonstration as to anything concerning them [these perplexities]."

Argument Summary: In this important argument, Maimonides makes several points: that the function of the astronomer is to convey knowledge of a theoretical system and not empirical reality (1.1); that only God has a complete knowledge of the supralunar spheres (1.6); that God's existence cannot be proved based on supralunar evidence (1.8); and that most (or all?) humans are unable to grasp the knowledge of the heavens with certainty (1.11, 1.12).

Can we conclude on the basis of this extended argument that Maimonides denies that the heavens can be configured by humans? And if so, how does this reading correlate with the implication of the *Mishneh Torah* that the heavens *can* be configured? In other words, we must determine which of these accounts more accurately represents Maimonides' mature understanding of current astronomy. Does the *Mishneh Torah* account summarized earlier represent the absolute codification of physical and metaphysical truth or is it a provisional account of cosmology, subject to change?[25] Kellner has argued that the *Mishneh Torah* "does not represent the most perfect possible exposition of that science available to him and most certainly does not represent the highest stage that astronomy can reach."[26] On this reading the *Mishneh Torah* represents the conventional Ptolemaic wisdom of the time, as contrasted with the anti-Ptolemaic account given in the *Guide.*

Underlying Kellner's query lie two deeper, related issues. The first has to do with Maimonides' attitude toward the alleged incompatibility of astronomical models with Aristotelian physics. For the epicycles utilized by astronomers violate Aristotle's principles that the motions of the heavens be uniform, circular, and around a fixed center. Nevertheless, as Langermann has pointed out, the results they achieve are quite precise.[27] The second issue raises a broader epistemological concern, namely whether Maimonides believes that the heavens can be configured. The description provided in *Mishneh Torah* would imply that human beings can have the sort of astronomical sophistication required to achieve knowledge of the celestial order; on the other hand, (1.1), (1.4), (1.6), (1.7), and (1.11) appear to undercut these implications. My own reading is an amplification of Langermann's point that in contradistinction to implicit suggestions in 2.24, "Maimonides does regard the true configuration of the heavens as something humanly attainable."[28] On this reading, Maimonides' point is epistemological rather than ontological: it is not that the heavenly spheres are per se inaccessible to the human intellect, but rather that nobody yet has determined their true configuration.[29] The key passages for this interpretation are (1.7), (1.9), and (1.11), all of which emphasize the epistemological limits of human intellect: (1.7) tells us that the heavens are ontologically beyond human knowledge; (1.9) warns against pushing human intellect beyond its dispositional limits; and (1.11) suggests that it is not inconceivable that *some* mind may find a demonstration of these matters.

In order to appreciate these points more fully, let us distinguish four different ways in which humans can be said *not to know* an entity X.[30]

2.1 There are certain *per se* features of X such that humans cannot know X.

2.2 There are certain *accidental* features of X such that humans cannot know X.

2.3 There are certain *per se* features of the knower such that s/he cannot know X.

2.4 There are certain *accidental* features of the knower such that s/he cannot know X.

The real question, then, is in which of these senses the heavens can be said to be unknowable. (2.1) and (2.3) support the contention that the heavens are per se unknowable, either because of the essential configuration of the heavens, or because of certain inherent features of all humans *qua* human being. (2.2) and (2.4), on the other hand, emphasize the accidental nature of this lack of knowledge; presumably this lack can be surmounted with requisite training, and so on. When we return to Maimonides' passage, we see that only (1.6) comes close to reflecting (2.3), namely that the heavens are "too far away," and unknowable to humans (but not to God). The very recognition that the heavens are knowable to God suggests that it is only due to human limitation that we don't have full knowledge of the heavens. The other statements all are compatible with (2.2) or (2.4): (1.4) and (1.7) are entirely ambiguous, whereas (1.10), (1.11), and (1.12) clearly are compatible with (2.4). That is, these latter support a reading whereby it is only due to the *accidental* features of *some* knowers that the heavens are not fully known and comprehended. There is nothing in the nature of heavenly configurations per se that precludes their being known. In other words, (2.1) is not satisfied by any of Maimonides' statements, and hence there is nothing in the science of astronomy which is per se beyond human grasp.

Creation Models in Maimonides

Perhaps no topic has received as much attention in recent years as has Maimonides' theory of creation. In part because Maimonides himself encouraged the attribution of ambiguity with respect to this topic, scholars have expended enormous amounts of energy decoding the central passages in the *Guide* that deal with creation. Maimonides characterizes the doctrine of creation as an extremely challenging as well as volatile topic, precisely the sort of issue appropriate to an esoteric presentation. Readers of the *Guide* who turn to chapters 2.13–30, which are devoted to creation, have thus been forewarned by the author to expect a modicum of ambiguity at best, or outright deception at worst.

In *Guide* 2.13 Maimonides describes three opinions on creation, and then in *Guide* 2.32 he describes three opinions on prophecy, stating that "the opinions of people concerning prophecy are like their opinions concerning the eternity of the world or its creation in time."[31] Is the word "like" supposed to posit a one-to-one correspondence between the two sets of opinions? If so, can Maimonides' own position be linked with any one set of correspondences, or is his allegiance split? In answer to these questions, interpreters have suggested every possible combination of opinions, and have offered every possible strategy for determining which is Maimonides' own view.[32] In this chapter I shall not enter the Maimonidean taxonomy controversy per se. My main concern, rather, is to elucidate the theory of temporality that evolves out of his discussion of creation. I shall, in the course of my discussion, however, offer evidence from Maimonides' discussion of temporality to support the contention that Maimonides' doctrine of creation of the world incorporates important elements of Aristotelian eternity.

In arguing thus, I clearly align myself with scholars who see in the *Guide* an esoteric text addressed to the intellectual elite. It is important, however, that we fully appreciate the import of claiming that Maimonides has incorporated an esoteric doctrine into his work. Ravitsky has distinguished at least three ways scholars over the centuries have interpreted and presented Maimonides' esotericism.[33] The works of Samuel ibn Tibbon in the thirteenth century and his followers (characterized by Ravitsky as radical exegetes) emphasized the isomorphism between the science of the Law and Aristotelian philosophy, that is, between Maimonides' understanding of Rabbinic doctrines and Aristotle's physics and metaphysics. On this reading, Maimonides' secret doctrine on creation consists in claiming that both Aristotle and Mosaic law propound the same truth, and that both must be accepted equally. More specifically, the doctrine of Aristotelian eternity can be used to elucidate and uncover Aristotelian wisdom that already appears hidden in Genesis. A more moderate harmonistically-minded camp claimed that whereas in many areas Maimonides adopted a rationalist approach reflective of Aristotle's influence, nevertheless in areas of theological dogma (e.g., creation, divine will, miracles) wherein Aristotle contravened Jewish belief, Maimonides abandoned Aristotle in favor of Mosaic Law. On this reading Maimonides' esotericism consists in accepting Aristotle in some, but not all, areas of intellectual and theological concern, hence giving rise to attempts at harmonization when appropriate. A third camp, consisting primarily of Maimonides' opponents, saw the *Guide* as intentionally attempting to undermine traditional beliefs by propounding a secret doctrine inimical to the Law. On Ravitsky's taxonomy, Strauss and his twentieth-century followers (although not "opponents" of Maimonides) fall into the third camp. Unlike ibn Tibbon, who saw the secret of the *Guide* lying in the close similarity between the Law and Aristotle, Strauss saw the secret as consisting in "the vast discrepancy between Scripture and philosophy, in the profound gap between law and science."[34] On this reading, there exists an unbridgeable gap between the intellectual and the religious, between law and philosophy, between faith and science. To claim therefore that Maimonides adopts an Aristotelian view of eternity is tantamount, on this reading, to claiming that Maimonides has rejected the veracity and authority of Scripture. My own interpretation is that Maimonides recognizes internal difficulties with the view of Scripture having to do with issues of time, but that in the absence of definitive demonstrative argument in support of Aristotle, he is forced to a position of epistemological skepticism. This skepticism, however, represents a tacit rejection of the unwavering faith in scriptural account of creation.

In *Guide* 2.13 Maimonides states the three standard views on creation. The main features of these three views, characterized as the Law of Moses (scriptural), Platonic, and Aristotelian, can be summarized as follows:

> *3.1 The Scriptural view:* that the universe was brought into existence by God after "having been purely and absolutely nonexistent;" through his will and his volition, God brought into "existence out of nothing all the beings as they are, time itself being one of the created things,"[35]
>
> *3.2 The Platonic view:* that inasmuch as even God cannot create matter and form out of absolute nonexistence (since this constitutes an ontological impossibility and so does not impute impotence to God), there "exists a certain

> matter that is eternal as the deity is eternal. . . . He is the cause of its existence . . . and that He creates in it whatever he wishes."[36]
>
> *3.3 The Aristotelian view:* that matter cannot be created from absolute nonexistence, concluding that the heaven is not subject to generation/corruption and that "time and motion are perpetual and everlasting and not subject to generation and passing-away."[37]

Each of these positions carries with it both metaphysical and theological implications. View (3.1) clearly postulates creation after absolute nonexistence (Arabic: *baʿd al-ʿadam al-maḥḍ al-muṭlaq*). In contradistinction to Wolfson, Harvey has pointed out the importance of distinguishing this theory from creation *ex nihilo* (Arabic: *min al-ʿadam*). Whereas creation *ex nihilo* may be interpreted to signify continuous creation out of nonexistence, (3.1) states that before creation there was sheer nonexistence. It is thus incompatible with the eternity of time, whereas creation *ex nihilo* can in theory be rendered compatible with the eternity of time.[38] And so (3.1) incorporates four distinct propositions: that God brought the world into existence after absolute nonexistence; that he did so through his will and volition; that he did so not from anything (Arabic: *lā min shay;* Hebrew*: lo midavar*); and that time is created.[39] View (3.3) can be seen as postulating an eternally beginningless universe. A tantalizing but separate question, to which I shall return at the end of this study, is whether adopting (3.3) would commit Maimonides to the eternity of the universe *a parte post.*[40] Finally, (3.2) postulates both a creator as well as an eternal substance out of which the universe is created. That is, it represents a version of eternal creation, adopting features of both (3.1) and (3.3).

Maimonides specifies several observations concerning the relations among these three characterizations. First, contrary to those who "imagine that our opinion and his [Plato's] opinion are identical,"[41] Maimonides is quick to disabuse those who are tempted to posit a connection between (3.1) and (3.2). The Platonic view, he states, cannot be substituted for Mosaic doctrine, even though there appear to be superficial similarities—most notably the postulation of a creator—between the two. Secondly, Maimonides' attitude toward the relation between (3.2) and (3.3) is ambiguous. He first contrasts them on the grounds that the Platonists believe that the entire heaven is subject to generation and passing-away, whereas the Aristotelians believe that only the sublunar sphere is subject to such generation and passing-away.[42] But he then dismisses [3.2] as not worthy of serious consideration on the grounds that

> [both] believe in eternity; and there is, in our opinion, no difference between those who believe that heaven must of necessity be generated from and pass away into a thing or the belief of Aristotle who believed that it is not subject to generation and corruption.[43]

In other words, he dismisses the original grounds for contrast between (3.2) and (3.3) and then argues that if Aristotle can be refuted, so too can Plato's theory be disqualified as a justifiable creation theory. In short, Maimonides appears to equate the positing of eternal pre-existent matter with the positing of an eternally beginningless universe. Having dismissed (3.2) as a weaker version of (3.3), he argues that the scriptural account is no more flawed than is the Aristotelian account. Then, pointing to the possibility of (3.1), coupled with its Mosaic (and Abrahamic) sanction, Maimonides argues that the very plausibility of Scripture suggests the non-necessity of Aristotle. In other words, Maimonides employs the stance of what Kretzmann has termed the "bold" philosopher by claiming that, on the basis of the veridical nature of Scripture, Aristotle's theory of eternity must be abandoned.

In Chapter 25 Maimonides lays out several pragmatic reasons as well for supporting (3.1) over (3.2) and (3.3). The most important of these is that (3.3) would destroy belief not only in the Law but in miracles and prophecy as well:

> the belief in the way Aristotle sees it—that is, the belief according to which the world exists in virtue of necessity, that no nature changes at all, and that the customary course of events cannot be modified with regard to anything—destroys the Law in its principle, necessarily gives the lie to every miracle, and reduces to inanity all the hopes and threats that the Law has held out.[44]

Maimonides is quick to point out, however, that (3.2) is not nearly as devastating: the opinion of Plato would "not destroy the foundations of the Law and would be followed not by the lie being given to miracles, but by their becoming admissible."[45] Why, then, does Maimonides not accept the authority of (3.2)? The main reason, he tells us, is that the Platonic view has not been demonstrated: "In view of the fact that it has not been demonstrated, we shall not favor this [Plato's] opinion, nor shall we at all heed that other opinion [Aristotle's], but rather shall take the texts according to their external sense."[46] It would appear, then, that (3.1), the scriptural account of creation of the universe out of absolute nonexistence, is Maimonides' final view.

Let us return, however, to Maimonides' original demarcation between an exoteric and esoteric reading of controversial issues. Maimonides has given readers ample ammunition to interpret his espousal of (3.1) as an exoteric ploy and to search for the underlying, or concealed, theory of creation. And as commentators working through the text have demonstrated, textual evidence abounds to support either (3.2) or (3.3) as an alternative expression of Maimonides' esoteric view of creation. As Ravitsky has so aptly demonstrated, these modern readings of Maimonides had already appeared in their myriad forms in the thirteenth to sixteenth century commentators upon the *Guide*, starting with Maimonides' translator Samuel Ibn Tibbon himself.[47] Moreover, some recent scholars have suggested an even more radical reading of the *Guide*, namely that Maimondes did not adopt a definitive position with respect to creation. Klein-Braslavy, for example, following the suggestion of Pines, argues that ultimately Maimonides upholds a skeptical stance in light of the evidence and does not ascribe to any of the three positions.[48] Inasmuch as Maimonides has clearly questioned the demonstrability of each of these views, she suggests that Maimonides' ultimate position is one of epistemological skepticism: the human intellect is simply unable to resolve the issue. Disagreement in matters of metaphysics occurs in cases when demonstrative arguments are not available.[49] So too, when the evidence is conflicting and unsupported by sound Aristotelian demonstration, the only justifiably rational stance, on this reading, is to withhold ones belief until such time as adequate demonstration becomes possible.

The importance of Aristotelian demonstration to this skeptical contention cannot be overemphasized. According to Maimonides, Aristotle did not claim to have a demonstrative proof for the eternity thesis; Aristotle himself considered his proofs in support of eternity to be "mere arguments" as opposed to logical demonstrations.[50] Although he is quick to point out that Aristotle "does not affirm categorically that the arguments he put forward in its favor constitute a demonstration" for the eternity thesis, Maimonides clearly disagrees with the Mutakallimūn who attempted to demonstrate the impossibility of such a claim. Rather, Maimonides states that "it seems that the premise in question is possible—that is, neither necessary . . . nor impossible . . ."[51] Aristotle himself, he points out, only considered his theory to be probable and not necessary: "Now to me it seems that he [Aristotle] does not affirm categorically that the arguments he put forward in its favor constitute a demonstration. The premise in question is rather, in his opinion, the most fitting and the most probable."[52] It is because he was lacking demonstrative arguments that Aristotle had to "buttress his opinion by means of the fact that the physicists who preceded him had the same belief as he."[53] For Maimonides, Aristotle's proofs for eternity constitute not "a cogent demonstration" but rather dialectical arguments, and so cannot be regarded as indubitable support for the eternity thesis.[54] However, it is important to point out that Maimonides has offered no demonstrative arguments in favor of (3.1), the view of Scripture. Demonstration cuts both ways: Maimonides has offered his reader no indubitable foundation upon which to accept either (3.1) or (3.3). Although epistemological skepticism would not be quite as heretical as espousing either (3.2) or (3.3), it nevertheless constitutes a provisional rejection of (3.1), which is tantamount to a rejection of the Mosaic theory found in Scripture.

In what follows I shall argue that, based on considerations of the nature of time, Maimonides recognizes that (3.1) is untenable. He is inclined to accept (3.2) on the grounds that it offers the possibility of reconciling theories of creation and eternity. However, Maimonides has already intimated that a stringent reading of (3.2) is tantamount to an acceptance of (3.3). Although (3.3) would greatly reduce the need for a Creator of the universe, and would eliminate the emphasis upon will and volition, it would accord with Maimonides' own views on time and enable him to reconcile a theory of creation with an Aristotelian theory of time. This position, while similar to (3.2), employs aspects of (3.1) as well; it is closest in temperament to a Neoplatonic version of eternal creation.[55] In support of this reading I shall offer several considerations based on the nature of time and temporality. For as Hyman has pointed out, the atemporality of creation is "one of the main features of Maimonides' account."[56]

The definition of time plays an important role in several contexts. The first has to do with how time and motion can be created independently and draws directly upon Aristotle's theory of time. As we have seen, in both the *Physics* and *Metaphysics* Aristotle develops the notion of the instant (*to nûn*) as a basic feature of time. The instant is defined as the middle point between the beginning and end of time. Since it is a boundary or limit, it has no size and hence cannot be considered to exist. We have also seen that according to Aristotle, inasmuch as time is defined in terms of motion, there can be no time without motion. In the introduction to part 2 of the *Guide*, Maimonides lists twenty-five propositions drawn from Aristotle that purportedly he accepts. That Maimonides is sympathetic to an Aristotelian theory of time is evidenced by the following definition he adopts in the fifteenth proposition:

> (Proposition 15) Time is an accident consequent upon motion and is necessarily attached to it. Neither of them exists without the other. Motion does not exist except in time, and time cannot be conceived by the intellect except together with motion. And all that with regard to which no motion can be found, does not fall under time.[57]

In this formulation, Maimonides is clearly following the Aristotelian definition of time as the "measure of motion." Klein-Braslavy points out that this definition can mean one of two things: it can mean either that time is the measure of all motion whatever, or that time is the measure primarily of the celestial sphere.[58] Maimonides himself is ambiguous as to which usage he endorses. Nevertheless Klein-Braslavy suggests that Maimonides seems to accept the latter reading while not ruling out the possibility of the former. For if, as Maimonides has already stated in proposition fifteen, time is consequent upon motion, then time and moving things must be created simultaneously, since neither has any ontological status without the other. But Maimonides does not posit simultaneous creation. Rather, he suggests that what is moved is itself created and came to be after not having been. This statement suggests that first God created time, and then he created moving things in time. We shall return to this point below, when we examine the initial instant of creation.

A second critical passage is *Guide* 2.17 in which Maimonides wishes to show that Aristotle's arguments for eternity are not demonstrative. The exoteric conclusion of 2.17 is that (3.1) is preferable to (3.3) because Aristotle has offered no acceptable demonstrations for (3.3). But commentators have noted that 2.17 could in fact be read as supporting (3.3). Maimonides ostensibly would like to show that Aristotle's arguments for eternity rest on a fundamental assumption that can be shown to be false. Maimonides' rejection of this assumption is based on the presumed fact that the nature of the world after it exists "does not resemble in anything the state it was in while in the state of being generated."[59] Just as a grown adult does not resemble a fetus, and so inferring from the nature of the adult the nature of the fetus would be absurd, so too Maimonides argues that inferring from the state of the present existence of the world to the initial instance of the world is absurd. Both Ivry and Malino have discussed the obvious disanalogies between the world and a fetus, the most notable being that in the case of a fetus (unlike the world), we can in fact come to know the nature of the fetus observationally.[60] Furthermore, as Malino aptly observes, Maimonides' rejection of this assumption depends upon notions of temporality that are inapplicable when discussing the creation of the world.

A third context occurs in *Guide* 2.18 wherein Maimonides examines three arguments upheld by the followers of Aristotelians who support the doctrine of eternity. These post-Aristotelians first argue that those who claim that God created the world "are obliged to admit that the deity passed from potentiality to actuality inasmuch as he acted at a certain time and did not act at another time."[61] The thrust of the argument depends upon the Aristotelian conception of action as change from potentiality to actuality. If God is construed as Pure Act, however, he cannot be said to act at an instant. Secondly, the post-Aristotelians claim that "eternity is shown to be necessary because there do not subsist for him, may he be exalted, any incentives, supervening accidents, and impediments."[62] This argument actually comprises two main subarguments, which, following Sorabji, we can term versions of the "why not sooner argument" and the "willing a change vs. changing one's will" argument.[63] The second subargument is really a restatement of the first but this time in terms of will. There is an additional subargument that is drawn from the nature of God's wisdom.

Maimonides' response to these arguments draws upon the equivocal nature of God, as well as upon the homonymous nature of divine predicates.

In response to the first argument, he distinguishes two senses of the term "act": only with respect to material beings does "act" imply a move from potentiality to actuality. With God, or an immaterial being, 'act' does not imply such a move, and hence does not imply change. Similarly, his response to both the second and third arguments is to specify the ways in which divine will is unlike human will. With respect to the second, he argues that God has no need of special incentives to will; that is, God's will does not function like human will in that it is not activated at a particular instant. With respect to the third argument, Maimonides' point is that, unlike human acts of willing, when God wills a change there is no change in his willing nature.[64] Clearly, Maimonides has not confronted any of these arguments head on, nor has he challenged the underlying notions of time, temporality and acting at an instant that are assumed by these arguments. He prefers to undermine their underlying presuppositions concerning action and the Deity. His main contention has been that these post-Aristotelian arguments have relied upon a mistaken conception of the divine predicates.

The most pointed discussion of time occurs in *Guide* 2.13, in the context of delineating (3.1). Having stated that (3.1) involves the creation of all existence, including time, "time itself being one of the created things,"[65] Maimonides raises several puzzles concerning creation and time, raising the issue of the relation between God's actions and the domain of temporality. Surely, he claims, no temporal predicates can be used to describe God's activities or nature before the creation, since before creation there is no time:

> Accordingly, ones' saying: God "was" before he created the world—where the word "was" is indicative of time—and similarly all the thoughts that are carried along in the mind regarding the infinite duration of his existence before the creation of the world, are all of them due to a supposition regarding time or to an imagining of time and not due to the true reality of time.[66]

Several points are worth noting in this passage. First, inasmuch as God transcends the temporal sphere and does not operate in a temporal context, Maimonides describes the duration or eternity of the Deity in atemporal terms, so as to preclude any temporal predications of God. To predicate of God infinite duration has no temporal meaning. Secondly, and more important for our purposes, Maimonides' use of the terms "supposition" or "imagining" of time (*demut zeman*) brings to mind his dismissal of the Mutakallimūn on the grounds that they were unable to distinguish between imagination and intellect.[67] Maimonides is suggesting that (3.1) involves one in a crude or vulgar understanding of time based on imagination, one which is contrasted with the "true reality of time." This true reality, of course, is consistent with an Aristotelian theory of time. But as we have seen, an Aristotelian theory of time lends credence to the eternity theory of the universe.

Let us summarize our discussion to this point. With respect to creation, Maimonides maintains that "God's bringing the world into existence does not have a temporal beginning, for time is one of the created things."[68] Maimonides does not want to suggest that time itself is eternal, for "if you affirm as true the existence of time prior to the world, you are necessarily bound to believe in the eternity [of the world]."[69] But neither will he claim that the creation of the world is a temporally specifiable action, for on the Aristotelian definition of time, the world must be beginningless in the sense that it has no temporal beginning. While supporting on an exoteric level the scriptural reading of creation, on an esoteric level Maimonides is suggesting that an Aristotelian theory of time (which he accepts) is more consistent with an eternity model of the universe. This reading is reinforced by Maimonides' analysis of the term *be-re'ishit,* to which we shall return below.

Creation, Time, and the Instant in Gersonides

We have seen that Maimonides tended to compartmentalize physics and astronomy, a major implication being that the astronomer is free to use any mathematical model that best suits his theoretical purposes. Gersonides, however, draws a clear connection between the two. Writing in fourteenth-century France, Gersonides spent several years in the papal court in Avignon, and may at that time have come into contact with the views of Ockham and other fourteenth-century scholastics. His major work *Milḥamot ha-Shem* is a sustained examination of the major philosophical issues of the day: theory of knowledge, divine omniscience and free will, providence and the creation of the universe.[70] As we shall see, inasmuch as Gersonides will want to argue that both time and motion are finite (and created), he must eliminate Aristotle's notion of infinitely extended time altogether.

In contradistinction to Maimonides, for Gersonides natural philosophy and mathematics are mutually reinforcing disciplines which come together in astronomy: "In its perfection the investigation [astronomy] belongs to both sciences—to mathematics because of the geometric proofs, and to natural philosophy because of the physics and philosophical proofs."[71] As Freudenthal has cogently stated, all the sciences for Gersonides ultimately form a coherent whole. Adhering to the ability of human beings to attain to an overarching truth comprising all of reality, Gersonides presents a unified cosmology rooted in a thorough-going epistemological realism.[72]

This realist stance is stated in the context of examining al-Bitrūji's astronomical proposals. Gersonides' contention is that "no argument can nullify the reality that is perceived by the senses, for true opinion must follow reality but reality need not conform to opinion."[73] That Gersonides clearly considered his own observations to be the ultimate test of his system is explicit from his attitude toward Ptolemy. The importance of empirical observation cannot be underestimated, he claims, and he values his own observations over those of others. "We did not find among our predecessors from Ptolemy to the present day observations that are helpful for this investigation except our own,"[74] he says in describing his method of collecting astronomical data. Often his observations do not agree with those of Ptolemy, and in those cases he tells us explicitly that he prefers his own. Gersonides lists the many inaccuracies he has found trying to follow Ptolemy's calculations.[75] Having investigated the positions of the planets, for example, Gersonides encountered "confusion and disorder" that led him to deny several of Ptolemy's planetary principles.[76] He does warn his colleagues, however, to dissent from Ptolemy only after great diligence and scrutiny.

Gersonides' attitude toward previous astronomers, coupled with his faith in human reason, are reflected in his discussion of creation. Maimonides, as we have seen, goes to great length to maintain that the topic of creation is beyond rational demonstration. Gersonides, on the other hand, devotes many chapters in *Milḥamot* 6 to proving that the Platonic theory of creation out of an eternal formless matter is rationally demonstrable. Further, the two disagree over the relation between the superlunar and sublunar spheres. As we have seen, Maimonides has claimed that no valid inference can be drawn from the nature of the sublunar sphere to that of the super-lunar sphere. Gersonides, however, rejects the metaphysical bite to the distinction and argues that inasmuch as both spheres contain material elements, what we know about creation is based on astronomy, and astronomy is fundamentally no different a human science than physics.[77] Astronomy can only be pursued as a science by "one who is both a mathematician and a natural philosopher, for he can be aided by both of these sciences and take from them whatever is needed to perfect his work."[78] Gersonides sees the ultimate function of astronomy to understand God. Astronomy, he tells us, is instructive not only by virtue of its exalted subject matter, but also because of its utility to the other sciences. By studying the orbs and stars, we are led ineluctably to a fuller knowledge and appreciation of God.[79]

Like Maimonides, Gersonides is concerned with whether time is finite or infinite, as well as with whether the creation of the world can be said to have occurred at an instant. Unlike Maimonides, however, Gersonides' discussion includes the physical manifestation of time and matter, as well as the theoretical implications of temporality. Having posited that the world was created at an initial instant of time by a freely willing agent, Gersonides must decide whether the world was engendered out of absolute nothing or out of a pre-existent matter. His examination, therefore, focuses upon the concept of matter underlying creation.

In *Milḥamot* 6.1.2 Gersonides lists three views of his predecessors who discussed the creation of the world. The first, that the world comes into existence and passes away an infinite number of times, has been associated with the rabbis as well as with certain ancient philosophers.[80] The second view, that the world was generated only one time, is associated with two sets of proponents: first is the version of Plato that the world was created one time out of some thing (*nitḥadesh mi-davar*), and second is the view attributed to the Kalām and to Maimonides, that the world was created out of absolute nothing (*nitḥadesh mi-lo' davar be-muḥlat*).[81] The third view is the eternity thesis of Aristotle, that the world is eternal and hence has not been created.

Gersonides' critical refutation of Aristotle's eternity thesis introduces the motif of time and its relation to motion. In contradistinction to Aristotle who postulated the eternity of time and motion, Gersonides will insist that both time and motion are finite. Gersonides hopes to refute Aristotle's eternity of the world by showing that the infinity of time and motion fail as exceptions to Aristotle's own finitistic universe.[82] According to Gersonides, Aristotle offered at least nine arguments in support of the eternity thesis: of these, three have to do with temporality. Aristotle's first argument has to do with the nature of time in general, the second is based on the nature of the "instant," and the third is based on the nature of temporal language. Gersonides' statement of Aristotle's first argument can be summarized as follows: If time came to be, it would have come to be in time. But this would imply a time before the original time. Since time is inseparable from motion and motion is connected to the moved object that moves in a single, continuously circular motion, it follows that time too is continuous. Hence time must be eternal.[83] This argument is based on those aforementioned passages in which Aristotle argues that since time is defined in terms of motion, there can be no time without motion.[84]

Aristotle's second argument, as stated by Gersonides, is based on his definition of the instant as the middle point between the "before" (*ha-qodem*) and "after" (*ha-mit'aḥer*) and goes as follows: If time came to be, there would have to be an actual instant (*'atah*) at which it came to be. But this would entail there being a potential instant before the present instant was actualized. But every part of time has only potential existence, and so no such instant could exist. Hence time could not come to be.[85] The main thrust of this argument, as presented by Gersonides, is that in order to account for the coming into existence of any present instant, there must exist a prior actual instant; but in the case of the first instant, there could be no prior instant, actual or potential. The third argument, based upon the second, takes into consideration the nature of temporal language. Whoever tries to speak about the coming into existence of time must utilize words that imply temporality before the existence of time itself. But these references would imply that time exists prior to its coming into existence. Time must therefore be eternal.[86]

Other arguments in favor of eternity are taken from Aristotle and center for the most part on the nature of motion, the impossibility of a first motion, the incorruptibility of the heavens, and the impossibility of a void.[87] Gersonides concludes his summary of Aristotle's arguments with two general comments which link the metaphysical considerations to those of a more theological nature. Gersonides offers the suggestion that ultimately what may have motivated Aristotle to support the eternity thesis were theological considerations based on the nature of the deity. First, echoing Maimonides' "Why Not Sooner Argument" mentioned above, he argues that it would be inappropriate to suggest that the deity causes at one time rather than at another. Furthermore, it is not appropriate that the deity exist independently of the world, which functions as the object of God's self-conception. And finally, Gersonides reminds us, as did Maimonides, that Aristotle himself did not regard his arguments in favor of eternity as demonstrations (*lo' ba' 'aleihem mofet*) but rather as containing fewer doubts than other arguments.[88]

In order to reject Aristotle's eternity thesis, Gersonides must demonstrate the finitude of time. To this end he first makes a number of observations pertaining to the general characteristics of time that will affect his argument. Time, Gersonides argues, falls in the category of continuous quantity. We speak, for example, of the parts of time as being equal or unequal; time itself is measured by convention (*be-hanaḥa*) as opposed to by nature; and its limit is the "instant," which itself is indivisible.[89] Echoing Aristotle, Gersonides points out that time cannot be comprised of "instants" because the instant measures time, but is not a part of time. Unlike time which is divisible, the instant is indivisible.[90] Further, Gersonides claims that time can be construed both as separate from its substratum and as residing in it. That time resides in its substratum is demonstrated from the fact that it has distinguishable parts: that is, present time is distinguished from both past and future time. Were past, present and future not distinguishable, argues Gersonides, then any part of time would equal the whole of time. But this is not possible. However, time must also be separate from its substratum. For if it were identical to its substratum, there would be as many times as there are substrata. But we know that there is only one mode of time, and not a multiplicity of times. Hence time is separate from what it measures.[91]

According to Gersonides time is partly potential and partly actual. Aristotle had argued that the past, in being a potency, was infinite. Gersonides however claims that potency refers only to the future and not to the past.[92] If the past were potential, then, Gersonides argues, contrary possibilities would inhere in the past as well as in the future; this however is absurd, since we know that the past has already occurred.[93] Hence, Gersonides concludes that only future time carries within itself potency. As we will discuss further in the next chapter, these comments are consistent with his statements elsewhere regarding the nature of future contingents.[94]

Having laid out these general characteristics of time, Gersonides now demonstrates that time must have been generated. We have seen that time is contained in the category of quantity. Gersonides will argue that just as quantity is finite, so too is time. A number of arguments are based on the nature of body, the spheres, and the regularity of the eclipse.[95] These arguments reflect the influence of Islamic Kalām thought and center on the Aristotelian principle that no infinite can be greater than another.[96]

Gersonides does, however, base a number of arguments on his concept of time. One argument for the finitude of time utilizes the nature of the "when" (*matai*) and can be restated as follows:

- 4.1 No "when" is infinite, since every part of time is finite and the "when" measures the finite.
- 4.2 Since no "when" is infinite, no part of past time is an infinite distance to the present now.
- 4.3 But if time were infinite, given any past "when," the relation of the time before it to the time between it and the now would be the relation of infinite to finite.
- 4.4 But time is homogeneous; that is, all its parts have the same ontological composition.
- 4.5 So (4.3) is impossible, since it implies that time is both finite and infinite.
- 4.6 Hence time must be finite.[97]

The main point of this argument is to show that time must be finite (and not infinite) on the grounds that no one part of time (or "when") is itself infinite. It should be noted that in this argument Gersonides uses the notion of a part of time to correspond to the Aristotelian sense of a "span." Medieval Aristotelians generally reflected Aristotle's distinction in the *Categories* between two senses of the term "part:" the first is an interval of time (i.e., instant), whereas the second is a stretch of time (i.e., span).[98] Gersonides uses the second of these two meanings in his argument.

Additional arguments are based on the nature of the past and future. If past time were infinite, then the motion it has measured would be actually infinite. But inasmuch as one infinite cannot be greater than another, the motion which measures the present could not exceed that infinite past motion. Hence time would exist without motion, which is not possible.[99] Furthermore, we have already seen that potency resides in the future alone. Gersonides now argues that if time were infinite, it would have to occur with respect to the future because only the future is potential. But the future cannot become infinite in quantity. And if the future cannot become infinite, the past can surely not become infinite. And so Gersonides concludes that time must be finite.[100]

Having shown to his satisfaction that time is finite, Gersonides must now refute the original arguments proffered by Aristotle in support of the infinity of time. Aristotle's first argument was that since all generation must take place in time, there can be no beginning to time. Hence phrases like "beginning of time" have no referent. Gersonides, however, refutes this argument by distinguishing two types of generation. The first he bases on Aristotle's notion of a change from contrary to contrary and takes place in time. The second, however, is what Gersonides terms absolute generation and is atemporal; that is, it is instantaneous and does not take place in time. It is in this second sense that Gersonides argues that time was generated. Before this absolute generation, there was no time. Phrases such as "beginning of time," Gersonides argues, must be understood in an equivocal sense.[101]

Aristotle's second objection centered around the notion of the instant as the limit between the past and future. In answer to this argument, Gersonides makes a number of points. To Aristotle's objection that we cannot imagine an "instant" before which there is no time, Gersonides claims that there are many truths that we cannot imagine (just as many imaginable things are not true). Reminiscent of the Kalām controversy over the doctrine of admissibility, Gersonides' point is that the non-imaginability of a claim is not a sufficient condition for rejecting its truth.[102]

A final argument concerns the potentiality of the instant. Gersonides attributes to Aristotle the idea that the instant can only have potential existence since it has only a hypothetical beginning and end. Because the instant divides time hypothetically, it turns out that past and future time are relative given their position with respect to other times. Gersonides however claims that the instant is only potential in the sense that it does not exist in the present; when I allude to the present instant it has already receded into the past. The instant is not potential in the sense of a point, namely, that it is simultaneously both the beginning and end of the measure in question. Hence we can posit the potential instant as the beginning of time.[103]

But Gersonides' rejection of Aristotle's eternity thesis, and his support of creation, do not commit Gersonides to a theory of creation *ex nihilo*. Arguing that creation out of nothing is incompatible with the facts of physical reality, Gersonides adopts a version of the second view, adopting a Platonic model of matter drawn ultimately from the *Timaeus*. The opening verses of Genesis 1 are used to distinguish two types of material reality: *geshem* and *ḥomer ri'shon*.[104] Totally devoid of form, *geshem* is the primordial matter out of which the universe was created. Since it is not informed, it is not capable of motion or rest; and since it is characterized by negation, *geshem* is inert and chaotic.[105] This primordial matter is identified with the "primeval waters" described in Genesis 1.2 *(tohu, tehom* and *mayim)*. However, Gersonides points out that *geshem* does not itself exemplify absolute nonbeing, but rather is an intermediary between being and nonbeing.[106]

In contrast to *geshem, ḥomer ri'shon* is the second type of reality. *Ḥomer ri'shon* is understood in the Aristotelian sense as a substratum that is allied to form. *Ḥomer ri'shon,* or matter, is inferior to form and hence cannot be known in itself. It contains within itself the potentiality to receive forms, yet has no actuality of its own.[107] Inasmuch as it does not contain its own actuality, *ḥomer ri'shon* is not an ontologically independent entity. Gersonides refers to this *ḥomer ri'shon* as "the matter that does not keep its shape (*ha-geshem ha-bilti shomer temunato*)."[108] In *Milḥamot* 6.2.7 Gersonides compares this matter to darkness, for just as darkness is the absence of light, so too this matter represents the absence of form or shape.

This cosmological principle plays a crucial role in Gersonides' astronomy. In an important study of the chronology of Gersonides' works, Glasner has argued that the notion of a "matter that does not keep its shape" is essential for Gersonides' "eccentric cosmology" in a number of ways: it explains why the movements of the different planets do not interfere with each other; it solves the problem of the contact between an eccentric orb and the orbs above or below it; and it solves the problem of the center of motion.[109] More specifically, Gersonides agrees with his predecessors that each of the astral bodies occupies its own celestial sphere, with the stars all residing in one sphere.[110] He further rejects as untenable the view that there exists a diurnal sphere deprived of stars which carries the world in one simple movement.[111] But how do we account for the interaction between these spheres? The celestial spheres are concentric and contiguous, and in principle we might expect that they move in unison. However, as Gersonides has already pointed out, each has its own proper movement, and often these movements are in conflict. In order to account for the independence of each sphere, Gersonides argues that between each sphere there exists a fluid body that determines the movements of the spheres.[112] This body, uninformed and without movement of its own, is described as "the matter that does not keep its shape" and lies between the spheres of the astral bodies.[113] Its function is to assure the motion of the celestial bodies and precludes the possibility of one planetary motion affecting another: "it is clearly appropriate that there be [enough] fluid between the spheres of one planet and the spheres of another such that a motionless layer may remain in its midst to make sure that the motions are not confused."[114] Gersonides compares this diaphanous body to the bones of an organic animal that make possible movement of the animal.[115] In *Milḥamot* 5.1.130–31 Gersonides tries to compute the thickness of these fluid layers.[116]

Gersonides' major objection, however, centers on Aristotle's formulation of the notion of the instant. In contradistinction to Aristotle, Gersonides distinguishes two roles of the instant: an initial instant that does not yet constitute time, and subsequent instants that demarcate "before" from "after." According to Gersonides, these two notions of the instant serve different functions. The first delimits a particular portion of time, namely continuous quantity, and is characterized in terms of duration. The latter, on the other hand, reflects the Aristotelian function of the instant as characterizing division. Gersonides claims that if there were no difference between these two functions of the instant, we could not distinguish between any two sets of fractions of time, for example three hours and three days, because our measure of the two sets would be identical. Since each period of time would be divided by the same kind of instant, there would be no way of distinguishing three days from three hours.[117] On the basis of distinction Gersonides therefore defines time as "the measure of motion (*ha-zeman hu' ma'aseh ha-tenu'ah bi-khelallah*) as a whole according to the instants which form the boundaries of motion but not according to the instants which only distinguish the before from the after."[118]

Figure 2.1 reflects what Gersonides has in mind. If we construe the instant in the second sense, then the intervals on AB would be identical to those on CD and EF, since all we have to go by are the number of instants and their corresponding intervals. Further, the instants on AB would be interchangeable with those on CD and EF. But we know that a year is not equivalent to a day or an hour. Gersonides points out that, technically speaking, I_1 on AB is not identical to that on CD, but rather takes on the characteristics of its temporal span. Understanding the instant in the first sense, then, allows us to see that the duration of these sets of intervals differs in each case.

But how does this distinction resolve the original problem? Gersonides' point is that Aristotle's original objections to the finitude of time obtain only if the instant is understood in the second sense. When the instant is taken as an initial instant of a temporal span, we see that there can

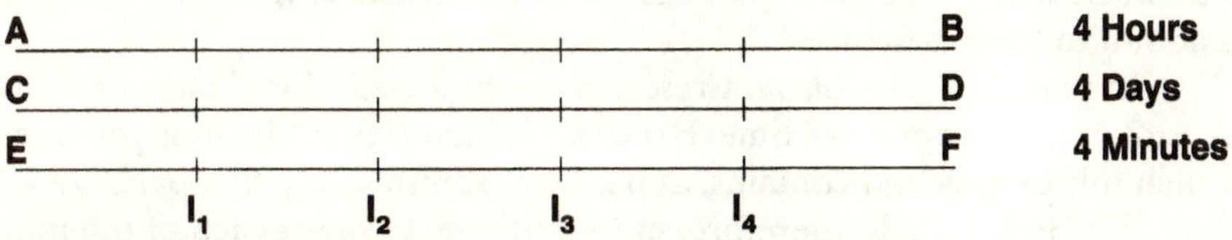

Figure 2.1. Intervals of the Instant

be a "first instant" without contradiction. Gersonides' point is that the instant taken in the sense of duration need not be preceded by a past time.[119] Hence his philosophical arguments lend credence to the theological position.

Creation, Time, and Duration in Crescas

Gersonides has presented an ambitious account of the finitude of time, one which attempts to refute the eternity thesis while at the same retain a sense of "initial instant" that remains true to Aristotle. Let us turn now to two critics of Gersonides, namely Crescas and Albo. Crescas' characterization of time occurs in *'Or 'Adonai,* in the context of elaborating Maimonides' summary of Aristotle's twenty-five metaphysical propositions.[120] Crescas first summarizes arguments on behalf of each proposition, and then critically assesses these arguments. Crescas' purpose in this exercise is to restore faith in those Jewish tenets which have been eroded by Aristotelian theory.

Proposition fifteen, as was introduced and discussed already by Maimonides, pertains to time and is summarized by Crescas as follows:

> Proof of the fifteenth proposition which reads: "Time is an accident that is consequent on motion and is conjoined with it. Neither one of them exists without the other. Motion does not exist except in time, and time cannot be conceived except with motion, and whatsoever is not in motion does not fall under the category of time."[121]

This statement, which is taken directly from Maimonides, is then contrasted with Aristotle's own definition: "Aristotle defines time as the number of priority and posteriority of motion."[122] According to Crescas, Maimonides' restatement of Aristotle contains four premises: that time is an accident; that time is conjoined with motion in such a manner that neither one exists without the other; that time cannot be conceived except with motion; and that whatever is not in motion does not fall under the category of time. Crescas then summarizes how each of these premises is proved.[123] His argument follows for the most part that of Aristotle, and is based on the notion that time is divisible into past and future parts by the present instant; inasmuch as none of these parts has present existence, time as a whole is said not to have existence.[124]

In part 2 of *'Or 'Adonai* Crescas turns to a critical evaluation of this Aristotelian conception of time. His own position is that "the four premises which this proposition contains, as has been shown in the first part, are all false."[125] He proceeds, therefore, in his critique, to reject each of the four arguments. This rejection will enable him to replace Aristotle's definition of time with his own. Crescas' own characterization of time is intimated in the context of his critique of the first premise. Rather than agreeing with Aristotle that time measures primarily motion, Crescas wants to claim that time can measure rest as well. His contention is based on the following argument:

5.1 Rest can be described as long when an object remains at rest for a long time, and short when it remains so for a short time.

5.2 Hence time is measured by rest without the presence of actual motion.

5.3 Even we allow the possibility of potential motion in the characterization of rest (that is "we measure rest only by supposing a corresponding measure of the motion of an object moved during the same interval"), it still follows that actual motion is not necessary in the conception of time.

5.4 Since rest can be characterized as long or short, it follows that time can be measured by rest alone, without a corresponding motion.[126]

On the basis of this argument, Crescas will want to make several points. The first is that time can measure rest as well as motion. Secondly, time can be measured by rest as well as by motion. And finally, time exists only in the soul. The first two points are captured in Crescas' revised definition of time: "the correct definition of time is that it is the measure of the continuity of motion or of rest (*she'ur hitdabequt ha-tenu'ah 'o ha-menuḥah*) between two instants."[127] In this definition Crescas retains Aristotle's and Maimonides' notion of time as a "measure" or "number." However, Crescas adds the important qualification that time is the measure not only of motion or change, but of rest as well. We should remember in this context that although Aristotle did allow that time could measure rest, he did not amplify this suggestion. By *hitdabequt* Crescas means to emphasize that time is not to be identified with physical motion or bodies, but with the duration of the life of the thinking soul. Wolfson suggests that Crescas in effect combines two characteristics in his definition of time: the notion of pure duration that is independent of motion, and the more conventionally Aristotelian notion of time as a measure of motion between instants.[128] In this way Crescas is able to account for something's being in time yet not capable of motion.

Crescas then goes on to say that the genus most appropriate to time is magnitude. Inasmuch as time belongs to continuous quantity and number to discrete quantity, if we describe time as number, we describe it by a genus that is not essential to it. Time is "indeed measured by both motion and rest, because it is our conception (*tziyurenu*) of the measure of their continuity that is time."[129] On this basis Crescas concludes that "the existence of time is only in the soul."[130] It is because humans have a mental conception of this measure that time even exists. The continuity of time depends only upon a thinking mind, and is indefinite, becoming definite only by being measured by motion. Were we not to conceive of it, there would be no time. Hearkening back to Aristotle's original query, whether time would exist if there were no souls, it is clear that Crescas' answer is negative: without human cognizing, there can be no time.

It is in this context that Crescas comes closest to reflecting his near scholastic contemporaries Peter Aureol and William Ockham. According to Peter Aureol, for example, time exists only in the mind.[131] And William Ockham develops an even more subjective view, according to which time is a "cosmic clock" which measures the duration of temporal events and things. Like Crescas who denies the real existence of time as an accident of substance, Ockham claims that time and instants of time are not really existent Aristotelian accidents.[132]

Returning, finally, to the original four premises upon which Aristotle's conception of time was based, Crescas is able to refute all four. The first premise, that time is an accident, is true only if we mean by it that time is not a substance; it should not be taken to mean, however, that time is an accident outside of the soul:

> for time depends as much upon rest as upon motion, and rest is the privation of motion and privation has no existence. It thus follows that time depends upon our supposition of the measure of the duration of either motion or rest inasmuch as either of them may be described as great and small.[133]

With respect to the second premise, he claims that time may exist without motion, i.e. that time is measured by rest or by the supposition of motion without its actual existence. With respect to premise three, he again claims that "to say that the idea of time cannot be conceived except if it be connected with motion must be denied."[134] And finally, with respect to the last premise, Crescas wants to argue that "the Intelligences, though immovable, may still have existence in time, inasmuch as it can be demonstrated that time existed prior to their creation on the ground that time does not require the actual existence of motion, but only the supposition of the measure of motion or rest."[135] That is, Crescas argues that on his theory, creation and time are two separate issues. Inasmuch as time is separate from motion, time can exist prior to the actual existence of motion.

The implications of Crescas' theory of time are apparent when we turn briefly to his discussion of creation in *'Or 'Adonai* 3.1. Without entering into the intricacies of this technical discussion, several important points can be made. Crescas takes as his point of departure the doctrine of creation *ex nihilo*. First he summarizes the three arguments Gersonides gave in behalf of his thesis that the world was created out of a pre-existent matter. In direct response to Gersonides, Crescas is adamant that both the matter and form of the universe derive directly from God.[136] Unlike for his predecessors, however, creation *ex nihilo* for Crescas is a nontemporal concept. Crescas tries to show that eternal creation is a plausible doctrine even in the context of creation *ex nihilo* by exploring the notion of divine omnipotence. God's power is infinite in the sense that God's acts are not temporally limited. Inasmuch as God acts under no constraints, when God creates the world, he is able to create something that is infinite in duration, or eternal. It is in this sense that Crescas claims that the world is both eternal and created.[137]

Crescas first summarizes various arguments that were given in support of eternal creation. The first argument reiterates the connection between time and motion, and states that if time exists it must be eternal, for otherwise it would be existent prior to its existence as the measure of motion, which is absurd.[138] The second argument pertains to the nature of the instant (*'ata*). It claims that if time came into existence, it would follow that the instant would have no "before," which is an absurdity. Just as a point divides a line into "prior" and "posterior," so too the instant divides time into "before" and "after."[139] In response to this argument, Crescas follows the precedent of Gersonides, who distinguished two types of instant. Not every instant divides past from future; just as a point can serve as the beginning of a line, Crescas argues that an "initial instant" can serve as the absolute beginning of time without implying a prior temporal unit. On the basis of this distinction, Crescas is able to posit an initial instant that marks the creation of time. Crescas continues and claims that "God created and brought forth the universe at a definite instant (*be-'et yedu'ah*)."[140] That is, that the universe has a temporal beginning.

Although Crescas posits an initial instant to creation, this does not mean that he rejects the doctrine of eternity altogether. In 3.4 he rejects Maimonides' contention that the world has a temporal beginning, claiming that it is based on the mistaken Aristotelian equation of time and motion. Because he has already abandoned this Aristotelian conception, Crescas is able to argue that the notion of "creation" of the world does not refer to a temporal beginning. Rather, for Crescas, the world is both eternal and created: because time and motion are not interconnected, Crescas is able to adopt a position which on Aristotelian grounds appears to be self-contradictory.[141] Scholars have worked hard to offer an interpretation of Crescas that preserves both. Feldman, for example, offers an alternative reading of Crescas that interprets eternal creation as "the continuous and limitless creation by God of an infinite series of worlds, each of finite duration. Thus interpreted, eternal creation implies eternal creativity: it is not the product that is eternal but the activity of creation."[142] We are not to understand creation as a single event in which time begins "after" not exsiting. Eternity does not pertain to temporality but to the neverendingness of time. The world is both eternal and created.[143]

The Subjectivity of Time according to Albo

The author of *Sefer ha-ʿIkkarim* [*The Book of Principles*], Albo incorporated Maimonides' discussion of pre-existent matter into his own examination of creation, which is couched in the context of developing a theory of time. He is one of the first Jewish philosophers to espouse the view that time is a phenomenon of the imagination, a motif introduced by Crescas and recurring in Spinoza. Albo's discussion of time occurs in the context of demonstrating that God is independent of time. For Albo, God's independence of time comprises both eternity (*ha-qadmut*) and perpetuity (*ha-nitzḥiyut*) and is upheld as a basic principle:[144]

> The third dogma is that God is independent of time. This means that God existed before time, and will exist after time ceases, therefore his power is infinite. For everyone who is dependent upon time is necessarily limited in power, which ends with time. Since, therefore God is not dependent upon time, his power is infinite.[145]

That God is prior to all existing things, including time, was a commonplace of Aristotelian thought. But inasmuch as this view was not held by his teacher Crescas (who argued that time is independent of the physical world and had existed prior to it), so Albo must be ready to explain to Crescas' followers his own position vis-a-vis primordial time.[146] By God's priority Albo means that nothing was prior to God, not even non-existence; God has always existed "in the same way without change."[147] Similarly God's eternality means that nothing is posterior to God, not even time. For if time outlasted God either *a parte ante* or *a parte post,* then God would exist at one instant of time and not at another; this, of course, would undermine God's necessary existence.

These comments lead Albo to examine the nature of time and creation more closely. God's eternality holds, he claims,

> even if by time we mean unmeasured duration (*ha-meshekh ha-bilti meshoʿar*) conceived only in thought, existing always, both before the creation of the world and after its cessation, but without the order apparent from the motion of the sphere, since the sphere was then neither in motion nor existent.[148]

Only measured time cannot exist without motion. Time itself, according to Albo is not dependent upon motion and even pre-existed the world. This non-Aristotelian motif is developed more fully. In another context Albo compares the commandments to time inasmuch as both time and commandments are not actual existents (*bilti nimtzaʾim be-foʿal*):

> [just as] time is not an actual existent, for the past is no longer here, the future is not yet, and the present is merely the now which binds the past to the future. The now itself is not real time (*zeman ʿal derekh ha-ʾemet*), since it is not divisible, whereas time is divisible, pertaining as it does to continuous quantity (*ha-kamah ha-mitdabeq*). The now is related to time as the point is related to the line. Time is therefore not an actual existent, and yet it gives perfection of existence to all things existing in time.[149]

Albo then distinguishes between "plain time" and "the order of time" as follows:

> Our Rabbis are of the opinion that time in the abstract (*ha-zeman be-shiluaḥ*) is such a duration. Time measured or numbered through the motion of the sphere they call "order of times" (*seder zemanim*), not simply time (*zeman stam*). According to this there are two species of time, the one is numbered and measured by the motion of the sphere, to which are applicable the terms prior and posterior, equal and unequal. The other is not numbered or measured but is a duration (*hemshekh*) existing prior to the sphere, to which the words equal and unequal do not apply.[150]

Whereas plain time is neither numbered nor measured, the order of times is numbered and measured by the motion of the diurnal sphere. In contrast to ordered time, plain time is eternal duration. Albo then raises two perplexities pertaining to time. The first puzzle is whether time originates in time or not. The solution is that although time has no origin and does not come to be in time, the "order of time" originates in time.[151] The second puzzle concerns the instant: "The now (*ha-ʿatah*), it is said, divides the past from the future. There is therefore a time before the first now, and hence time and the sphere are eternal."[152] Albo's answer, relying on his twofold notion of time, is that Aristotle's argument refers only to the "order of times" and not to "plain time": plain time in which there is no motion "has not the elements prior and posterior, and it is not subject to measure because measure cannot apply to time without motion. The terms prior and posterior apply to it [plain time] only figuratively and loosely."[153]

Scripture, Philosophy, and the First Instant of Creation

Let me now turn to the theological implications of these philosophical discussions. I suggested in the introduction to this chapter that the phrase *be-reiʾshit* designates the fact that there was a beginning. But positing a beginning instant carries with it assumptions about what preceded this instant, and how this instant is reflected in Scripture. The theological implications of these issues become apparent when we turn to the following text from *Genesis Rabbah.*

> ["And there was evening and there was morning," (Gen. 1.3)] Said R. Judah b. R. Simon, "'Let there be evening' is not what is written here, but rather 'and there was evening.' On the basis of that formulation we learn that the sequence of time had already been laid out." Said R. Abbahu, "on the basis of that same formulation we learn that the Holy One, Blessed be he, had been engaged in creating worlds and destroying them prior to the moment at which he created this one." Then he said, "This is the one that pleases me, but those did not please me."[154]

This paragraph became the focus of profound philosophical speculation among Jewish philosophers wishing to ground Scripture in scientific legitimacy. A number of pressing questions emerged from this passage: First, how does ones' understanding of Rabbi Judah's phrase *seder zemanim* affect ones' interpretation of the temporality of creation in Genesis 1.1? Second, what is the status of the existence of time before day four when the temporal markers were created? And finally, how is time measured in these first three days of creation? In short, these questions crystallize the attempt of Jewish philosophers to reconcile rabbinic dicta and sentiments with an Aristotelian theory of time.[155] Inasmuch as their interpretations of Rabbi Judah's words serve as a revealing litmus test for the theological implications of their metaphysics of time, let me end this chapter with a brief examination of their discussions.

The first question is the subject of Maimonides' commentary on the phrase *be-rei'shit,* the first words of Genesis 1.1. Why does Scripture start with the phrase "In the beginning God created . . ."? In order to explain the sense of "beginning" being used in this context, Maimonides turns in Guide 2.30 to an interpretation of the two terms *teḥilah* and *re'shit,* both of which can mean "start" or "beginning." Maimonides distinguishes between the two on the basis of causal priority.[156] An event *E* can specify an event/state of affairs *A* as being causally prior to *B* in one of two ways: first, When *A* is a part of *B,* and second, when *A* is not a part of *B* but rather appears simultaneously with it. In both cases the term *teḥilah,* or causal beginning, can be used. In contradistinction, the term *re'shit* refers not to a temporal priority of *A* to *B,* but rather to its ontological genesis.[157] On this basis Maimonides is able to allow for an interpretation of the phrase *be-rei'shit* in such a way as to accord with eternal creation. The proposition *bet* in the phrase *be-rei'shit* is not, on this reading, a temporal indicator, but rather fixes the event in question ontologically: it refers not to a temporal beginning but to an underlying ontological state. So that when we read the statement in Genesis 1.1 ("*be-*

rei'shit bara' 'Elohim," "in the beginning God created"), we should understand it to describe a nontemporal event, one which specifies that God is the creator of the universe, that is, its ontological ground of Being, or what Harvey describes as the continuous ontic dependence of creation on the creator. On this interpretation, *be-rei'shit* refers to a "principle" in the sense of the Greek *archē.*[158]

Having postulated the nontemporal mode of creation, Maimonides is able to interpret those Rabbis who understood the Creation account in Genesis to postulate a domain of temporality before the creation event. That is, in order to explain how there can be "one day," at the beginning of creation, when the temporal indicators, that is, the sun and moon, were not created until the fourth day, Maimonides quotes both Judah ben Simon and Abahu to support his contention that "time existed prior to the existence of this sun."[159] Recognizing that their statements support an eternity thesis, Maimonides adopts two separate strategies. The first is simply to recognize that their comments imply that "the order of time necessarily exists eternally *a parte ante.* That, however, is the belief in the eternity *a parte ante* of the world, and all who adhere to the Law should reject it."[160] In other words, one strategy is simply to recognize that these sages were supporting a version of eternal creation and hence to reject their interpretation.

Maimonides' second strategy is to subsume their comments as corollaries of those of Rabbi Eliezer. In 2.13 Maimonides refers to Rabbi Eliezer, whose commentary on creation postulates creation by means of pre-existent matter. Maimonides depicts this commentary as admitting "the eternity of the world, if only as it is conceived according to Plato's opinion."[161] Interestingly enough, Maimonides does not respond directly to Eliezer's statement; his only response to it is to claim that it may "confuse very much indeed the belief of a learned man who adheres to the Law. No persuasive figurative interpretation with regard to it has become clear to me."[162] Uttered by an individual who is generally not at a loss for interpretative prowess, for whom the "gates of interpretation" are rarely if ever closed, Maimonides' stance suggests that he himself is not as uncomfortable with Eliezer's statements as one might expect.[163] If so, then Maimonides' second strategy with respect to Rabbis Judah and Abahu is similar to his attitude toward Eliezer. Maimonides claims that their comments are "only the counterpart of the passage in which Rabbi Eliezer says, 'Wherefrom were the heavens created.' "[164] Inasmuch as Maimonides is not bothered by the latter, so too can it be inferred that he is not bothered by the former. Note that Maimonides seemingly discredits these rabbinic comments altogether by questioning the authority of the speakers: "To sum up: you should not, in considering these points, take into account the statements made by this or that one."[165] Here Maimonides seems to be suggesting that in considering the issues of time and creation, one ought not be misled by the opinions of sundry Rabbis. If this is so, what sense, then, should we make of Maimonides' overt espousal of (3.1) on the basis of Mosaic authority? At least (3.2) and (3.3) have other considerations in its favor. But if the sole basis for (3.1) is authority, then Maimonides seems to be undermining its very plausibility.

In an apparent attempt to deflect the strength of the eternity thesis, Maimonides suggests that the worry underlying Rabbi Judah's passage really can be understood in terms of the metaphysics of time, namely the difficulty that "the notion that time existed prior to the existence of this sun."[166] In order to account for this worry, Maimonides gives an allegorical explanation, suggesting that the term "heavens" (*shamayim*) is sometimes called "firmament" (*raqiʿa*) which comprises the spheres and the celestial lighters. On this interpretation the luminaries were created with the entire *shamayim* on the first day but not suspended until the fourth day. The stars as well as the sun and moon are situated within the firmament of the heaven. Presumably, then, days one, two and three could be measured by these temporal markers. Technically, however, Maimonides interprets the creative act as a simultaneous creation of heaven and earth, and so the demarcation of days subsequent to creation represents a gradual unfolding of differentiation within creation.[167]

Gersonides deals with Rabbi Judah's dicta both in *Milḥamot* 6.2.8, and in his Torah commentary. Gersonides first turns to the term *be-rei'shit,* which is seen to have many meanings, only one of which refers specifically to temporality. *Be-rei'shit* can mean priority in order, priority in degree, or priority in cause and in nature.[168] Accepting the precept that time existed from the instant of creation and not before—"the coming to be of the world from the Lord, may he be blessed, was without time (*be-zulat zeman*)"[169]—Gersonides reflects Maimonides and states that everything described in the first six days of creation was "actually created together," that is, simultaneously.[170] On Gersonides' interpretation, the term "day" does not reflect temporal priority but rather causal priority: the term "day" is used "to indicate the priority of some existent things over others."[171] Gersonides then reiterates the point that, inasmuch as all things were created simultaneously, the supralunar spheres were actually created on the first day, and so their light was available to account for the demarcation of days one through three.[172]

In part because they reject the Aristotelian equation of time and motion, Crescas and Albo have less trouble accounting for Rabbi Judah's dicta. We have seen that for Maimonides, time as well as the celestial spheres were created. If however, as Crescas believes, time is independent of motion, and exists prior to the creation of the world, then the spheres can be in time even before the creation.[173] This is the import of Crescas' interpretation of

Rabbi Judah's statement. In contradistinction to Maimonides who had to interpret this statement figuratively, Crescas is able to adopt its literal meaning: ". . . the passage of Rabbi Judah, son of Rabbi Simon, which reads: 'It teaches us that the order of time had existed previous to that,' may be taken in its literal sense."[174] To Rabbi Judah ben Simon's related dictum that the order of time pre-existed creation, Albo is quick to point out that Rabbi Judah really meant that the time that is measured by the motion of the sphere is called "order of times" and is contrasted to time simply (*ha-zeman be-shiluaḥ*), which has no priority, posteriority, or order.[175] Because the phrase "order of times" applies only to time that is correlated to motion, Albo argues that Rabbi Judah's statement implies that time existed prior to the fourth day of creation. But Albo recognizes that ultimately this solution does not resolve the issue. Quoting the famous rabbinic passage that one must not explore what is above, below, before, and behind, Albo accedes that he has not really explained "how there can be a duration before the creation of the world which has in it neither prior nor posterior."[176] Or as Harvey has succinctly argued, Albo has not succeeded in explaining how, *within* the "order of times," there can be a first instant that has no "before."[177]

Let us turn finally to Isaac Arama who in his sermons combines philosophical and theological motifs with creative ingenuity. In dealing with Genesis 1, Arama addresses all three of our original concerns. Starting with the question how could there have been a day, that is a unit of time measurement, before the fourth day when the luminaries were created, Arama first notes the two main views of time, that of Aristotle and that of Maimonides. He agrees with Maimonides that "the beginning of time and the beginning of creation exist as one and the same."[178] Given this position, he must explain how time could possibly pre-exist the fourth day of creation. Arama resolves the conundrum by introducing a notion of time that is not tied to spherical motion, but rather to motions that already existed prior to the fourth day.[179] Arama argues that the spheres were created on the second day but started their movement on the fourth day when the sun and moon were introduced. Although according to Arama the sun and moon were created on the first day, they were not actually suspended (*nitlu*) until the fourth day, which demarcates the commencement of motion.[180]

But then Arama offers another solution to the problem of time before the fourth day in the context of Crescas' theory. This second solution is based upon Maimonides' basic contention that there is a basic connection between time and movement of the spheres. We have seen that Arama is not willing to accept Maimonides' position that time exists from the first day on because the sphere was already created on the first day. According to Arama, spherical motion begins on day four. So he says that the time of days one to four can be construed as "complete time" only in the mind of God. Time turns out to be an image in God's soul.[181] "Although it is known that time cannot be created at all without the motion of the spheres, nevertheless Scripture will not, because of this, refrain from describing it in terms of the future. For God speaks and Moses writes. . . . For insofar as complete time exists in God's wisdom. . . . Scripture mentions it and defines it even before it becomes into existence in its final description."[182] Arama's point here is that Scripture has recorded the creation of time retrospectively.

On this basis, then Arama distinguishes between two types of time: complete, or natural, time (*zeman shalem*), which is connected to spherical motion; and incomplete time (*zeman bilti shalem*), which is not tied to spherical motion but rather connected to other forms of motion.[183] In an interesting analogy drawn from Ibn Rushd, Arama claims that it does not make sense to argue that those who do not experience spherical movement do not experience time: imagine, for example, individuals imprisoned in a cave since childhood, who have not experienced natural time since they have not experienced spherical motion of the heavens. Arama argues that nevertheless, they experience other motions and hence time, by way of inference.[184] Using this example, then, Arama uses this distinction to resolve the original dilemma. Arama sees the existence of time as a distinction between before and after. When this distinction between before and after is made, time in the incomplete sense exists.

How, then, do we even recognize the first day as day number one that is, as the *first* day? Arama claims that light was created after a measure of dark, and then continues this analogy to other days as well. That is, the existence of time is not dependent upon the existence of the motion of the spheres, but upon the existence of any motion in general. On the first three days there is only incomplete time, and only on day four is there complete time, which is time predicated upon day and night.[185]

Arama finally turns to Rabbi Judah's dicta. According to Arama, the order of times precedes not the creation of the world but only the fourth day, preceding the creation of the luminaries. The order of times preceding the luminaries was that of "incomplete time" which is not tied to planetary motion.[186] Arama tries to find support for these two solutions in *Bereshit Rabba* 3 and offers the following interpretation of Rabbi Judah's words: According to Arama, Rabbi Judah teaches that time can be estimated without the motion of the spheres, hence supporting the position that time exists after creation but before the fourth day. Creation brings with it the concept of prior and posterior, and so it contains the "order of time". The term "day" refers not to an exact measured time, but rather to a concept of a definite duration of time, which is independent of the movement of the spheres. While rejecting the views of Maimonides, Crescas, and Albo, Arama continues to postulate a concept of time before the fourth day.

Conclusion

We see then that theological speculation has both reinforced and been reinforced by philosophical considerations of time and creation. More pointedly, I submit that concerns of temporality are central to understanding medieval Jewish views of creation. Much of Maimonides' effort has been aimed at showing that the scriptural view of creation is inconsistent with an Aristotelian theory. And yet, that Maimonides is unwilling to support Scripture altogether is reflected in his comments on Rabbi Judah. Gersonides is less willing ostensibly to compromise the temporal beginning of the universe, and so he creatively reinterprets Aristotle's notion of the instant in such a way as to allow for a temporal beginning to creation out of a preexistent matter. In this way he has retained a first instant to the act of creation in light of the notion of an eternally beginningless substance whose existence ontologically proceeds that of created composites. Hence according to Gersonides the world is both created at an instant and eternally existent.

In contradistinction to Maimonides and Gersonides, both of whom adhere fairly closely to Aristotle's characterization of time as the measure of motion, Crescas and Albo deviate from this Aristotelian depiction of time in terms of motion. Employing elements that are implicitly embedded in Aristotle, Crescas emphasizes the discontinuity of time and motion. From this discontinuity Crescas develops two implications: the first has to do with the subjectivity of time, while the second emphasizes the dissociation of time from creation. Hence Crescas, drawing upon Gersonides, is able to claim that God did create the universe at an instant in a way that accords with *Genesis Rabbah*. The subjectivity of time is then posited by Albo, who emphasizes the role played by imagination in perceiving time, and developed more fully by Arama who uses the distinction between complete and incomplete time to account for the existence of time.

Thus both Platonic and Aristotelian theories of time are important in shaping medieval Jewish modes of discourse pertaining to creation: the two theories are interwoven along with Rabbinic material to create an intertextual garment that takes on a life of its own. In the next chapter, I shall examine even more closely the individual threads of this garment; the warp thread uniting the threads of this garment is that of infinity.

CHAPTER THREE

Time, Motion, and the Instant: Jewish Philosophers Confront Zeno

Introduction

In chapter 2 we examined the arguments in favor of an initial instant of creation. In this chapter I turn to a specific subset of these arguments, namely those arguments that are devoted to refuting the eternity of the world by incorporating issues pertaining to traversal of the infinite. The issue of whether or not the infinite can be traversed is descended from Zeno's Paradoxes of Motion; these paradoxes have inspired an enormous body of literature devoted to issues connected with infinity, the continuum, and the divisibility of both magnitude and time.[1] To this day philosophers of science and mathematics address the underlying problems spawned by this vast literature.

That a predominantly mathematical notion such as the infinite might in fact provide the setting for theological issues has already been noted by H. Thijssen who points out that Neo-Thomist theologians of the late nineteenth century, for example, found common ground with Cantor's theory of transfinite set theory.[2] So too were medieval discussions concerning the eternity of the world couched in considerations having to do with traversal of the infinite, God's knowledge of the infinite, and God's own infinity. In this chapter I examine Jewish philosophical responses to the set of issues raised by Zeno's Paradoxes. I concentrate upon the following questions that arise directly from these paradoxes: Is any temporal or spatial spread infinitely divisible? Is the infinite divisibility of space analogous to that of time? Can either infinitely divisible spread be traversed? Can an extended duration be comprised of durationless points? And finally, is time composed of instants which themselves have no duration? I shall argue in this chapter that these technical issues have direct theological implications. More specifically, Jewish philosophers, although not particularly concerned with Zeno's paradoxes per se, nevertheless were very much interested in the theological and cosmological implications of the paradoxes, especially as they pertained to the question of creation. Before turning to these questions, however, let us start with a brief look at the most influential statement and transmission of these paradoxes.

Traversing the Infinite: Zeno, Aristotle, and John Philoponus

The most trenchant of Zeno's four paradoxes is the Dichotomy, or the "Paradox of Half-distances," which can be summarized as follows. Imagine Achilles attempting to traverse the distance AB. In order to reach B, he must first traverse half that distance, or AC. But in order to reach C, he must traverse half that distance, or AD. Assuming an infinite number of subdivisions between A and B, Achilles will never be able to reach his goal B.[3]

In fact we can distinguish two ways of reading this paradox, which are expressed in Figure 3.1.

On the first reading (3.1.1) no moving body can ever complete its journey—it can never take the last step to B since there is no last step which will complete the journey. On the second reading (3.1.2), no body can ever start its journey—it can never take its first step since there is no first step to take. On both readings, Zeno means to show that no moving body can traverse an infinite distance in a finite stretch of time.[4]

These paradoxes are part of a continuing dialogue between what Fred Miller has termed the nihilistic and atomistic strands of the continuum debate.[5] The main question in this debate concerns the characterization of the smallest units of space and time: are they composed of divisible or indivisible units? Another way of stating this question is to ask whether a magnitude is divisible everywhere, that is, perpetually divisible into smaller units, or divisible only down to some atomic magnitude, beyond which subdivision is no longer possible. In answer to this query, there were two schools of Greek thought: the nihilists believed in the infinite divisibility of space and time, whereas the atomists believed that time and matter were made up of smallest basic units. Further, the nihilists denied the existence of a vacuum and argued that matter is continuous, whereas the atomists

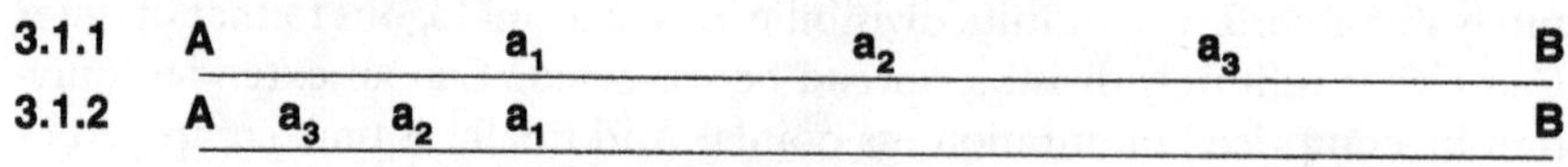

Figure 3.1. Zeno's Paradoxes of Motion

supported the notion of a vacuum, which supplies the space in which indivisible atoms move.[6]

Zeno's purpose (although unstated in the paradoxes themselves) is to demonstrate the untenability of both atomism and nihilism. His main contention is that if time is construed as infinitely divisible, then change and motion are impossible. But if atomism is adopted, then notions of place and change become untenable. In short, Zeno constructs these paradoxes in order to demonstrate that pluralism suffers "even greater absurdities" than monism.[7]

In fact, however, there are two underlying principles Zeno employs in order for the dichotomy paradox to work: first he assumes that if anything moves, it performs infinitely many tasks; and second, he assumes that nothing can perform infinitely many tasks. From these two statements Zeno concludes that nothing moves. As Barnes points out, however, the first statement is only true if "space is continuous or infinitely divisible; for it rests upon the assumption that any stretch of space AB, contains infinitely many spatial points a_i."[8] But this, of course, is precisely the issue at stake.

When Aristotle turns his attention to solving Zeno's paradoxes, his main concern is to refute the atomistic horn of the dilemma. In the *De Generatione et Corruptione,* Aristotle poses the dilemma as follows: Are the basic units that grow, things that are indivisible or divisible?[9] The problem as he sees it arises with respect to the second possibility, for if things are infinitely divisible, there is nothing that will ultimately survive the division. Assume, says Aristotle, that body is everywhere divisible. Each of these parts is then everywhere divisible. But then body would be comprised either of indivisible points, or of nothing, neither of which Aristotle allows. As Furley has pointed out, the resulting segments of an infinite division could neither have any magnitude nor be without magnitude.[10] Aristotle's reason for saying they could not have any magnitude is simply that it would contradict the hypothesis that a division everywhere had been completed. Aristotle reiterates this point by claiming that the number of points in a line is finite as well. Why? because a more than finite number will yield the absurd result that an extended line is composed of unextended points.[11] Hence it would appear that body must be comprised of indivisibles, that is, of ultimately irreducible atoms.[12]

Aristotle, however, conceives of reality as a continuous plenum in which space, time, and matter all have the quality of continuity. Hence he will want to maintain that no continuum can be made up of indivisibles. In *De Generatione et Corruptione* he resolves the issue by claiming that magnitude is infinitely divisible only in a restricted sense.[13] This discussion is reiterated in the *Physics* when he defines a continuum as "those limiting extremes of the two things in virtue of which they touch each other, become one and the same thing, and (as the very name indicates) are 'held together', which can only be if the two limits do not remain two but become one and the same."[14] In *Physics* 6.1 Aristotle offers an argument to the effect that "everything continuous is divisible into divisibles that are always divisible."[15]

How then does Aristotle avoid the two horns of Zeno's dilemma? In *Physics* 6.2 Aristotle distinguishes two senses of infinity (*apeiron*): infinity with respect to divisibility and infinity with respect to extent.[16] He argues that time is infinite in the first sense, and then claims that this distinction allows us to maintain that Achilles has an infinite amount of time in which to traverse an infinite distance. But Aristotle soon sees that what's at stake is whether it is possible to traverse an infinite number of anything.[17] His first line of defense, that an infinite amount of time is required to touch infinitely many points, appears to solve the paradox only because we are, as Sorabji put it, "all hemmed in by the finitude of our lives. For Aristotle time, like space, is continuous, whereas for Zeno time is granular while space is smooth."[18] Sorabji's claim that Aristotle cannot, on his own principles, afford to make this concession, is recognized by Aristotle as well.

> For suppose the distance to be left out of account and the question asked to be no longer whether it is possible in a finite time to traverse an infinite number of distances, and suppose that the inquiry is made to refer to the time itself (for the time contains an infinite number of divisions): then this solution will no longer be adequate.[19]

In *Physics* 3.6 and 8.8 Aristotle argues that those who had posited indivisible magnitudes in order to respond to Zeno had confused two senses of indivisibility. These two senses can be distinguished as follows: infinite$_1$ refers to a complete divided state, whereas infinite$_2$ refers to the process itself of dividing. The first Aristotle terms actual infinity, and the latter he terms potential infinity. The atomist argument in favor of indivisibles relies upon the notion of actual infinity. According to Aristotle, however, a solution to Zeno's paradox depends upon the latter sense of infinity. That is, according to Aristotle, the infinite with respect to division exists potentially but not actually; infinite divisibility is a continuing process and so the potency in question is actualized whenever the process is in effect.[20] In *Physics* 8.8 Aristotle maintains that he had formerly misconstrued Zeno's challenge, and now claims that the notion of potential division is crucial for understanding Zeno: "to the question whether it is possible to go through an infinite number of units, either of time or of distance, we must reply that in a sense it is and in a sense it is not. If the units are actual, it is not possible, if they are potential, it is possible."[21] Aristotle's point, intimated already in *De Generatione et Corruptione,* and elaborated in *Physics* 3.6, is that no continuum can be divided everywhere simultaneously. Once we realize that the continuum can be only potentially and not actually divided, we can accept the infinite divisibility of the continuum.[22]

Aristotle, then, allows infinity to exist only in a restricted sense. "For generally the infinite has this mode of existence: one thing is always being taken after another, and each thing that is taken is always finite, but always different."[23] What this means is that Aristotle takes infinity in terms of the finite: however large a finite number is taken, even more can be taken. In an important sense the infinite "in respect of addition" is the same as the infinite "in respect of division."[24] Infinity is "extendible finitude" and hence correlated with a process.[25] In short, Aristotle has claimed that despite the fact that nothing can perform infinitely many tasks, nevertheless Zeno's paradoxes can be dissolved.

The two implications of this view are first, that the infinite cannot be traversed, and secondly, that infinity is merely potential and not actual. Aristotle uses the principle that the infinite cannot be traversed to assert that there cannot be infinite number: number is countable, and so if there were infinite number, then one could count, and traverse infinity: "nor can number taken in abstraction be infinite; for number or that which has number is numerable. If then the numerable can be numbered, it would also be possible to go through the infinite."[26] But are these implications convincing? Barnes, for one, thinks not. In fact he finds all the arguments in support of the claim that nothing can perform infinitely many tasks to be unconvincing.[27] This is an important point, the implications of which will emerge more fully in our discussion of Jewish texts. For once we superimpose a theological framework onto the discussions of Zeno and Aristotle, the paradoxes take on additional complexity. For example, can it be said that God knows the infinite divisions in a line? Can the universe be infinite and yet contain finite existent creatures? Can there be a first instant of creation?

The individual who enabled medieval Jewish philosophers to formulate these questions in a theological context is John Philoponus, whose works contain a refutation of Aristotle's theory of the eternity of the world from the perspective of theories of the infinite. John Philoponus' major work *Contra Aristotelem* has been lost and survives only in quotations from Simplicius' commentaries on Aristotle's *De Caelo* and *Physics.*[28] In this work, Philoponus hopes to demonstrate the creation of the world by arguing that Aristotle's assumption of eternal motion leads to untenable conclusions. Philoponus' works were known to Arabic philosophers, and were transmitted by the Kalām school, through Saadia Gaon, to eleventh- and twelfth-century Jewish and Christian philosophers.[29] For this reason his arguments are of crucial importance to theological attempts to refute Aristotle's arguments for the eternity of the universe.

In his arguments, Philoponus takes Aristotle's definition of the infinite (which Aristotle used to prove eternity) and turns it against itself, showing that in fact the definition of the infinite precludes eternity. The three specific sets of arguments offered by Philoponus can be summarized as follows:

2.1 If the universe were eternal, the generation of any object in the sublunar universe would be preceded by an infinite series of generation. But an infinite cannot be traversed, and so no objects would be generated.

2.2 The eternity of the universe would imply an infinite number of past motions that is continually being increased. But an infinite cannot be added to.

2.3 The numbers of the revolutions of the heavenly bodies are multiples of one another and thus eternity would imply infinite numbers of past motions in varying multiplicities. But infinite numbers cannot be multiplied.[30]

In *Contra Aristotelem* Philoponus presents two sets of arguments in support of creation, both of which are directed against Aristotle's eternity thesis. According to Simplicius, he assumes as axiomatic that "it is impossible for an infinite number to exist in actuality or for anyone to traverse the infinite in counting, and that it is also impossible that anything should be greater than the infinite, or that the infinite should be increased."[31] From this axiom he argues as follows. Imagine an infinite series of transformations that has taken place among the four elements. In an eternal world, these transformations would constitute an infinite series. But, using Aristotle's characterization of infinity, it is clear that an infinite number cannot exist actually or be traversed. So in an eternal world, the infinite series of transformations could never be completed and the particle now known to exist could never in fact have come into existence. "If then, the motion of the particular fire came to be, an infinite number of motions surely did not exist first."[32] Further, imagine that the scenario were expanded to the spheres. If the motion of the heavens is without a beginning, and if spheres revolve at unequal periods of revolution, then it is necessary that the sphere of Saturn has rotated with an infinite number of revolutions; but on this celestial model, the sphere of Jupiter must have rotated with nearly three times more revolutions, the sun with thirty times more revolutions than Saturn, and that of the fixed stars more than ten thousand times greater. But, Philoponus, argues, "if it is not [even] possible to traverse the infinite once, is it not beyond all absurdity to assume ten thousand times the infinite, or rather the infinite an infinite number of times?"[33] Hence he concludes that the circular motion of the heavens is not eternal but must have had a beginning.

Now whether Philoponus' insights can be upheld as philosophically cogent in light of modern understandings of infinity (not to speak of heliocentrism itself) is doubtful. As Sorabji has argued, Philoponus has not demonstrably shown that the universe must have had a beginning. In fact, it is unlikely that this thesis can ever be supported with the absolute certainty required by Aristotelian demonstration.[34] But I agree with Sorabji that the importance of Philoponus' arguments lies in their providing medieval Jewish and Christian philosophers with the ammunition they so desperately needed to refute the eternity thesis.[35] It is precisely this ammunition, for example, which Saadia Gaon utilizes in his support of creation *ex nihilo.* For as a result of Philoponus' critique, medieval Aristotelians had to confront a number of problems that arose with the proofs from the impossibility of an infinite. For example, if Aristotle denied the possibility of an infinite number, why did he advocate the eternity of the world, hence implying the existence of an infinite number of objects and motions? Further, if Aristotle denied the possibility of an infinite magnitude, why did he advocate the existence of an infinite time continuum?[36] And finally, if Aristotle denied the possibility of an infinite series of causes and effects, why did he affirm the existence of eternal series in which each prior member is construed as the cause of the following one? These are the issues confronted by Jewish philosophers, and it is to these questions that we now turn.

Jewish Neoplatonic Considerations of Infinite Divisibility

Turning first to Neoplatonic texts, we see that consideration of the metaphysical and physical constitution of reality forms an important component of Jewish Neoplatonic thought. Whether or not matter, space and time are infinitely divisible becomes a topic of considerable interest, as evidenced in the writings of Isaac Israeli and Ibn Gabirol. Both the Aristotelian and atomistic positions were available to Jewish thinkers, who for the most part sided with Aristotle against atomism. By the time of Israeli, in the ninth century, Kalām atomism is as influential as the atomism of Democritus. I shall return to the details of Kalām thought more fully below. Suffice it to say at this point that Kalām atomism differs from its Greek precursors in two important respects. First, atoms in Kalām thought are construed as unextended, sizeless points, as opposed to Greek atoms that were thought to be indivisible yet with extension.[37] Second, the Kalām theologians introduced an occasionalist tinge to atomism, arguing that these atoms are destroyed and recreated at every instant by the deity.

When we turn to Isaac Israeli (c. 855– c. 955), we find operative many of these cosmological ingredients superimposed onto an emanation scheme derived ultimately from Plotinus. A physician and philosopher, Isaac Israeli is considered to be the first Jewish Neoplatonist. He was born in Egypt and began his career as an oculist.[38] At about age fifty, he emigrated to Tunisia to study medicine under the Muslim Ishāq ibn-Irām. Later, he was appointed court physician by 'Ubāyd Allah al-Māhdi, the founder of the Fatamid dynasty, and became known as one of the great physicians of the Middle Ages. His works were widely circulated and translated into Arabic, Latin, and Hebrew.[39] Of his many surviving works, the *Book of Definitions* and the *Book of Substances* are the main sources of Israeli's philosophical ideas. His best-known work, the *Book of Definitions* deals with definitions of philosophical, logical, and other terms.[40] The *Book of Substances* has survived only in incomplete fragments of the original Arabic.[41] It appears to be a general treatise on philosophy aimed at a general audience rather than being a specifically Jewish work. Finally, the *Chapter on the Elements* (the *Mantua Text*) exists only in manuscript, at Mantua.[42] From this text we learn that Israeli based his view of creation and the series of emanations on an earlier text known as *Ibn Ḥasdai's Neoplatonist.*[43]

Israeli's philosophical cosmology describes the various stages of being as a series of emanations, or hypostases, from the intellect. The intellect itself is constituted by the union of first matter and first form, which are created by the power and will of God. Israeli thus upholds the notion of creation *ex nihilo* in the case of the first three hypostases, while adopting the Plotinian concept of emanation for the others. But, unlike Plotinus, in the *Book of Definitions* each hypostasis acquires more shadows and darkness, out of which the next hypostasis emanates. The shadow accounts for its loss of strength. In the *Book of Substances,* he tries to combine both metaphors by claiming that the shadow is the new substance; the essence or light is not what emanates. Unfortunately, Israeli is seemingly unaware of the contradiction between maintaining that emanation is both the passing of the essence and the passing of a shadow.[44]

Whereas Plotinus describes intellect as emanating directly from the One, Israeli, following his pseudo-Aristotelian source, interposes two simple substances—first matter and first form or wisdom—between the Creator and intellect as representing the first hypostatis.[45] Soul follows intellect in this triad of hypostases and is divided into a higher phase and a lower one, which Israeli calls nature.[46] The doctrine of atomism is clearly expressed in Israeli's *Book on the Elements.* In this work Israeli is concerned to define the term "element," out of which the sublunar world is composed. In the context of this discussion he combats the atomistic theory of the Mu'tāzilites and of Democritus, and attempts to prove that a line is not composed of points. Israeli's extant writings do not reveal a systematic thinker who developed his key concepts. However, from fragments scattered throughout his works, we can pull together a number of definitions and concepts which pertain to his view of time and infinity. Israeli introduces the discussion in the context of discussing Galen's definition of the term element. According to Israeli, by "element" Galen means "the minimum part of a thing,"[47] echoing Aristotle's use of the term. He then introduces a fictitious interlocutor who suggests that by "part" he means "those parts into which a body is divided naturally and of which it is composed, just as a body is divided into surfaces and surfaces into lines and lines into points."[48] In support of his interpretation, the fictitious interlocutor quotes Democritus as well as the Mu'tāzalites:

> And I will adduce a proof from the words of Democritus since the wise man Democritus said that a body is composed of surfaces and surfaces are composed of lines and lines are composed of points. And he mentioned as well the agreement of Abraham ibn Nazzām and his followers from among the Mu'tāzalites, from which is understood from their words that body is composed of parts which are not divided [except] into points.[49]

According to Israeli, then, Democritus maintained that matter is composed of spaceless atoms analogous to mathematical points.[50] But Israeli rejects the mathematization of atoms by arguing that the union of two points can be conceived in two ways: either the totality of the one unites with the totality of the other, or a part of one comes in touch with a part of the other:

> you may say either that the point when it is connected to another point is a conjunction of whole with whole, or a part with a part. But if you say that the conjunction of one point with another is a conjunction of whole with whole, then the two points will form one point and their place will be one place, for the whole of the one is the whole of the other, seeing that neither of them has extension nor is there between them any difference. And the same will hold true of the third point and the fourth and the others after it to infinity. And if the conjunction of one point with another is as part with part of the other, then the division and dissection of the point would necessarily follow, but this would do away with his claim that the point is indivisible.[51]

In the first case, there is no distance between the two, and so the result would be a point; in the second case, a contradiction results—a partial union of atoms that are ex hypothesi spaceless and devoid of parts. In either case mathematical points cannot produce an extended body.[52] Israeli therefore concludes that inasmuch as bodies cannot be composed of atoms that are both indivisible and unextended, nor can they be composed of atoms that are indivisible and extended, it is not appropriate to postulate the existence of indivisible magnitudes.

Echoes of Israeli's Neoplatonic cosmology can be found in Ibn Gabirol's works as well. Representing the flourishing of Jewish intellectual life in Andalusia under the enlightened reign of the Umayyad caliphate, Gabirol was one of the first Jewish philosophers in Spain to benefit from the intellectual ferment of this Golden Age. Although Gabirol lived barely forty years, he is known primarily for his metaphysical writings; his major philosophical work *Meqor Ḥayyim* is a purely metaphysical treatise that presents a rigorously defined Neoplatonic cosmology.[53]

Without characterizing the many details of Gabirol's Neoplatonic cosmology, let us concentrate upon several of his own specific modifications which pertain to our topic. Gabirol's most creative contribution centers around his hylomorphic conception of existence. All substances in the world, both spiritual and corporeal, are composed of matter and form. Types of matter are ordered in a hierarchy that corresponds to a criterion of simplicity: general spiritual matter, general corporeal matter; general celestial matter; general natural matter; and particular natural matter. Individual matter is associated with prime matter, which lies at the periphery of the hierarchy, thus epitomizing the very limits of being.[54]

How are form and matter interrelated? Gabirol is ambivalent toward this question and presents two alternatives. On the one hand he argues that form and matter are mutually interdefined and are differentiated only according to our perspective of them at a particular time; accordingly both are aspects of simple substance. On the other hand, he emphasizes the complete opposition between matter and form, suggesting that each possesses mutually exclusive properties that render a reduction of one to the other an impossibility.[55]

The importance of these discrepancies is reflected in Gabirol's discussion of creation. When describing the yearnings of matter, Gabirol argues that inasmuch as matter was created bereft of form, it now yearns for fulfillment.[56] However, in other contexts, he asserts that matter subsists not even for an instant without form.[57] In this latter case, matter is and always was united with form. Additionally, Gabirol offers two accounts of the actual process of creation. According to *Meqor Ḥayyim* 5.42, universal matter comes from the essence of God, and form from the divine will, whereas other texts suggest that both were created by the divine will.[58]

Gabirol raises the issue of the infinite divisibility of matter and substance in treatise two of *Meqor Ḥayyim,* in the context of working out his ontologies of matter and form. Although he does not mention Zeno by name, his analysis pertains to the ultimate divisibility of the parts of substance and reflects the issues raised by Zeno. Having just maintained that each composite of substance is composed of that of which it was put together, and since the parts of the quantity of the substance in question are all similar, Gabirol raises the following question: Are the ultimate constituents of reality divisible or indivisible? In other words, can the parts of substance be divided or not?

> Intelligence divides the quantity of the substance of the world and reduces it into its parts. Thus the quantity of the substance of the world must be composed of its parts. And since the parts of the quantity of substance of the world are similar in themselves, they do not possess difference among themselves in the sense of quantity. Let us therefore pose one of the parts of which we shall speak. Can you say of this part, which is one of the parts of the quantity of substance, that it can be divided in potency inasmuch as it is indivisible, or rather is it indivisible in potency?[59]

In posing this question, Gabirol reflects the concern of the Mutakallimūn, who had argued for the ultimate indivisibility of matter. His aim will be to show that quantity exists only with substance.

Gabirol first characterizes the terms 'point' and 'part' as follows:

> In effect, if we say that the point is a part of body and that the body is divided of itself, this is only by accident and not by essence or nature, because the point is an accident which subsists in body and does not pertain to it by its nature. The same follows for color: we say that it is a part of body in the sense that body is composed of surface, color, figure and other accidents which are carried in it; but they do not participate in the nature of body. But, for the part between the parts of the quantity of the substance of the world of which we have spoken, it does not follow, for it is a part of the nature of quantity and resembles it, just as a part of water is not external to the nature of water.[60]

Based on this distinction, Gabirol presents a number of arguments in 2.17 designed to support the divisibility of parts and concludes that " the part in question between the parts of the quantity of the world is divisible, and it is clear to me that it is divided into substance and accident."[61] Gabirol's disciple, however, is reluctant to agree so soon and pushes further, presenting an argument in favor of infinite divisibility: "But what would you say if, changing my opinion, I said that this part is not further divisible into other parts smaller than it in quantity. . . . After we have posed that this part is the smallest of the parts, how is this smallest unit divided?"[62] Gabirol responds by distinguishing between two types of divisibility, epistemological and ontological. "It is not impossible that this part be the smallest with respect to the senses but not in itself. It is divisible, in effect, in that it is impossible that it be a part of quantity and not be divisible."[63] Gabirol's point is that we may not be able to detect divisible parts beyond our senses, but nevertheless they do exist. The following extended argument is presented in support of his position.

3.1 Assume the existence of a smallest part *p* which is not divisible.

3.2 *p* is either part of another or not.
3.3 Imagine that *p* is part of another part *p**.
3.3.1 If *p* is a part of *p**, when we add *p* to *p**, *p** is greater than when it was alone.
3.3.2 But is *p** divisible or indivisible?
3.3.2.i If you say that *p** is indivisible and that *p* was already indivisible, then *p** cannot be divisible.
3.3.2.ii But if two united parts do not form a divisible part, these parts and one single part would be equal, and two would be equivalent to one, which is absurd. If, for example, an ensemble of parts were equal to one part, the body of the entire world would be equal to one of its indivisible parts.
3.3.2.iii Hence a second part *p* added to a first *p** must be divisible.
3.3.2.iv And if *p* is divisible, it follows that *p** must be divisible as well.
3.3.2.v And, similarly, it follows that all the parts of body be divisible, since it is impossible to find a part which is indivisible.
3.4 If, on the other hand, *p* is not a part, we could not find another part *p*** larger than *p*.
3.5 If we found *p*** larger than *p*, it would be necessary that *p* be a part of *p***.
3.6 But when we add any part of one body to another, the ensemble will be greater than each of its simple parts which compose it.
3.7 So, there does not exist a part *p*** greater than *p*.
3.8 *p* then, is the smallest of parts.
3.9 But then *p* would be both part and non-part.
3.10 Therefore *p* must be a part.
3.11 And being a part, *p* is necessarily divisible.[64]

Argument Summary: What Gabirol does in this argument is show that any smallest part p must be further divisible into smaller parts because if not, then its being a "part" of another entity is ontologically indefensible. He does so by envisioning two possibilities, and showing that on either scenario, the part 'p' must be further divisible.

In this argument, however, it is clear that Gabirol is equivocating with respect to the term "part." In short, Gabirol concludes that "the smallest part in question is not nondivisible, for we cannot find an indivisible part; and it is clear as well that the part in question between the parts of the quantity of the substance of the world is composed of substance and accident."[65] In conclusion, Gabirol has posited the infinite divisibility of substance on the grounds that there is quantity only where there is substance. "I concede now in a clear and certain manner the continuity of the totality of the substance with the totality of quantity."[66] Gabirol's contention is that extension and indivisibility pertain to two different kinds of being: the former is associated with matter, and the latter with spirit. It is impossible to reduce the one to the other. Hence matter cannot be composed of indivisible, spaceless atoms (*minimae partes*). Inasmuch as any indivisible unit must be of a spiritual nature, once we begin to speak of spiritual matter, we leave the issues of quantity and matter behind. Gabirol therefore envisions the possibility that all of the world might exist in a point and that extension is not essential to matter.[67]

Having seen that matter is infinitely divisible, let us turn to Gabirol's arguments for the divisibility of form. Gabirol clearly asserts that both finitude and divisibility pertain to form. That finitude is the result of form is evinced in the following passages. In answer to the question why the intellect only seizes things that are comprised of matter and form, the Master states that

> [another reason] is the fact that the intellect, when it perceives a thing, understands it; and it can only comprehend what is finite with respect to it; in effect, a thing is finite only by virtue of form, for an infinite thing does not possess a form by which it becomes one and differs from another; so the eternal essence is infinite for it does not possess a form.[68]

And in another context, Gabirol claims that "the form is seized by the senses as finite. In effect, the form is finite for it is the limit of the body . . ."[69] Form is the principle of divisibility as well. Clearly what distinguishes the finitude of both matter and form is the fact that they are mutually interdependent: in this context finitude signifies not so much the sense of spacial limitation as ontological dependence. Having characterized the finitude of matter and form, we are now in a position to characterize more fully the notion of infinity used by Gabirol to describe God. By infinite in the qualitative, or substantive, sense, Gabirol means a totally independently existing entity, one which requires no ontological support. An infinite being possesses no form (4.6), is not divisible (3.3) and is not subject to change (3.6).

Interestingly enough, Gabirol says little about infinity itself, but rather devotes considerably more time to Divine will, which resides in the intermediary sphere between finitude and infinity: the finite and infinite intersect in the will. Speaking of the intelligible substance, the disciple asks, "Tell me whether the forms of these substances are finite or infinite; if they are finite, how they can have the being of an infinite force; if they are infinite, how something finite in act can issue from them."[70] Gabirol's response requires aligning form with the creative will: in and of itself, form is identical with will. It is only when it enters into creative act with matter that it becomes finite. In other words, both form and will, that is to say the force that produces these substances, are finite by virtue of their effect and infinite by virtue of their essence. But the will is not finite by virtue of its effect except "because the action has a beginning and so follows the will; and it is infinite by virtue of its essence for it does not possess a beginning. And inversely, we say of the intelligible substance that it has a beginning because it is caused, and that it has no end for it is simple and not temporal."[71] This point is emphasized even further in the following passage. In answer to a number of difficulties concerning creation, Gabirol points out that the will is both finite and infinite.

> Your doubt arises from the fact that you have assumed that the will is infinite; but whereas it is infinite when we consider it by its essence, without its action, it is not infinite by virtue of its form which emanates from it; this is because its action is finite, in that it possesses a beginning and its form commences from it. It follows therefore that it must be finite in the beginning of its action and in the beginning of the form which emanates from it. The will is therefore intermediate between the elevated essence and the form which emanates from the will.[72]

Hence the process of creation is seen as the projection of infinite form upon finite matter, and the retention on the part of matter of a part of this infinite form. Theoretically, were form able to exist independently of matter, it would be infinite and not finite. An even more interesting question concerns the finitude of matter: if matter were able to exist independently of form, would it be infinite as well? No, for it contains within itself the grounds for finitude. So that although form is allied with finitude, Gabirol reserves the possibility of speaking of the infinity of form.

Meeting the Kalām Challenge: Kalām Atomism Described

As intimated above, Kalām atomism differed from its Greek counterpart both metaphysically and theologically. Wolfson suggests that theological considerations, connected in particular with God's causal powers, may have led Kalām theologians to the metaphysical insistence upon an atomist theory that was historically associated with a denial of causality.[73] These considerations are manifested in Maimonides' presentation of Kalām thought in *Guide* 1.71–76. In these chapters, Maimonides sets out to present only those views that are common to both the Mu'tāzilites and the Ash'ārites, and that pertain to God's existence, unity, and incorporeality, as well as the existence of the world.[74] Although I shall not deal with Maimonides' discussion of the history and transmission of these teachings, it will be useful to examine his statement and understanding of the actual teachings of Kalām atomism as it relates to issues of creation and infinite divisibility of the continuum.[75]

According to Maimonides, Kalām teaching can be summarized in twelve premises as follows: premises 1, 2, 3, and 11 pertain to atoms, premises 4, 5, 6, 7, 8, and 9 pertain to the nature of substance and accidents, and premises 10 and 12 pertain to the epistemological implications of atomism. Inasmuch as the first group deals with the issue of atomism and infinite divisibility, let us examine them more closely. Maimonides' first premise posits the existence of indivisible atoms which are unextended and yet when combined form extended bodies. Premise 1 (P.1), that atoms exist, is expressed in the following set of characteristics:

- The world as a whole is composed of very small particles that, because of their subtlety, are not subject to division.
- The individual particle does not possess quantity in any respect. However, when several are aggregated, their aggregate possesses quantity and becomes a body.
- These particles are alike and similar to one another, there being no difference between them in any respect whatever.
- These particles are not restricted in their existence for they are constantly created by God whenever He wishes, and their annihilation is likewise possible.[76]

What the first premise tells us is that on Maimonides' understanding of Kalām atomism, atoms are indivisible, indistinguishable from one another, and unextended entities which are constantly created by God. Premise 1 (P.1) also points explicitly to the connection between atomism and God's omnipotence in that God can create and destroy these atoms at will. This latter point will have important implications with respect to a metaphysical theory of causality, as well as to a theological theory of predestination.

The second premise (P.2) postulates the existence of a vacuum and states that according to the Mutakallimūn, "vacuum exists and that it is a certain space or spaces in which there is nothing at all, being accordingly empty of all bodies, devoid of all substances."[77] The unstated Kalām argument, already intimated in Aristotle, is that in order for atoms to be individuated one from the other, there must be space between them which is "atom-less," or void of body. In order to account for motion, the void must therefore exist. The third premise applies the metaphysics of space to that of time and introduces the existence of time atoms. Not surprisingly, these time atoms are then correlated with arguments concerning infinite traversal hearkening back to the arguments of John Philoponus.

Premise 3 (P.3) is characterized as follows:

- Time is composed of instants which, because of the shortness of their duration, are not divisible.

- If time were continuous and infinitely divisible, it would follow of necessity that all particles would be divisible.
- Similarly if distance were supposed to be continuous, it would follow of necessity that the instant that had been supposed to be indivisible would be divisible.
- Therefore distance is not continuous, but composed of parts at which divisibility comes to an end; so too the division of time ends with the instants that are not divisible.

(P.3) postulates that time-atoms are indivisible and without duration. Maimonides uses the following analogy to exemplify what the Mutakallimūn have in mind:

> For example, an hour consists of sixty minutes, a minute of sixty seconds, and a second of sixty thirds. And thus this division of time ends up accordingly with parts constituting, for instance, tenths or something even briefer, which cannot in any respect be separated in their turn into parts and are not subject to division, just as extension is not subject to it. . . . Consequently time becomes endowed with position and order.[78]

Maimonides' point in this passage is to demonstrate that according to the Mutakallimūn, just as distance is indivisible, so too is the temporal spread indivisible. Just as the world is composed of indivisible spatial aggregates, so too must there exist individual temporal aggregates.

The eleventh premise (P.11) pertains directly to the problem of infinite divisibility and states that the existence of the infinite in any mode is impossible. Maimonides first reminds us that four kinds of actually existing infinites have been shown by Aristotle to be impossible: the existence of an infinite magnitude; the existence of an infinite number of causes; what is infinite *in potentia* or accidentally; and the existence of what is infinite by way of succession.[79] These four types of infinite are conflated by the Mutakallimūn, who according to Maimonides believe that "there is no difference, in their opinion, between saying that a certain infinite magnitude exists and saying that bodies and time are liable to be divided to infinity."[80] So too the Mutakallimūn deny any qualitative difference between asserting the "simultaneous existence of things infinite in number, arranged [at the same time] in orderly fashion," and asserting that "things infinite in number came into existence, but passed away one after the other."[81] In short, the Kalām theologians reject all types of infinite, some on metaphysical grounds, and others on epistemological grounds.[82]

From these premises, the Kalām atomists deduce a number of consequences that deny the possibility of motion and which hearken back to Zeno's paradoxes of motion. These consequences apply as well to arguments for the existence of the world. The first Kalām consequence (C.1) postulates that motion is illusory and is presented by Maimonides as follows:

> C.1.a They [the Mutakallimūn] said that motion is the passage of an atom belonging to these particles from one atom to another that is contiguous to it.
>
> C.1.b It follows that no movement can be more rapid than another movement.
>
> C.1.c In accordance with this assumption, they said that when you see that two things in motion traverse two different distances in the same time, the cause of this phenomenon does not lie in the greater rapidity of the motion of the body traversing the longer distance; but the cause of this lies in the motion that we call slower being interrupted by a greater number of units of rest than is the case with regard to the motion we call more rapid, which is interrupted by fewer units of rest.
>
> C.1.d And when the example of an arrow shot from a strong bow was alleged as an objection against them, they said that the motions of the arrow were also interrupted by units of rest.

The Kalām theologians are asserting in this example that because what we call motion is nothing more than the passage of one atom from one void to another, such passages all occur at the same rate. What accounts for apparent differences in speed is the number of rest units which must be traversed.

The example of the arrow alludes, of course, to Zeno's third paradox of motion according to which an arrow, which appears to be moving through the sky, is actually at rest in each of the units it occupies. To the example of the arrow, Maimonides raises the counterexample of a millstone whose complete revolution demonstrates that the circumference travels more quickly than the inner circle at the center, and hence that body is in continuous motion. In this example, two concentric circles with different radii move in such a way that each of them rolls along a straight line (fig. 3.2). But how can these two lines be equal in length if they are produced by circumferences of different radii?

To this the Mutakallimūn reply as follows:

> C.1.e Their answer to this objection is that the various portions of the millstone become separated from one another in the course of its revolution and that the units of rest that interrupt the motion of all the revolving portions that are near the center are more numerous than the units of rest that interrupt the motion of the parts that are farther off from the center.[83]

In other words, the Kalām atomists respond to the millstone example, which is adduced against Zeno's paradox of the arrow, by arguing that the units of rest in the center of the stone exceed those at the ends; hence neither circumference can be said to move more or less slowly than the other.

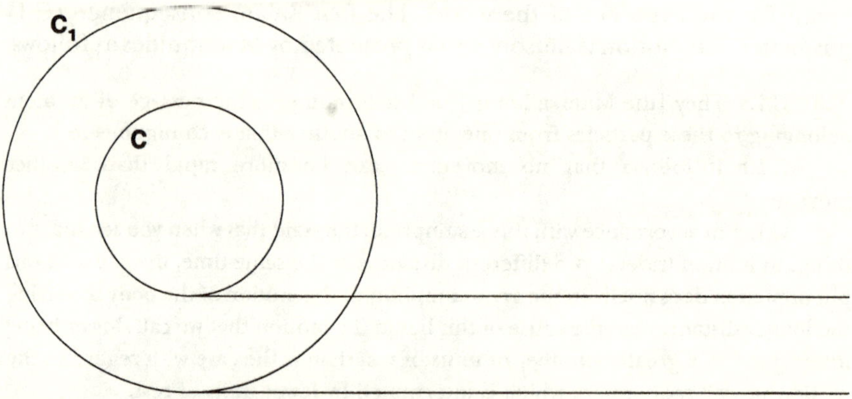

Figure 3.2. Maimonides' Millstone Example

Other consequences have direct application to the issue of creation and draw upon Philoponus' first two premises, (2.1) and (2.2). (C.2) is a restatement of (2.1) and offers a Kalām argument that the whole has been created in time on the grounds that each individual must also be created in time:

> Once upon a time this Zāyd was not here; then he is. Accordingly if he is, it is impossible in every respect that he should not derive from his father Umar. Now his father Umar is likewise produced in time. Accordingly if his father is, it is impossible that he, Umar should not be generated from Zayd's grandfather Khalid. But Zayd's grandfather Khalid is also produced in time. Thus this goes on to infinity. However, they have assumed, as we have made clear in the eleventh of their premises, that the existence of an infinity of this kind is impossible.[84]

(C.3) draws upon the createdness of the atoms to postulate the existence of the world. It is summarized by Maimonides as follows:

> They say: the world in its entirety is composed of substances and accidents. Now no substance can be exempt from one or several accidents. And all accidents are produced in time. Accordingly it follows necessarily that the substance that serves as a substratum for them is also produced in time. . . . Accordingly the world in its entirety is produced in time.[85]

Both consequences clearly draw upon the impossibility of an infinite in succession and reflect the influence of John Philoponus. In both the second and third consequence, Maimonides emphasizes how the Kalām doctrine of atomism, combined with the denial of an infinite, leads ineluctably to postulating the beginning of the world. While other Kalām arguments were incorporated into Jewish thought as well, these are the arguments that pertain directly to issues of infinite divisibility and that reappear in Jewish discussions of Kalām thought. As we shall see in the next section, both Maimonides and Saadia reject these arguments on both metaphysical and epistemological grounds.[86]

Rejection of Kalām Atomism: Saadia Gaon, Halevi, Ibn Daud, and Maimonides

Writing in tenth-century Egypt, Saadia Gaon (882–942), incorporated Kalām influences into his major philosophical work *'Emunot ve-De'ot* (*The Book of Doctrines and Beliefs*).[87] The ostensible purpose of this work is to show that the truths of scientific inquiry can be reconciled with the tradition of Torah; to this end he examines a series of philosophical conundra, many of which are drawn from Kalām writings. In his chapter on creation, Saadia presents eight arguments for the creation of the world that can be divided into two groups of four arguments each: the first group proves that the world must be finite (i.e., not eternal), and the second group that the world was created *ex nihilo* and not out of a pre-existent matter.

Saadia Gaon's fourth proof of creation "from time" draws upon John Philoponus' first proof of creation and is based on Philoponus' premise (2.1) that no infinite can be traversed. Saadia's argument proceeds as follows:

- 4.1 We know that time is threefold: past, present, and future.
- 4.2 Although the present is shorter than any instant, let us take the instant as one takes a point.
- 4.3 By (2.1) above, we know that it is not possible to traverse the infinite.
- 4.4 If we assume that time is infinite, it is "impossible for thought to penetrate to the furthest point of that which is infinite."
- 4.5 Hence if an individual should try in his thought to ascend from that present point in time to the "uppermost points," it would be impossible for him to do so.
- 4.6 On the same reasoning it is impossible that the process of generation should traverse an infinite period down to the lowest point so as ultimately to reach us.
- 4.7 But if the process of generation did not reach us, we would not be generated, and the beings now existent would not be existent.
- 4.8 However I find myself existent.
- 4.9 I therefore know that the process of generation has traversed time until it has reached me.

4.10 We therefore conclude that time must be finite.[88]

Argument Summary: Saadia's argument is based on John Philoponus' premise that the infinite cannot be traversed. He argues that if the present instant is infinite, then it is never possible to traverse the very instant in which we find ourselves. So too, it would not be possible to traverse any other infinite instant in order to reach the point we now inhabit. But we know that we have in fact reached this present point in time. It follows, therefore, that the time we have to traverse is finite.

Having argued that (4.9), proof of the traversal of past time, supports his postulating the finitude of time, Saadia then applies the argument to the traversal of future time as well. But he fully recognizes that (4.3) needs further examination. It is in the context of supporting (4.3) that Saadia raises a possible objection to the argument. Saadia attributes to an anonymous heretic a variation of Zeno's celebrated paradoxes of motion:

4.11 The following objection has been made by "a certain heretic in conversation with one of the Believers in the Unity of God."
4.12 We know that any distance which an individual walks can be "divided into an infinite number of parts."
4.13 But we also know that an individual can in fact cover the distance between these two points.
4.14 Therefore it is possible to traverse an infinite distance.[89]

Argument Summary: In Saadia's statement of Zeno's paradox, the heretic claims that inasmuch as any distance is infinitely divisible, the fact that a person can travel from one point to another demonstrates that the infinite can be traversed.

The heretic's point in this argument is that (4.14) contravenes (2.1) and hence (4.3) must be rejected. How, then, can Saadia account for (4.14), traversing an infinite distance, without abandoning his argument for the finitude of time? Aristotle's original solution was that the individual has an infinite amount of time in which to traverse the infinitely divisible distance in question. Saadia, however, is more interested in Kalām solutions to the problem according to which one might apply the notion of the leap, or atomism. The Kalām philosopher al-Nazzām, for example, introduced the notion of the leap as a response to Zeno.[90] Believing in infinite divisibility, but eschewing atomism, al-Nazzām adopted the idea of infinitely divisible leaps in order to explain how we can traverse an infinity of subdistances. On this theory any journey involves a finite number of variably short leaps.[91]

Rejecting this Kalām theory as untenable, Saadia proposes his own solution, which is a reflection of Aristotle's distinction between actual and potential infinity. As we have seen above, Aristotle argued that Zeno had confused two senses of indivisibility: a complete divided state as opposed to the process itself of dividing. Aristotle then argued that a solution to Zeno's paradoxes relies upon this latter sense of infinity: because infinite divisibility is a continuing process, the potency in question is thus actualized whenever the process is in effect.[92] This distinction between actual and potential infinity, mediated through the works of John Philoponus, is reflected in Saadia's discussion.[93] Saadia argues that Zeno's paradox is sophistical in that it fails to note that "the infinite divisibility of a thing is only a matter of imagination (*maḥshavah*), but not a matter of reality (*po'al*)."[94] If, Saadia argues, the infinite traversal had occurred in the past in the imagination alone, the paradox would be valid. But since the process of generation has traversed real time and reached us, (4.12) "cannot invalidate our proof, because infinite divisibility exists only in the imagination."[95] In answer to this paradox, then, Saadia distinguishes between actual and potential traversal. Traversing a finite spatial distance is not the same as traversing infinity, because in this case there *is* no actual infinity, but only an infinitely divisible finite distance. That is, reminiscent of Aristotle's distinction, Saadia's point is that the infinite exists potentially and not actually.

Aristotle has shown that a magnitude is potentially infinite, by way of division, while number and time are potentially infinite by way of addition. He needs this point to demonstrate the eternity of the universe.[96] Saadia's contemporary al-Kindi rejected this view of the world by rejecting the notion of potential infinity as a mere "imaginative fancy"; he is thus left with an actual finite world which is dependent upon God. Saadia accepts, like al-Kindi, the Aristotelian idea of a potentially infinite divisibility but minimizes its significance.[97] The key to John Philoponus' argument was that the infinite cannot be traversed and the same is true of Saadia's argument. But as Davidson points out, Saadia's emphasis differs from that of John Philoponus in that Philoponus considered transformations whereas Saadia considers time.[98] That is, Philoponus' infinite is an infinite series, whereas that of Saadia's is an infinite continuum. This move simplifies the argument considerably, so that Saadia does not have to extend the argument separately to the sublunar and superlunar spheres.

But did Saadia even recognize the difference between an infinite series and infinite continuum? Evidently not, for otherwise he would not have moved so seamlessly between time and space. As Davidson points out, Saadia's response to this paradox is that the objection misleadingly adduces the traversing not of an "actually" existing infinite, but of an infinite existing solely in "imagination," whereas the actual proof relies on the fact that an actual infinite cannot be traversed.[99] That is, the objection views distance not as an infinite continuum, but rather as an infinite series of discrete parts. By accepting the cogency of the objection, Saadia is either "consciously drawing an analogy between an infinite series of discrete parts and an infinite continuum, or else he did not detect any distinction between the two."[100]

Saadia then presents four arguments in chapter 2 to show that the world was not created out of a pre-existent matter. Of these the very last pertains to traversing the infinite.

> I went still further, arguing that if we fail to admit the existence of something which has nothing prior to it, it is impossible for us to accept the fact that there exists anything at all. For if we consider in our mind that one thing comes from another thing, we have to predicate the same thing of the second as of the first, and say that it could only have come into being from a third thing; the same predicate again must be made of the third thing, namely that it could only have come into being from a fourth thing, and so *ad infinitum.* Since, however, an infinite series cannot be completed, it follows that we are not in existence. But, behold, we are in existence, and unless the things which preceded us were finite (in number), they could not have been completed so as to reach us.[101]

Note that in this argument, Saadia does not consider the time continuum, but rather the series of transformations leading up to what exists at the present moment. A second point worth nothing is that in this argument, Saadia feels obliged to append a statement to the effect that if the things preceding us were not infinite, they would not have been completed so that we might exist. This point was absent in the first argument, and reflects a similar point in John Philoponus.[102]

Before turning to Maimonides' critique of Kalām atomist theory, let us briefly note in passing the way in which two other philosophers deal with the themes and arguments so carefully adduced by Saadia. In Halevi's *Kuzari,* Saadia's arguments are conflated into one garbled series. Attempting to describe to the King of the Khazars the metaphysics of the Mutakallimūn, Halevi offers the following hodgepodge of arguments, set out in two sections, drawn from Kalām thought:

> [1] The first axiom deals with the creation of the world, with the object of making it an established fact, and it denies the theory that it is without beginning. If time had no beginning, the number of individuals existing in the past down to our own age would be endless. That which is endless cannot be actual. How could those individuals have become actual, being so many as to be without number? There is no doubt, however, that the past had a beginning, and that the existing individuals are limited by a number. It is within the power of the human mind to count thousands or millions multiplied without end, at least in theory, but this cannot be done in reality. For that which becomes actual and can be counted as one, is like the number which is both actual and

finite without doubt. How can the infinite become actual? The world has, therefore, a beginning, and the revolutions of the spheres are subject to a finite number.[103]

[2] Further, that which is infinite can neither be halved nor doubled, nor subjected to any arithmetical calculation. We are aware that the revolutions of the sun are one-twelfth of those of the moon, and that the other movements of spheres stand in similar relation to each other, one being the divisor of the other. The infinite, however, has no divisor. How could the one be like the other, which is infinite, being either below or above it, I mean larger or smaller in number? How could the infinite come to us? If an infinite number of things existed before us, how could the [idea of] number come to us? If a thing has an end, it must also have had a beginning, otherwise each individual object must have waited for the [prior] existence of an infinite number of others; so none would ever come into existence.[104]

In this garbled rendition, we can recognize several of John Philoponus' contributions, overlaid with Kalām rhetoric. Axiom 1 is based on Philoponus' premise (2.1) that no infinite can be traversed and reflects Saadia's fourth proof of creation from time (4.1–4.10). Halevi has conflated three motifs in this first rendition: the premise that if time is infinite there would be an endless series of individuals existing from the past to the present; the notion that an actual infinite is not countable; and the fact that the revolutions of the spheres are subject to a finite number, a separate argument of Philoponus which is tacked onto Halevi's summary. In his second set of arguments (axiom 2), Halevi conflates the divisibility and augmentability of the infinite with the revolutions of the spheres, as well as with the very existence of number. Here he combines Philoponus' second proof of creation that the infinite cannot be increased (2.2) with his third proof from the impossibility of an infinite number based on the movement and speed of the planets (2.3).[105]

Ibn Daud's argument for the impossibility of an infinite is more focused that that of Halevi. While reflecting the influences of Aristotle, John Philoponus, and the Kalām philosophers, Ibn Daud's discussion is brief and limited to demonstrating that no actually existing, ordered quantity can be infinite. Occurring in the context of an extended analysis of Aristotle's categories, his argument focuses upon infinite lines and runs as follows. Imagine two lines AB and CD which extend infinitely (see fig. 3.3). If, in line CD we remove a finite length (*ba'alat takhlit*) CE from it, and then lengthen the line ED until E is where C used to be, then ED is either equal to AB or shorter than AB. If the two are equal, then CD is equal to ED, which is impossible. If ED is not equal to, but shorter than AB, this too is impossible since ED was hypothesized to be infinite, and no infinite can be smaller than another. ED must therefore be finite. If we then replace the finite part of

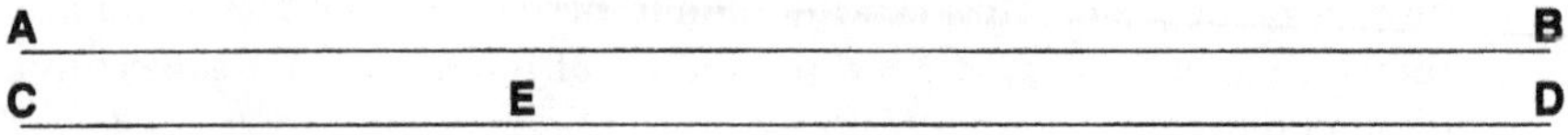

Figure 3.3. Ibn Daud's Argument for the Impossibility of the Infinite

CE, then the whole must be finite as well. By hypothesis then, "since line CD is equal to line AB, therefore both AB and CD are finite lines of a known magnitude, that is, it is possible to know their magnitude. Therefore, it is impossible that an infinite line exists."[106]

From this point Ibn Daud then argues that neither infinite surfaces nor infinite bodies exist. He then turns to the question of infinite series of counted things, and concludes that it too cannot be infinite since a series of counted things is by definition countable and hence finite. In other words, an actually existent infinite series of counted things is a contradiction in terms.[107] Interestingly enough, Ibn Daud does not mention the Aristotelian view that infinity exists in potentiality, perhaps, as Fontaine suggests, to avoid any implications that might ensue with respect to the doctrine of the eternity of the world.[108]

We have already presented Maimonides' formulation of Kalām atomism above. Let us now look briefly at his refutation of those arguments that make use of Kalām assumptions. As we have seen, Maimonides' summary incorporated several Kalām premises and consequences, most notably (P.1) the existence of indivisible and unextended atoms, (P.2) the existence of a vacuum, (P.3) the existence of indivisible time-atoms, (P.11) the denial of an infinite, (C.1) the illusion of motion, (C.2) the necessary creation of the world based on the necessary creation of individuals in time, and (C.3) the necessary creation of the world based on the creation of accidents in time.

With respect to (P.1) and (P.2), Maimonides has little of his own to add; these premises are exegetical and Maimonides simply presents the underlying Kalām considerations that led to them. The third premise is more problematic, however, and Maimonides is quick to point out that the Mutakallimūn, even more than "the cleverest philosophers," have "no knowledge at all of the true reality of time."[109] He does not present a counterargument, however, and is content to rest with an ad hominem statement which will be amplified in *Guide* 2.13, in the context of discussing creation. Nor does Maimonides have much to say with respect to (C.1), except to dismiss it as "abhorrent," and resulting in consequences even more abhorrent. For example, Maimonides objects to (C.1) on the grounds that it precludes the study of geometry altogether; geometrical demonstrations lose their cogency as evidenced by the fact that their theory precludes the division of a line composed of an odd number of atoms: "As for the other [Kalām positions] the demonstrations proving them are not cogent, as when we say we want to divide a line into two equal halves. For in the case in which the number of its atoms is odd, the division of the line into two equal parts is impossible according to their assumption."[110]

The eleventh premise is discussed in *Guide* 1.74.3. Maimonides points that whoever wishes to demonstrate the creation of the world in time must use (P.11) in order to rule out infinite regress. For from the hypothesis that the world is eternal, it follows that many imaginary infinite series can be postulated. Those who postulate the eternity of the universe believe both "that an infinite may be greater in number than another infinite," and congruously that "an infinite number of revolutions may be greater than another infinite number of revolutions."[111] Having rejected the infinite in all its guises, the Mutakallimūn are able to reject the hypotheses of the eternalists. Maimonides' refutation is embedded in (P.11). Against the first argument, Maimonides emphasizes the distinction made by Aristotle between actual and potential, and between essential and accidental infinite. The second argument is dismissed by his saying that "all these things are mere fictions and have no reality."[112]

Inasmuch as consequences two and three pertain to the issue of creation, Maimonides has much more to say about these arguments. Maimonides criticizes (C.3) by pointing out that it assumes three premises: "The first premise is that anything that is infinite through succession is impossible," while the second is that "all accidents come into being in time."[113] But Maimonides then presents Aristotle's arguments that the circular movement of the heavenly spheres contravenes both premises. The burden of proof, then, is upon the Mutakallimūn to demonstrate that, in contradistinction to Aristotle, the accident "circular movement" is produced in time.

The major thrust of Maimonides' rejection of the Kalām arguments, however, is contained in *Guide* 1.71. Maimonides' general contention is that these arguments "are derived from premises that run counter to the nature of existence that is perceived."[114] Maimonides himself adopts the theoretical stance of an eternalist and tries to argue for creation using Aristotle's own premises, rather than those of Kalām. In this way, he feels, the shaky metaphysical ground of Kalām metaphysics is obviated.

Gersonides on the Continuum

When Gersonides turns to the problem of infinite divisibility of space and time, he does so primarily against the backdrop of Aristotelian arguments. His discussion of the continuum is contained primarily in books 3 and 6 of *Milḥamot ha-Shem,* in the context of his discussions of divine omniscience and creation respectively. The issue of infinite divisibility is raised in a curious context, namely that of God's knowledge of particulars. According to Gersonides some modern philosophers used the claim that continuous quantity (*ha-kamut ha-mitdabeq*) is divisible into what is indivisible in order to demonstrate that God does not know particulars. The general argument of these philosophers was that if God knows particulars, then if he has knowledge of what is divisible, then he knows all its parts. But in accordance with what is entailed by divisibility, these parts would be infinitely divisible. Hence God would know those parts that are infinitely divisible. But by definition, God's knowledge is only of bounded objects. Hence he cannot know divisibles. But since there are no indivisible magnitudes, these philosophers concluded that God does not know particulars.[115]

From this argument it would seem to follow that either God does not know particulars, or that continuous quantity is divisible into indivisible parts. Gersonides, however, wants to claim both that God does know particulars and that there are no indivisibles. He therefore rejects the argument of these philosophers by arguing that God knows the division of body according to its nature: "He knows that whatever is divisible can be divided further insofar as it is a quantity (*ba'al kamah*). God has no knowledge of an end of division, which by its very nature has no end."[116] In other words, Gersonides claims that although God knows the nature of quantity *qua* quantity (*ha-kamut be-mah she-hu' kamah*), namely that it is infinitely divisible, nevertheless he does not know of each particular instance of quantity what its infinite divisibility comprises.

This point is further emphasized in Gersonides' rejection of attempts to explain God's knowledge by claiming that God knows the possible division of magnitude into parts, but not its actual division. Again, Gersonides argues that God knows that every part into which a quantity is divided is capable of division, but he does not know the limit of what is limitless by nature.[117] It is here that Gersonides is already sowing the seeds for his subsequent rejection of any notion of potential infinity.

Gersonides now turns to the problem of the divisibility of the continuum. He first states what he takes to be Aristotle's formulation of the argument on behalf of the atomists, showing that body must be divisible into indivisibles.[118] We have already seen that Aristotle himself resolved this dilemma by distinguishing two senses of divisibility: actual and potential divisibility. Averroës, however, understood Aristotle's solution somewhat differently. In his commentary on *De Generatione et Corruptione,* Averroës argues that the actual division of a body is impossible. He emphasizes the impossibility of simultaneous division and argues that just as a person cannot acquire knowledge of all the sciences simultaneously but must do so successively, so too a body cannot be divided at all of its points simultaneously. Simultaneous division, according to Averroës, could occur only if the points in question were contiguous.[119] The main thrust of Averroës' argument is summarized by Gersonides in *Milḥamot* as follows: "Thus when we have chosen a point at any desired place where the magnitude can be divided, once we have divided the magnitude at that point, it is impossible to divide it at any second point at a place we might choose, for it is impossible to divide it at a point consecutive to the first point."[120] Gersonides, however, disagrees with Averroës on several counts. First he points out that Averroës must assume the very thing he sought to deny: for if, as Averroës thought, division is not possible at any point A_1 contiguous to A, and if body is infinitely divisible, then there is an infinite number of points at which division is not possible. Hence, on Averroës' argument, it follows that body is comprised of indivisibles—the very condition that he sought to deny. Second, Gersonides claims that divisibility at each point is possible. That is, according to Gersonides, division at A_1 is possible even if division has already occurred at A. Even were he to grant Averroës his assumption, Gersonides claims, the dilemma would not be resolved.[121]

Gersonides therefore rejects Averroës' solution and offers his own solution to Aristotle's original dilemma, arguing that "when we posit a body actually divided at every place where it can be divided, we have postulated as existent something that is impossible, for it cannot be divided except into that which is [capable] of [further] division."[122] In other words, Gersonides is distinguishing here between two senses of infinite divisibility: the claim that a continuum is divisible into parts which are themselves further divisible, and the claim that a continuum is actually divisible into indivisible parts. The main thrust of Gersonides' discussion is to distinguish between what I shall call infinite and endless division: a continuum is not actually infinitely divisible, but rather endlessly, or potentially, divisible. The first refers to a complete divided state, whereas the second refers to the process itself of endlessly dividing, a process that is not completable.[123]

Gersonides' distinction between infinite and endless division is brought out even more strikingly in his discussion of number. This discussion occurs in two contexts. In *Milḥamot* 6 and 3 Gersonides has characterized matter in terms of continuity (*hitdabbequt*), that is, that by virtue of which matter can be infinitely divided and its parts still retain their continuity.[124] In *Milḥamot* 6 Gersonides poses a possible objection to his characterization as follows: Suppose one were to argue that inasmuch as number is augmentable, so too is quantity augmentable, or infinite. Gersonides' response is that the endlessness we find in number is not an endlessness of quantity, but rather of the *process* of division and augmentation. That is, Gersonides distinguishes between quantity itself and the act of increase/diminution, which is based on quantity. Although the act of augmenting is never exhausted, quantity itself remains finite.[125] This is my understanding of Gersonides' statement that "quantity is infinitely finite."[126]

In *Milḥamot* 3 the question is posed in terms of whether God can have knowledge of a non-augmentable number. Gersonides summarizes the argument of those who argue that there might exist a non-augmentable number that is known to God, and claims that such a position is absurd.[127] The fallacy is that it assumes an "existent that is logically impossible [i.e., the greatest number]."[128]

The issue that emerges from this is how number and divisibility of the continuum are related. Averroës had argued that in a body there is an infinite number of points that exists potentially, and division is possible at each point.[129] Gersonides dismisses Averroës' position as absurd on the grounds that if a body did contain finite dimensions, and these dimensions were potentially infinite, then they could not be in contact. Why not? According to *Physics* 6.1, between any two points there would be another point. Since there is an infinite number of points between any two points, the measure between any pair of dimensions would be infinite in number. But Gersonides points out that this would result in a finite body being infinite, for the reason that if all the distances within the finite body were totalled, the result would be an infinite number.[130]

Going back to our original diagram of Zeno's paradoxes, we can see what Gersonides has in mind (fig. 3.4). Imagine the finite body, AB which is divided at C and D. The distance from AB = f, which is a finite number. The distance CD = i, which is infinite, however, since there is an infinite number of points between C and D. So the distance of CD, which is a part of AB and hence smaller than it, would nevertheless be infinitely large and greater than the finite distance of AB. In order to avoid positing the existence of the infinitely large, Gersonides concludes therefore that "a continuous magnitude does not have an infinite number of points, neither potentially nor actually."[131] So how does he account for the fact that the continuum can be infinitely divided, or number infinitely augmented? He claims that infinity with respect to divisibility is found not in the number of divisions undertaken but in the very act of dividing. The number of divisions in any continuum is always actually finite, although this number may be augmented indefinitely. For example, Gersonides claims, take a continuum that is divided into two parts, each of which is divided into two more parts, and

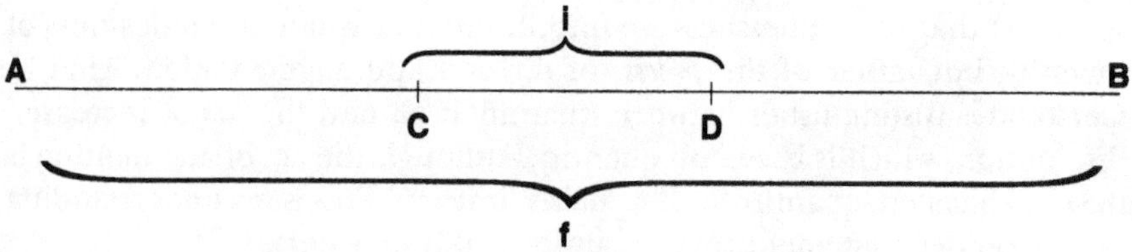

Figure 3.4. Gersonides' Conception of Zeno's Paradoxes

each in turn divided so that there are now eight parts. Gersonides states that "the process can be carried on indefinitely, although the number of parts is always finite."[132]

The application to number is explicitly made: "Just as a number is not augmentable, either potentially or actually, into that which is infinite in number [i.e., an infinite aggregate or infinite number], so, too, is this the case in the division of a continuous magnitude."[133] The former is upheld on the basis of two principles. The first states that inasmuch as every number is even or odd, and these numbers are all finite, it follows that every number is finite. The second principle is a generalization of the first and states that "in augmentation a number has no potentiality for becoming a non-number but the potentiality for becoming a greater number than it is."[134] In other words, if a number could become infinite it would no longer be a number, since by the first principle, a number is by definition either odd or even, and hence finite.

In short, just as number always remains finite and cannot become infinite, so too Gersonides concludes that "continuous quantity" cannot contain infinite parts: "A continuous magnitude does not potentially have an infinite number of parts; hence it is evident that it does not potentially have an infinite number of points."[135] We have seen that according to Gersonides quantity by its very nature is finite: "The nature of quantity necessitates that it be finite."[136] If, Gersonides argues, quantity had an infinite number of parts, it would follow that a finite quantity would be infinite, for each of its potential parts would be comprised of quantity. In order to avoid this conclusion, Gersonides therefore characterizes infinite divisibility of the continuum in terms of the possibility of subdividing its parts. Thus, just as no number can be infinite, so too no continuum can contain either parts or dimensions that are infinite in number.

What we have seen here is a modification of Aristotle. Gersonides has argued that the notion of potential infinite divisibility is misleading at best, and leads to serious paradoxes of its own. For if, as Aristotle has argued (*De Caelo* 1.12), all potentialities are eventually actualized, then at one time all the possible divisions of a magnitude will have been completed and an actually existing infinite will exist. In order to avoid this conclusion, Gersonides replaces the notion of "potential infinity" with his own notion of "indefinitely divisible" infinity.[137]

In answer then to the question how to account for the infinite divisibility of the continuum, Gersonides is unwilling to deny infinite divisibility altogether, as evidenced in his characterization of matter. And he is aware of its importance with respect to the addition/division of number. Hence his solution is to deny the reality of the infinite while at the same time affirming the possibility of the process of infinite division/addition. That is, according to Gersonides, what we mean by the infinite is a process that by nature is never-ending; the members of the process, however, are finite. Thus, for example, the natural number series is infinite in the sense that its members can be ever-augmented; but each member of the series is finite.

Crescas on Infinity, Space, and the Vacuum

In the process of upholding the basic dogmas of Judaism, Crescas subjects Aristotle's physics and metaphysics to a trenchant critique. His rejection of Aristotle's theories of place and the infinite forms part of an extended attempt to weaken Aristotle's hold upon Jewish philosophy. In Aristotle's *Physics* 4.1, space (discussed originally in *De Caelo*) is identified with place (*topos*) and forms an integral part of Aristotle's theory of motion, which is defined as "change of place."[138] Any definition of the place of a body must satisfy certain requirements: it must serve as the container of a thing and not its part; it must be neither greater nor smaller than the thing contained; it must be separable from what it contains; and place is distinguishable with respect to "up" and "down." Place is then properly defined by Aristotle as "the boundary of the containing body at which it is in contact with the contained body."[139] On the basis of this characterization, Aristotle proceeds in *Physics* 4.6 to reject the possibility of a vacuum, for in a theory that does not allow for a place not correlated to any body, there can be no "empty space" or void.

But one problem that immediately asserts itself has to do with the place and subsequent motion of the outermost celestial sphere. Aristotle had maintained in other contexts that the outermost celestial sphere could not be in a place, since body, place, void, or time did not exist beyond the limit of the world; and yet, as we have seen already in chapter 2, that this sphere moved was a vital piece of Aristotle's cosmology and astronomy.[140] How, then, could this outermost sphere move (i.e., change its place) with constant velocity if it was itself without place?[141] Aristotle's own solution, that the sphere "somehow" contained itself, that the soul and the world are "in a way, in place, for all its parts are; for on the circle one part contains another,"[142] proved most unsatisfactory to his followers. Either, they maintained, Aristotle's theory of place had to be modified or the dogma of the motion of the outermost sphere had to be abandoned.[143] Crescas adopts the former strategy, rejecting Aristotle's theory of place by pointing out the many absurdities to which it leads.

One important implication of Crescas' alternative conception of place and infinity has to do with his postulating the existence of the vacuum. According to Crescas, place is prior to bodies: in contradistinction to Aristotle's conception of place, space for Crescas is not a mere relationship of bodies but is the "interval between the limits of that which surrounds."[144]

Space is seen by Crescas as an infinite continuum ready to receive matter. Because this place or extension of bodies is identified with space, there is no contradiction in postulating the existence of space not filled with body, that is, the vacuum.[145] Crescas, in fact, assumes that place is identical with the void, on the grounds that "place must be equal to the whole of its occupant as well as to [the sum of] its parts."[146]

Crescas' theory of space has ramifications with respect to his conception of the infinite as well. In order to postulate the infinity of space, time, and number, Crescas must refute Aristotle's theory of the infinite. Crescas' general contention is that Aristotle's arguments are all victim to a common fallacy in that they assume that one can argue against the existence of the infinite from the analogy of the finite. Crescas, however, will want to maintain that the assumptions one makes about the finite are inapplicable to the infinite—the infinite and finite are qualitatively different and cannot be compared.[147] Let us turn, then, first to those arguments having to do with infinity and the vacuum, and then to their implications with respect to number and time.

Of the many arguments adduced against the infinite, several have relevance to our topic of infinite divisibility. Crescas' critique in *'Or 'Adonai* centers on the twenty-six Aristotelian propositions adduced by Maimonides in the *Guide*. The first three propositions deal with the infinite and reiterate the Aristotelian dicta that an infinite magnitude is impossible, that the coexistence of an infinite number of magnitudes is impossible, and that an infinite series of causes and effects is impossible.[148] In order to deny the existence of infinitude magnitude, Aristotle advanced arguments to show the impossibility of incorporeal infinite magnitudes, of corporeal infinite magnitudes, of the rectilinear or circular motion of infinite bodies, and the existence of an actual infinite.

Crescas immediately singles out Aristotle's denial of the vacuum as the underlying basis for his contention that incorporeal infinite magnitude cannot exist. Aristotle had argued that motion is a function of two forces, medium and motive force. The rarer the medium, the quicker the motion. And so if there were a medium of infinitely rare density, one would expect that the body in question would move in an infinitesimal time. Because a vacuum has no density, it would then follow that motion in a vacuum would happen in no time. But this scenario, according to Aristotle, is absurd: inasmuch as the distance moved is divisible, and a moving body must pass from one point to another, it follows that motion must take time.[149] Crescas' response to Aristotle is that every moving body has its own fundamental velocity: "even by eliminating the receptacle there will still remain an original time of motion."[150] And so a body moving in a vacuum will therefore move according to its own fundamental velocity.

Another argument has to do with the interpenetrability of bodies.

Aristotle had argued that filling immaterial infinite magnitude by an extended body would violate the impenetrability of bodies. For Crescas, however, matter alone does not account for impenetrability, but matter and dimensionality are together sufficient.[151] Extended space and void are not two kinds of space, but one—the former is mixed with matter whereas the latter is pure and hence invisible. The vacuum exists therefore and is not a physical or metaphysical contradiction. That the vacuum is infinite is presented in the following argument. Assume that an incorporeal magnitude is infinite by nature. It is also simple and homogenous. But an infinite extension, according to Aristotle, is divisible. If this infinite incorporeal magnitude is divisible, its parts are then infinite, which would imply that an infinite is composed of infinites. In order to avoid this difficulty, Crescas utilizes the analogy of a mathematical line:

> Examination of the argument which he has framed to prove the impossibility of an incorporeal infinite magnitude. We say that the argument is fallacious and a begging of the question. For he who assumes the existence of an incorporeal magnitude likewise affirms the existence of an incorporeal quantity. By the same token, it does not follow that the definition of the infinite would have to be applicable to all its parts, just as such reasoning does not follow in the case of mathematical line. Nor would there have to be any composition in it except of its own parts.[152]

In this analogy Crescas is making a distinction between two types of divisibility: that which comprises composition and that which does not. A syllable, for example, is divisible into letters and composed of letters, whereas a mathematical line is divisible into linear parts but not composed of these parts. In the latter case the linear parts are bounded by points, and so if the line were composed of parts, it would be composed of points. But we know (from Euclid) that a line is not composed of points. And so Crescas is able to conclude that when a thing is continuous and homogenous, it is divisible into parts but not composed of parts. Then Crescas applies this analogy to the vacuum: like the line, the vacuum is divisible into infinite parts but not composed of them, and so the infinite has no composition "except of parts of its own self."[153]

The infinity of the vacuum is reinforced by additional arguments having to do with the limits of infinity. Altabrizi, for example, offered what Crescas calls "the argument from application" (*mofet ha-devequt*), which is stated as follows:

> Suppose we have a line infinite in only one direction. To this line we apply an infinite line [which is likewise infinite in only one direction], having the finite end of the second line fall on some point near the finite end of the first line. It

> would then follow that one infinite line would be greater than the other. But this is impossible, for it is well known that one infinite cannot be greater than another.[154]

Crescas' reply is that terms like "greater than" and "smaller than," do not apply to infinite magnitudes.[155] This point is reinforced with respect to time and number as well. When we say that time and number are infinite, we understand that terms such as "many" and "few," "large" and "small," "equal" and "unequal" are simply not applicable. To suggest that one infinite is larger or smaller than another infinite is to commit a category error not unlike that described by Maimonides in his theory of divine predication.[156]

Having postulated the existence of an infinite vacuum, Crescas turns to Aristotle's arguments against the existence of an infinite revolving sphere on the grounds that an infinite has no first point and an infinite distance cannot be traversed in finite time. According to Aristotle the movement of an infinite revolving sphere would require traversing an infinite distance in a finite time. Crescas reconstructs Aristotle's argument as follows:

> If an infinite, spherical body moving in a circle existed, it would have to traverse an infinite distance in finite time. But this is impossible. Hence the existence of an infinite endowed with circular motion is impossible. The proposition which denies the consequent is self evident. As for the connection of the consequent with the antecedent, it may be made clear as follows: Let an infinite line emerge from the center; and let also a chord intersect the sphere. Since the sphere is assumed to be infinite, it is clear that the chord will have to be infinite. Let that chord be at rest. Now, if we suppose the radius to revolve on its center, it will at some time meet the chord and intersect it while at another time it will not meet it. As a spherical body rotating upon itself must complete its rotation in finite time, it follows that the radius would traverse an infinite distance, namely, the given chord, in finite time. But this is a flagrant absurdity, inasmuch as motion completed in finite time must take place over a finite distance.[157]

Crescas' response to this and similar arguments from circular motion is that "they are likewise inconclusive, being again based upon the analogy of a [finite] sensible body."[158] More specifically, he argues that motion has no absolute beginning, since it is infinitely divisible. The time of motion, as well, has no beginning. When two infinite lines (from a sphere) meet, they do not meet at an absolute first point or at an absolute first time. Therefore there is no infinite distance being passed through in finite time. And so any distance traversed by a sphere in finite time, and on a finite axis, will be finite.[159] Crescas' response to the first argument highlights the point that infinite increase and decrease still do not change the nature of a finite magnitude. For example, "it is possible for a distance infinitely to decrease and still never completely to disappear . . . if in the case of decrease, there is always a certain residual distance which does not disappear, *a fortiori* in the case of increase it should be possible for a distance, though infinitely increased, always to remain limited."[160] Hence, returning to our original query, Crescas is able to explain the motion of the outermost celestial sphere on the grounds that it rotates in an infinite vacuum; the sphere is no longer conceived as the final limit or boundary of the space of the universe.

In an interesting gambit, Crescas uses Aristotle's arguments against an absolute beginning to motion (in *Physics* 6.5.23a 236a ff.) in order to uphold an infinite series of causation. Appropriating Aristotle's dictum that there can be no first part of motion, because every object that is moved must have already been moved, Crescas maintains that "it is not inconceivable, therefore, that the infinite line [in question] should meet the other line in a finite distance with a finite motion, and this may be accounted for by the fact that the extreme beginning of motion must take place in no-time."[161] In his parting company with generations of Aristotelians who had used the denial of an infinite series of causes to postulate the necessary existence of a prime mover, Crescas therefore will have to resort to other arguments to postulate the existence of God.

Crescas' refutation of Aristotle's theory of infinity affects notions of time and number as well. In order to reject Aristotle's idea that every number is odd or even, and hence finite, Crescas distinguishes what is in theory numerable from what is actually numbered, allowing him to postulate the notion of infinite number: things that "have the capacity of being numbered but are not actually numbered (*asher mi-darkam she-yisapru 'aval 'einam sipurim be-foal*) . . . are not excluded from the possibility of being infinite."[162] To infinite numbers, ascription of even and odd are inapplicable.[163] What Crescas has in mind here is an infinite series of finite numbers which has no end. With respect to the infinity of time, Crecas takes seriously the notion that the infinite is not traversible, claiming that just as a numerical series can be infinite, so too can the temporal series of causes be infinite. Again, turning Aristotle against his own principles, Crescas argues that inasmuch as it is not impossible for an infinite number of effects to emanate from one single cause, it is not inconceivable to postulate an infinite series of intermediate causes and effects within a causal series emanating from and coexisting with an uncaused eternal cause.[164] Replying to the obvious counterargument encountered above that an infinite series cannot be traversed and so our own existence is proof of a finite causal series, Crescas responds that "that which cannot arrive except by the precedence of what is infinite does actually arrive."[165] More specifically, Crescas maintains that in the case of things that coexist in time, such as the infinite series of intermediate causes and effects, their nontraversibility must be proved and not simply assumed. Just as a causal series can be infinite, time as well can be conceived without a beginning.

Conclusion

In this chapter I have maintained that the philosophical thrust of Zeno's denial of the divisibility of the continuum reappears not only in Aristotle's direct reply to Zeno, but in John Philoponus' refutation of the eternity of the universe and then in Kalam atomism. The preoccupation of the Mutakallimūn with metaphysical atomism was transmitted to Jewish Neoplatonists, and inherited by Saadia Gaon and his successors. By the time of Gersonides and Crescas, arguments concerning the infinite divisibility of space, time, and continuum had developed a complexity that was then superimposed onto theological concerns having to do with the eternity of the universe and God's knowing the infinite.

Perhaps the best example of this integration occurs in the work of Gersonides. We have seen how Gersonides amplifies Maimonides' discussion of infinite divisibility. How can this solution be applied to Zeno's paradoxes? I submit that, given what Gersonides has said about infinite divisibility of the continuum, his solution in this work is surprisingly modern. In his capturing the spirit of modern mathematical notions of infinity, two points stand out. First, with respect to the infinite divisibility of a continuum, we have seen that Gersonides disagrees with Aristotle's distinction between the actual and potential infinite in order to resolve Zeno's paradox. Gersonides' solution is in fact closer to modern limit theory. According to this theory, mathematicians speak of an infinite converging series (i.e., the sum of subdivisions between 0 and 1) having a finite sum that approaches 1. It is important to note that the finite sum functions as a limit and is not part of the converging series itself. Gersonides is at pains to point out that infinite numbers cannot have a finite sum. However, his characterization of infinity as a process rather than as a state leads ultimately to a modern sensibility. For what limit theory espouses is that the most we can say is that an infinite series converges to a finite sum. Limit theory by itself does not resolve Zeno's paradoxes. As Max Black has pointed out in his discussion of infinity machines, the operation of summing the infinite series will tell us where and when Achilles will catch the tortoise if he can catch the tortoise, but that is a big if.[166] This is the point of Gersonides' discussion as well.

The second point has to do with the augmentability of number. We have seen that in answer to the question "Can God have knowledge of a non-augmentable number?" Gersonides has argued that given the nature of

God's knowledge, such a state is impossible. At most God can know that the number series is indefinitely augmentable, but there is no "last" number in the series to know. In this discussion, Gersonides foreshadows more recent discussions of Cantor and others.[167] In light of his contention that God only knows that a continuum is infinitely divisible, but not its individual parts, Gersonides can be construed as arguing that the most God can know is a denumerably infinite set: he can count the members of the set, that is, he knows "how many" members of the set there are, but he does not know the "infinitieth" member of that set, since there is no such number to know. Of course, Gersonides differs from Cantor in that he would deny that such a set comprises an actual infinity, for as we have seen he denies the completability of an infinite series. But in terms of the implications of such a set for God's knowledge, Gersonides would accept the description of such a set. Let us turn, then, to the issue of God's knowledge in light of temporality.

CHAPTER FOUR

Temporality, Human Freedom, and Divine Omniscience

Introduction

The general problem of divine omniscience comprises a number of subsidiary problems: the problem of logical fatalism as introduced by Aristotle in his *De Interpretatione* and further developed by the Stoics; the problem of God's foreknowledge of human events and the relation of this knowledge to free will; and particular theological difficulties centering around the notions of prophecy, providence, and retribution.[1] For our thinkers, the question of whether God knows particulars arises out of a clash of two worldviews. On the one hand Jewish philosophers accept an epistemology that identifies knower with object known in certain respects; hence, to claim that God knows events in the sublunar world is to posit at the very least a likeness, if not an actual isomorphism, between God and his creation. On the other hand, as theologians, these thinkers are committed to certain theological presuppositions about the transcendent nature of God, which in many instances seems to preclude such a likeness. The primary worry underlying these questions is whether the claim that God's knowledge extends to particulars in the sublunar world is inconsistent with the doctrine of God's unity and immutability. In this chapter I examine the major philosophical moves made in medieval Jewish philosophy to address these issues. Before turning to the second and third clusters of issues surrounding God's foreknowledge, however, let me first explore the underlying logical dimension of the problem. For it is the logical dimension that underscores the relevance of God's omniscience to the domain of temporality.

The Problem Defined: Aristotle's Sea-Fight Paradox

The logical dimension of this issue can be traced back to Aristotle's *De Interpretatione.* In his celebrated "Sea-Fight" Paradox (*De Interpretatione* 9), Aristotle is concerned to safeguard the limits of the Law of the Excluded Middle with respect to statements about future particular events. Having claimed earlier that this law pertains to statements about past and present events, he now in chapter 9 limits it, arguing that any statement about a future contingent event is now neither true nor false. The argument proceeds as follows:

1.1 With respect to statements about past and present events, it is necessary for the affirmation or negation to be true or false.
1.2 It is therefore necessary for everything (regarding past and present events) to be the case or not to be the case.
1.3 There is, then, no chance *now*, with respect to past and present events.
1.4 Assume that with respect to statements about future events, it is also necessary for the affirmation or negation to be true or false.
1.5 If (1.4) is so, then it is necessary for statements about future events to be the case or not to be the case.
1.6 There is, then, no chance, with respect to future events.
1.7 If there is no chance with respect to future events, then all future events happen of necessity and there is no point for humans to deliberate with respect to future events.
1.8 But we see that "what will be has an origin both in deliberation and in action."[2]
1.9 Hence it is not the case that statements regarding future events are the case or are not the case.
1.10 With regard to future particular events, they necessarily will be or will not be; but neither disjunct is now necessary.
1.11 Affirmations or negations about future particulars are therefore not now true or false.
1.12 In this way the indeterminacy of the future is safeguarded, as is human deliberation and choice.

Argument Summary: Aristotle states that statements about past and present future events are now true or false. But with respect to future events, statements about these events are not now already true or false. If they did have a truth value, there would be no human freedom; but Aristotle postulates that we know that future events originate in our deliberation. Hence there is human deliberation, and future events are indeterminate with respect to truth value.

That is, the world at present (now) is indeterminate with respect to the existence or nonexistence of future contingents. As an example of a future contingent event, Aristotle discusses a sea-battle, stating that "it is necessary for there to be or not to be a sea-battle tomorrow; but it is not necessary for a sea-battle to take place tomorrow, nor for one not to take place—though it is necessary for one to take place or not to take place."[3] Distinguishing in this case between simple necessity and temporal necessity, Aristotle states that "what is, necessarily is, when it is; and what is not, necessarily is not, when it is not. But not everything that is, necessarily is; and not everything that is not, necessarily is not."[4] It is Aristotle who thus connects the relevance of future indeterminacy to human action: if future contingents have a truth-value now, it would follow that "nothing of what happens is as chance has it, but everything is and happens of necessity. So there would be no need to deliberate or to take trouble (thinking that if we do this, this will happen, but if we do not, it will not)."[5]

On the basis of this distinction between simple and temporal necessity stated in (1.10), Aristotle is thus able to safeguard the indeterminacy of the future (1.12) without sacrificing the truth-value of all other statements (1.1)–(1.3).[6] In this model, the future is ontologically distinct from both the past and the future; this distinction is rooted in human behavior as well as in our linguistic utterances.

The medieval commentary literature upon Aristotle's sea-battle demonstrates the many interpretations given to the ontological status of future contingent events.[7] The relevance of Aristotle's sea-fight battle to our topic of temporality becomes clear when we focus upon the ontological status of both the future and the past in God's knowing intellect. As Normore has so succinctly pointed out, the issues underlying divine omniscience "pose some of the clearest challenges to a philosophical theory of God's relation to time."[8] The main issue that concerns us in this chapter is whether past, present, and future, and God's knowledge of past, present, and future, are necessary or contingent. More specifically, if God is present to all of time, and knows all time in one act, what does this imply about the ontological status of the future? If what is future is as available to God as what is past, does it follow that the future is as fixed ontologically as the past?[9]

Medieval philosophers, concerned with safeguarding the freedom of human action, worried specifically whether God's foreknowledge of future contingent events entailed the necessary occurrence of these events. The general form of their worry can be construed in terms of the following argument:

> For any true substitute for *q*, where *q* stands for any future contingent state of affairs, there is a true substitute for *p*, where *p* is about the past, which makes these premises true:

2.1 Necessarily ($p \rightarrow q$)
2.2 Necessarily p
2.3 Necessarily ($p \rightarrow q$) $\rightarrow$ (Necessarily $p \rightarrow$ necessarily q)
2.4 Necessarily q[10]

When p is allowed to stand for "God foreknows q," we derive the precognitive form of the argument according to which God's foreknowing q entails q.

This argument is further strengthened once we unpack the notion of divine knowledge. Sorabji has suggested that an analysis of God's foreknowledge follows that of human knowledge in that both involve belief states.[11] Divine knowledge, however, unlike its human counterpart, is infallible. Hence the phrase "God foreknows p" can be unpacked as: God believes p in advance of the occurrence of p; p is true; and God is incapable of error. The crucial part of this analysis is the emphasis upon God's infallibility. It is God's infallibility, coupled with his prior knowledge, rather than the causal force of his knowledge, that carries with it the suggestion of the necessity of the objects of his knowledge. What concerns medieval philosophers in general, and Jewish philosophers in particular, is that if God is infallible, then the objects of his knowledge *cannot fail to be* what God already knows them to be. How to account for the ability of humans to contravene the prior infallible knowledge which God has of their actions becomes of paramount importance.

In much of medieval Jewish literature, two main solutions to the problem of divine foreknowledge presented themselves: compatibilism and incompatibilism. Most Jewish philosophers, along with their scholastic contemporaries, adopted a form of compatibilism, claiming that God's foreknowledge of future contingent events in no way impedes human freedom.[12] I will take compatibilism to be the view that God's knowledge is compatible with human freedom. More specifically, let us define compatibilism as follows:

> *Compatibilism* = God's knowing at t_1 that I do p at t is consistent with the possibility of my doing $-p$ at t.[13]

On this reading, a compatibilist position permits the following statements to cohere:

3.1 God knows at t_1 that I will do p at t.
3.2 At t I will do p.
3.3 At t I can refrain from doing p.
3.4 My doing p at t is a free action.
3.5 Hence God knows at t_1 that I will do p at t, and my doing p at t is a free action.

The compatibilist, therefore, has no problem with asserting both that God has foreknowledge that I will do a particular action (3.1) and that I do that action freely (3.4). That compatibilism is the dominant position in Jewish philosophy will become apparent as we examine the main views in this chapter. But compatibilism is not immune to logical difficulties, and incompatibilists are quick to point to discrepancies between upholding both foreknowledge and human freedom. Centering their critique around premises (3.1) and (3.3), incompatibilists maintain that a free action must be construed as follows:

> Action p is *free* at t = it is possible for me to do p at t and to refrain from p at t.

Incompatibilists disagree, however, over the implications of accepting this definition.[14] One form of incompatibilism, which I shall term *indeterminism*, is that God simply does not know future contingent events. The indeterminist will want to maintain that if (3.1) is true, then p cannot be construed as a free action. Starting with human freedom and the existence of contingency as a given, the indeterminist will deny (3.1) on the grounds that if p is truly indeterminate prior to its actualization, then it cannot be known by God:

> *Indeterminism* = my freely doing p at t precludes God's knowing p at t_1.

Clearly this position safeguards human freedom at the expense of divine omniscience. Another strand of incompatibilism, *determinism*, claims that if God knows the causal chain of events that unfolds from his knowledge, human actions are ultimately determined by this knowledge. The determinist in other words rejects both (3.3) and (3.4) in light of (3.1), arguing that given the definition of free action as the ability to do p and $-p$ at t, it follows that if God knows p at t, then the agent is not free:

> *Determinism* = God's knowing at t_1 that I do p at t entails that it is not possible for me to do $-p$ at t, and hence I am not free.

Both indeterminism and determinism had their adherents in Jewish philosophy, albeit in very few numbers. For as I shall argue below, both determinism and indeterminism undermine other basic tenets of Jewish belief, most notably prophecy and doctrines of reward and punishment. But before turning to these positions, we must turn to two ancillary issues. The first pertains to the concept of knowledge and how it functions in the context of divine omniscience. The second issue has to do with Jewish thinking about the influences of astrology upon human freedom.

Medieval philosophers in general are concerned with the extent and limits of God's knowledge of particulars in the sublunar universe. Their presuppositions result from adopting a monist ontology. In a monism, God is characterized as a unity, unchanging in essence and omniscient. That God knows at least his own essence is reflected in Aristotle's dictum that God is thought thinking itself.[15] Yet once it is claimed that God's knowledge extends beyond his own essence to his creation, a number of problems arise. If it is argued that God knows particular entities, does this compromise either the unity or the immutability of his knowledge, and thus, of his essence? The first would appear to follow from the following considerations. The doctrine of divine simplicity suggests the identity of God's essence and knowledge. If, however, God's knowledge constitutes a multiplicity, then God's essence would reflect this plurality. The second follows from the fact that if the objects of God's knowledge are subject to generation and corruption, then his essence must change along with changes in the objects of his knowledge. In both cases the underlying implication is that God's essence has been acted upon or otherwise affected by the objects of his knowledge in much the same way as the human intellect is acted upon by the objects of its knowledge. How to account for divine knowledge while denying, on the one hand, plurality, and on the other hand, the effects of causal activity upon God, becomes a major consideration for Jewish thinkers.

What emerges from these issues is that the term "knowledge" needs further clarification. Reflecting the doctrine of the ultimate unknowability of God's nature, Maimonides, for example, asserts the equivocal nature of the term "knowledge" when applied to God. That is, since God's act of knowing is totally unlike human acts of knowing, the term "knowledge" has one signification when applied to humans and another when applied to God. In Part 1:51–61 of the *Guide,* Maimonides develops an elaborate theory of divine predication, the purpose of which is to claim that human language is inadequate to predicate anything of God. In these chapters Maimonides argues that terms predicated of God must be understood in one of three ways. The first construes such terms as action predicates, descriptive of the "ways and the characteristics" of the Deity.[16] From these action predicates we infer corresponding mental states analogous to those states that humans experience when exhibiting those actions. For example, just as we attribute mercy to an individual who exhibits merciful actions, so too do we attribute to God the state of mercy when God exhibits actions that resemble our own merciful actions. Thus an extended analogy is drawn between God and humans as follows:

human action *m*, which is said to be merciful	analogous	Divine action m_1, which resembles human action m
human emotive state is "mercy"	to:	Divine emotive state is "mercy$_1$"

Maimonides, however, is quick to point out that the descriptive action terms "mercy" and "mercy$_1$" do not mean the same thing. This leads to his second theoretical point, namely that the four essential attributes of God—life, power, wisdom, and will—are of one simple essence; all other attributes are to be conceived either as descriptive of divine action, or as negative attributes. However, even these four attributes, when predicated of God, are used in a homonymous or equivocal sense.[17] The difference between human and divine predicates is qualitative: since the terms are applied by way of perfect homonymity, they admit of no comparison between God and his creatures.

Reaction to Maimonides' theory of homonymous predication was intense, both in scholastic and Jewish circles.[18] Gersonides, for example, attempts to salvage the ability of humans to talk meaningfully about God. Gersonides disagrees with Maimonides' doctrine, claiming that divine predicates are to be understood as *pros hen* equivocals rather than absolute equivocals (as Maimonides had argued). What this means is that according to Gersonides, predicates applied to God represent the prime instance or meaning of the term, whereas human predicates are derivative or inferior instances. So for example, knowledge when applied to God is perfect knowledge and constitutes the standard for human knowledge, which is less perfect than divine knowledge: "the term 'knowledge' is predicated of God (may he be blessed) *primarily* and of others *secondarily*."[19] Gersonides denies that terms have completely different meanings when predicated of God and of humans; it is only because of an underlying commonality of meaning that we can use language meaningfully at all.

In light of the linguistic implications of the doctrine of homonymous predication, Maimonides develops in *Guide* 1:58–60 his celebrated theory of negative predication, arguing that ultimately negative predication alone brings the human mind closer to an understanding of God: "Know that the description of God, may he be cherished and exalted, by means of negations is the correct description."[20] This third piece of Maimonides' theory of divine predication represents the logical culmination of his theory of language. Maimonides explicitly states that describing God by means of affirmations, by means of positive ascriptions, yields the paradoxical result of receding in knowledge from God. By ascribing to God terms that do not begin to capture his transcendent nature, humans are both insulting and denigrating God's true essence. Ultimately silence is the only appropriate linguistic response to divine predication: "Silence with regard to You is praise."[21]

What is most interesting about this discussion, however, is that the specific problems pertaining to the relation of divine omniscience to human freedom are independent of Maimonides' own analysis of divine predication. More pointedly, given the explicit disavowal of the ability of language to describe God's predicates, we would expect that Maimonides would have little to say about any of God's predicates, for example, omniscience. But, despite his theory of negative predication, Maimonides devotes several chapters in the *Guide* to divine omniscience, and finds quite a bit to say about it. By "omniscience" he refers to God's act of knowing everything that can be known. In fact it will emerge from the ensuing discussion that according to Maimonides, the two propositions "God knows *p*" and "Zaid knows *p*" can be parsed similarly. That is, although the term "knowledge" must be understood equivocally when applied to God and to Zaid, nevertheless the logical force of the term "knows" in both statements is the same.[22]

Astrological Determinism and Human Freedom

The second issue mentioned above concerns the role of astrology. Astrology has always occupied a contentious place in Jewish thought. Most Jewish philosophers supported natural astrology, the view that the celestial bodies affect sublunar life and existence to some extent. Aristotle had laid the groundwork for the theory that these bodies were responsible for the growth and sustenance of sublunar entities. That the sun and moon both affect natural cycles and events on earth is unequivocal and represents a classic paradigm of natural astrology.[23] The calculations of natural astrology overlapped those of astronomy, and could be utilized for practical purposes such as fixing the calendar.

The real question, then, concerns the coherence of judicial astrology, that is, the extent to which the stars and planets exerted an influence over human events in general, or more particularly, over those actions that entail human choice. On the one hand, judicial astrology was derisively dismissed in the Bible, identified with idolatry and pagan star worship. Two passages in Scripture explicitly prohibit the study of the stars: Deuteronomy 4:19 "Beware, when you look up into the heavens and see all the host of the heavens, the sun, moon and stars, that you do not let yourselves be allured into paying homage to them," and Deuteronomy 17:3 "[one who has] paid homage to them [i.e. alien gods] namely, the sun, or the moon, or the whole host of the heavens, which I prohibited." Other passages indirectly deplore the science of astrology.[24]

On the other hand, it was an accepted science that permeated ordinary life. In rabbinic texts, there are passing references to divination by means of planets. The Rabbis were especially ambivalent about the role played by astrology. Charlesworth, and more recently Kiener, have argued that just as as rabbinic Judaism became a repository for mathematical astronomy developed by the Babylonians, so too did the science of astrology develop, albeit in a benign form, among the Rabbis.[25] Most talmudic sages recognized the role played by the celestial bodies in determining human affairs, as evidenced by numerous talmudic passages that attribute astrological consultation in approving terms.[26] Every person was seen to have a patron star (*mazal*) which determined his destiny.[27] On occasion, natural events can be traced to astrological signs, as evidenced in the following passage:

> Samuel said: A vernal equinox which falls in [the hour of] Jupiter will surely cause trees to break, and a winter solstice which falls in [the hour of] Jupiter will surely dry out the seedlings . . . provided that the [previous] New Moon took place in [the hour of] the moon or in [the hour of] Jupiter.[28]

Further, individuals born under a specific planet are said to exhibit the qualities popularly assigned to that planet.[29] But most questionable is the status of Israel. In contradistinction to those who maintain that Israel, like other nations, is subject to celestial influence, a well-known rabbinic dictum asserts that "over Israel there is no sign (*'ein mazal le-Yisra'el*)."[30] The rabbinic examples in the text suggest that with sufficient prayer, Jews can counteract their astrological fate. Hence, ironically, the notion that a person's fate (*goral*) was determined by his constellation (*mazal*) led to the idea that this fate could be averted by attention to prayer and good deeds. As argued succinctly by Kiener, though, judicial astrology did not intrude into the decisions of Jewish law, nor was it to become "part of the arsenal of apologia in defense of Judaism against paganism. Astrology was a belief system foreign to the sacred realm of *halakhah*."[31]

The effect of these astrological motifs can be seen explicitly in perhaps one of the most influential works in medieval Jewish cosmology, the *Sefer Yetzirah*, a mystical commentary upon Genesis. Early Jewish philosophers, faced with a variety of philosophical cosmogonies—Kalām, Aristotelian, Neoplatonic—commented extensively upon *Sefer Yetzirah*, possibly because it offered what they considered an "authentic Jewish response, compatible with at least some of the philosophic theories, to the question of cosmogony."[32]

Contained in the *Sefer Yetzirah* is an elaborate cosmological scheme that can be summarized briefly as follows. The twenty-two letters of the Hebrew alphabet, along with the ten *Sefirot*, the numbers one to ten, comprise the foundation of the creation process. The first four *Sefirot* stand for the power of God, the transmutation of divine spirit into air, the emergence of water from air, and the emergence of fire from water. *Sefirot* five through ten describe the six possible dimensions or directions of the created universe. The twenty-two letters are the "foundation letters" (*'otiot ha-yesod*) out of which the universe is created by an intricate process of combination and permutation.

In order to actuate this creation process, the letters are grouped into "three books," that is, groups of letters grouped according to their special functions. The first group, the "mother letters" *'alef, mem* and *shin,* are the prime constituents of everything else in the universe. The "doubles," which include *beit, gimel, dalet, kaf, peh, resh,* and *tav,* become the basis for the creation of the planets. And the "simples," which comprise the remaining twelve letters, constitute the twelve months of the year, as well as the twelve signs of the zodiac, which are associated with the twelve vital organs of the human body.[33]

In the context of this cosmological description, then, we see that several astrological passages emerge as the basis for subsequent commentary. The first includes such topics as the relationship of the seven Hebrew consonants that take a *dagesh* to the seven planets and seven days of the week: "Seven Doubles: *BGD KPRT* of Foundation, he engraved them, he carved them, he permuted them, he weighed them, he transformed them, And with them he formed, Seven planets in the Universe, Seven days in the Year, Seven gates in the Soul, male and female."[34] Each of the letters corresponds to a specific heavenly sphere, day of the week, and psychological characteristic. The letter *kaf,* for example, is associated with the planet Venus, with Wednesday, and with the "left eye in the Soul."[35] Each letter, then, can be used to influence the part of the body with which it is associated. The second astrological section pertains to the twelve simple consonants with which "He formed twelve constellations in the Universe, twelve months in the Year, and twelve directors in the Soul, male and female."[36] The signs of the zodiac are associated with the twelve Hebrew lunar months (rather than with the position of the Sun as in Western astrology). A final astrological passage has to do with the power of the *teli.* The twenty-two letters form a triad that is represented by the dragon (*teli*) the sphere and the heart.[37] This dragon is depicted as having extraordinary astrological powers.[38]

These astrological passages in the *Sefer Yetzirah* have given rise to a rich commentary tradition. Both Saadia Gaon and Shabbetai Donnolo wrote extensive commentaries upon *Sefer Yetzirah,* emphasizing astrological passages in this work. Saadia Gaon wrote an Arabic commentary in the late ninth or early tenth century.[39] Another influential commentary, called *Sefer Ḥakhmoni,* was written by the Italian physician Shabbetai Donnolo (913–after 982).[40] While a full examination of this commentary tradition is beyond the scope of this study, let me emphasize one passage that especially interested philosophers. This is the passage mentioned above, having to do with the threefold reading of the term *sefer:* "He created his world by three *sefarim, s-f-r-, s-f-r,* and *s-f-r-.*"[41] This passage reappears in many philosophical contexts. Saadia's commentary canonizes the reading that is adapted by many subsequent thinkers.

> He created the world by three things, writing, number and speech. . . . I interpret *sefer* as writing as it says (Daniel 1:4): "To teach them the writing and language of the Chaldeans." *Sfar* means quantity and number, as it says about Solomon (2 Chronicles 2:16): "After the numbering which David his father has counted."[42]

The third would be *sippur*, which Saadia Gaon does not bother to mention, presumably because it was so obvious. Saadia then continues to explain the three in terms of speech, writing and thought.[43]

Interestingly enough, Halevi interprets the reference further; stripping it of mystical or poetical nuance, Halevi "was the first to interpret the three *sefarim* of our passage in light of a basic doctrine of Aristotelian rationalism: the unity of the subject, act, and object of intellection."[44] In *Kuzari* 4.25, Halevi refers to *Sefer Yetzirah*, "the 'Book of Creation' by the Patriarch Abraham," and explains the terms *s'far, sefer,* and *sipur* in terms of will, writing, and speaking. In the case of God, the three are a unity inasmuch as God's will, speech, and act are one and the same. With respect to humans, however, the three are nonidentical: "Man's will, writing and word are marks of the thing, but not the nature of the same."[45]

Ibn Gabirol refers to *Sefer Yetzirah* indirectly in *Meqor Ḥayyim* in the following passage: "And so we shall say that the composition of the world arose from the arrangement of (*ketivat*) number and letters in the air."[46] The allusion is not fleshed out however, and the implications of creation by number and letters are not developed further. In his poems, however, Gabirol is more explicit. In his poem *Shokhen 'Ad,* recited by some Sephardic communities on Rosh Hashanah, Ibn Gabirol refers to the *Sefirot* as well as to the *'ein-sof:* "He decided to reveal the set of ten *Sefirot*/ And he wrote Ten corresponding to them in the *'Ein-Sof.*"[47] He refers as well to the "three books":

> He who dwelleth forever, exalted is he alone from of yore
> Solitary in his royal grandeur is he, and there is none by his side
> From the light in which he is cloaked he fashioned the universe
> In the manner of the three sealed books.[48]

In these lines it is clear that the universe is an emanation from God, or more specifically from the light in which God is cloaked. The three sealed books may very well refer to the three books in *Sefer Yetzirah,* corresponding to the letters *s/f/r,* namely *sofer, sipur* and *sefer.* Zangwill suggests as well that they may refer to *Meqor Hayyim* 5.62 in which Gabirol compares Form to the script, and Matter to the tablet upon which the writing is engraved. On this reading the *sefarim* refer to entities that are accountable for the creation of the universe, namely, Will, Form, and Matter; these in turn may be compared to the three agencies involved in the writing of a book—the scribe, script, and scroll—again alluding to *Sefer Yetzirah.*[49]

Gabirol's major literary contribution comprises what we may term his "wisdom poetry." The best known and most elegant example of this philosophical poetry is Gabirol's masterpiece *Keter Malkhut* that comprises an elaborate description of the universe, incorporating the major elements of Ptolemaic cosmology along with astrological motifs. For this cosmology he turned to the works of the *Epistles of the Brethren of Purity* (*Rasāil ikhwān aṣ-ṣafā'*), and to the astronomical works of Al-Farghāni.[50]

Gabirol's poem incorporates the basic elements of Ptolemy's *Planetary Hypotheses:* a series of concentric spheres around the earth, with the five planets, moon and sun, the Zodiac, and a ninth diurnal sphere that imparts motion to all the others spheres. In *Cento* 10 the earth is described as an orb with the moon and four elements encircling it. The moon excites new events in our world every month, but Gabirol cautions that "Always her own Creator's will (*ratzon ha-Bore'*) she heeds," noting that astrological influences are subject to divine will.[51] *Cento* 12 describes the lunar eclipse, which again shows that "a Judge keeps them controlled/ Who raiseth one and brings another low." Gabirol then mentions the *Teli,* the mythological serpent that swallows heavenly bodies and causes their eclipse. The sun was created as a time-keeper:

> What mind could grasp thy greatness, how the Sun
> To be a timekeeper Thou hast designed
> To count the days, as into years they run,
> Each instant fixed, each period defined;
> And make the fruit-trees bud, beneath the bland
> Spell of Orion's band/And genial Pleiades, for richest yield[52]

After describing Jupiter, Mars and Saturn, Gabirol turns to the Zodiac, whose signs have a power to affect sublunar events:

> And power resides
> In those signs, whence that potency doth flow
> That each created thing can wield below
> Each after its own kind.[53]

The signs of the Zodiac function as "mansions palatine" to the planets. In all these passages Gabirol emphasizes that the influences that flow through the planets to the sublunar sphere do so at the will of their Creator.[54] In *Cento* 24 Gabirol introduces a tenth sphere, "the sphere of mind—Intelligence, the palace court most nigh/Unto Thyself." This tenth sphere, which serves as the location of intelligence, is Gabirol's own introduction. From this sphere the stuff of souls and angels ("psyches of highest rate") is made.[55]

As we move into the philosophical literature, astrological references proliferate. Many philosophers considered astrology as a genuine science. These included, in addition to the *Sefer Yetzirah* commentaries of Saadia Gaon, Shabbetai Donnolo, and Solomon Ibn Gabirol, the works of such thinkers as Abraham Ibn Ezra and Gersonides. Maimonides is the most vehement opponent of astrology, seconded by Joseph Albo in the fourteenth century. Before turning to the diatribes of Maimonides below, let us first attend to the seductive influence of the discipline among its supporters.

Abraham Ibn Ezra, born in Tudela in 1089, was a poet, grammarian, biblical exegete, philosopher, astronomer, astrologer and physician. He lived in Spain until 1140, where he was a friend of Ibn Tzaddik, Ibn Daud, and Judah Halevi. Most of his works were composed between 1140 and 1146. He died in 1164 in either Rome or Palestine.[56] Best known for his biblical commentaries, he also wrote many short treatises on grammar, astrology, and numbers, and translated into Hebrew the text *The Book on Eclipses,* which contains a theory of planetary conjunctions.[57] Although Ibn Ezra did not write any specifically philosophical works, he was strongly influenced by Ibn Gabirol. For example, he accepts Gabirol's doctrine that intelligible substances are composed of matter and form, and he uses Gabirol's descriptions of God as the source from which everything flows. Ibn Ezra's theory of soul reflects Neoplatonic motifs as well. The source of the rational soul is the universal soul and immortality is understood as reunification of the rational soul with the world soul.

In a recent article, Langermann has emphasized the importance of astrological themes in the biblical commentaries of Ibn Ezra. Recognizing the importance of stellar influences upon human destiny, Ibn Ezra nevertheless emphasizes the "salvation from astral decrees" that is available to humans.[58] It is the science of astrology that permits knowledge of impending disaster, and thus enables humans to take appropriate precautionary measures. Recognizing that one who is born in a defective stellar configuration cannot "become like the one who was born in a perfect configuration," Ibn Ezra nevertheless avers that "man can by means of his mind (*da'ato*) mitigate his misfortune somewhat."[59] The mitigation achieved by humans is fairly limited, however. Langermann's hypothesis is that a major key unifying the Neoplatonism of twelfth-century Hispano-Jewish thinkers is "the extension of naturalistic, specifically astrological, explanation to the phenomenon of religion as well."[60] More specifically, while the spiritual goal of all humanity is the same, diversity of religious practice is analogous to diversity in all natural phenomena and can be traced to stellar influences.

This tension is reflected in Halevi's *ha-Kuzari* as well. Halevi was ambivalent about the role played by the celestial orbs. Reflecting the importance of astrology and ascribing particular differences to the influence of climate, spheres, stars, and constellations, the philosopher downplays the importance of denominationalism, urging the King to adopt any religion, or even make up his own.[61] Halevi contravenes this position by emphasizing repeatedly the uniqueness of Judaism as a religion, and the Jews as a people. We cannot deny, he says, that the heavenly spheres do exert an influence upon the sublunar world. The particulars, however, are unknown to humans. "The astrologer boasts of knowing them [particulars] but we repudiate it and assert that no mortal can fathom them."[62] Astrology is seen as a deceptive science, which includes other pseudosciences as "conjuring, magic practices and other tricks which are rejected as completely by nature as by the Law."[63] With respect to astrology itself, he rejects naturalistic explanations of the uniqueness of Jews and Judaism.[64]

Perhaps one of the most outspoken philosophical supporters of astrology is Gersonides, whose astral determinism is explicitly developed in two contexts: In book 2 of *Milḥamot* he interweaves astrological motifs into his discussion of divine providence and prophecy, while in book 5 astrology occupies a central role in the context of his cosmological speculations. Let us look first at the discussions in *Milḥamot* 2, since they set the tone for the subsequent investigations of book 5. In *Milḥamot* 2 Gersonides is concerned to explain how divine knowledge operates, and to what extent divine foreknowledge of future contingents affects human choice.[65] His major thesis is that divine knowledge is predicated to a great extent upon knowledge of the heavenly bodies, which bodies are in turn "systematically directed toward his [man's] preservation and guidance so that all his activities and thoughts are ordered by them."[66] Because of the major role played by the heavenly bodies, it turns out that often astrologers, whose function it is to study these bodies, are able to predict human events correctly.

The technical aspects of this system are described more fully by Gersonides in *Milḥamot* 5, which is devoted to the workings of the heavens. However, in *Milḥamot* 2 he does allude to the role of the Active Intellect that imparts much of this information to the informed astrologer. And yet, as we all know, astrologers often err in their predictions. Astrological errors can be due to several factors. First, because of the "difficulty of obtaining the necessary positions of these bodies by observation,"[67] astrologers are often unable to verify their data. Furthermore, since the zodiacal position of a heavenly body at any given time is only repeated once in many thousands of years, astrologers have no access to the repeatability of those events that would be required to verify their knowledge. As a result, Gersonides claims, we know very little of the order of the heavenly bodies. Nevertheless, he assures us that whatever happens by chance is "determined and ordered according to this type of determinateness and order."[68] Outdoing even

Plato's hierarchical structuring in *Republic* 4, Gersonides argues that the ultimate perfection and ordering of society is due to astrological influence. People turn to particular crafts as a result of the "order emanating from the heavenly bodies."[69]

Having articulated the ordering power of the astral bodies, Gersonides describes in *Milḥamot* 5.3 the separate intellects and the spheres that they move. The movers emanate from God who is construed as the "First Separate Intellect."[70] They are ordered in a rational system that governs the sublunar domain. If there were no one first intellect, Gersonides argues, the rational order we see in the heavens would be the result of chance, which is unacceptable. The agent intellect thus functions as the link between these celestial bodies and human affairs. The kinds of information it transmits are of an astronomical type, as evidenced in the following example:

> it [the agent intellect] knows how many revolutions of the sun, or of the diurnal sphere, or of any other sphere [have transpired] from the time at which someone, who falls under a particular pattern, had a particular level of good or ill fortune . . .[71]

The agent intellect, then, serve as the repository for information communicated by the heavenly bodies. The patterns revealed in this communication between agent intellect and diviner (astrologer, prophet) are from the heavenly bodies that themselves are endowed with intellects and so "apprehend the pattern that derives from them."[72] Each mover apprehends the order deriving from the heavenly body it moves, and not patterns which emanate from other heavenly bodies. As a result, the imaginative faculty receives the "pattern inherent in the intellects of the heavenly bodies from the influence deriving from them."[73] This influence derives from the position of the heavenly bodies "by the ascendent degree or the dominant planet [in a particular zodiacal position]."[74] However, inasmuch as the heavenly bodies do not jointly cooperate with one another (*lo yishtatfu*) in this process, it is possible for the communication to be misconstrued.[75]

Why is it that certain communications are received more clearly than others? A constitutionally perfect imaginative faculty receives information from both dominant and weak heavenly bodies. By "weak," Gersonides means that certain bodies are too weak both to bring about events on earth as well as to transmit information about these events. Hence he concludes that information about the future emanates "from the dominant body in the particular proper face (*panim*) in which it has dominance but not from any of the attending planets (*ha-meshartim*)."[76] But to constitutionally imperfect imaginative faculties, the information received is only from the dominant heavenly bodies. Hence the overall quality of the information received will differ in the two cases.

Gersonides does admit that on occasion human choice is able to contravene the celestial bodies: "[Our intellect and choice] have the power to move us contrary to that which is determined by the heavenly bodies."[77] Nevertheless this intervention is rare, and true contingency is a rare state of affairs indeed in Gersonides' ontology.[78] Gersonides presents an argument to show that human choice guided by reason can subvert the celestial bodies despite their general ordering of our lives. The heavenly bodies can order human affairs either by virtue of their difference of position in the heavens, or from the difference of the bodies among themselves.[79] Astral bodies, however, will affect different individuals in different ways; they can also affect an individual differently at different times; and finally, two or more bodies can affect a single individual, resulting in multiple influences that can have contrary effects. Having noted these various ways that the astral bodies can affect human behavior, Gersonides notes that humans can contravene these effects: God has provided humans with "the intellectual capacity (*sekhel ba'al takhlit*) that enables us both to act contrary to what has been ordered by the heavenly bodies and to correct, as far as possible, the [astrally ordained] misfortunes that befall us."[80]

Both Maimonides and Gersonides disagree over the ultimate purpose of the celestial bodies. For Maimonides, as we shall see below, it is not possible that a greater entity, the heavens, exist for the sake of the sublunar universe. This point will constitute one of Maimonides' key arguments proffered against the acceptance of astrology. Gersonides disagrees, maintaining that it is not inappropriate that the more noble exist for the less noble. The stars, he argues, exist for the sake of things in the sublunar world.[81] More explicitly, they are designed for the benefit of sublunar existence, and they guarantee the perpetuation of life on earth.

The main characteristic of the astral bodies is their luminosity (*nitzutz*). This luminosity affects their actions and effects.[82] Gersonides is very much aware of the problem of accounting for how the astral bodies can affect actions at a distance. The sun, for example, functions as a paradigm for action at a distance. Once we understand, Gersonides claims, how the activity of heating reaches earth from the sun, we can understand how the particular activities of the other stars reach the sublunar realm as well. By explaining the efficient cause as the light or radiation of the stars, Gersonides can account for weak or strong effects. As Langermann has pointed out, Gersonides' account furnishes the basis for the introduction of astrological causation into natural philosophy.[83]

In *Milḥamot* 5.2.8 Gersonides lays out six astrological principles that affect his general cosmological scheme. These can be summarized as follows. First, each astral body exercises a different influence specific to it. Second, astral influence depends upon its position in the zodiac *(galgal ha-*

mazalot). Third, the longer a star stays in one place in the zodiac, the greater its effect because of the strength of its luminosity. Fourth, astral influence is dependent upon its inclination to the north or to the south; its effect will be strongest when it is in the middle, as evidenced by the sun, whose heat is strongest when it is at the Tropic of Cancer as opposed to being at the Tropic of Capricorn. Fifth, the greater the radiation or luminosity of a star, the stronger will be its influence. And finally, the closer to earth a star is, the stronger will be its influence.[84] These principles function as the underpinnings of his general astronomy as well.

Having looked at the arguments proffered in favor of astrology, let me turn now to its most vigorous opposition as represented by Albo. (I shall return to Maimonides' discussion of astrology in the context of his discussion of free will.) Joseph Albo is equally vehement in his condemnation of astrology. In his work *Sefer ha-ʿIkkarim* he introduces the topic in the context of prophecy. In order to ascertain daily events, the ancients (*ha-qadmonim*) would resort to the study of astrology (*yediʿat mishpaṭei ha-kokhavim*).[85] This science was distinct from prophecy. Albo distinguishes between natural and judicial astrology: the former, associated with Aristotle, asserts that "the heavenly bodies and the stars exert an influence upon the lower world by means of their various motions" and according to Albo applies only to the natural order and not to intellectual entities.[86] Judicial astrology, on the other hand, asserts that "all things that happen to a man, do so by the decree of the stars."[87] Albo is unequivocal in his rejection of judicial astrology. First he argues that it is prohibited by the Torah. Humans cannot come to have knowledge of the future for two reasons: First, the power of imagination far outstrips reality, and moreover, "God can destroy the power of the constellations and bring about the opposite of that which they determine."[88] Second, judicial astrology removes "the category of the contingent (*ṭevaʿ ha-ʾefshar*)," a category that is critical to Albo's upholding free will.[89] Furthermore, Albo argues, the actual constellations themselves are completely arbitrary and have no basis in reality: "They draw imaginary lines from star to star so as to produce the figure in question. . . . Why do they not draw other lines in another way and produce other forms?"[90] The principles themselves of astrology are often negated by real life, as for example when a number of passengers, each of whom presumably has a different horoscope, die at sea and receive the same fate.[91] Finally, Albo points that often astrologers are incapable of ascertaining the causes of past events; If so, how much harder is it to determine the causes of future events.[92] The ultimate difference between the astrologer and the prophet is that the latter "has a true knowledge of things, which cannot in any manner fail, because it emanates from God," whereas the knowledge of the astrologer can at any point be negated by God's will.[93]

Compatibilism in Jewish Kalam: Saadia Gaon and Halevi

It is against the backdrop of issues of astrological determinism that we must understand the concomitant attempt to reconcile human freedom and divine causal activity. The topic of human freedom became known to Jewish thinkers as a philosophical conundrum through their contact with Greek and Islamic writings, particularly with Kalām writings on the antinomies of free will.[94] The actual word for providence is not found in Jewish writings prior to Ibn Tibbon's introduction of the term in the process of translating Maimonides' appropriation of the Arabic term "*tādbir.*"[95] Nor is there a term in Hebrew philosophical language corresponding to the scholastic notion of free will, *liberum arbitrium.*[96] But this does not mean that the issue did not arise in the context of medieval Jewish thought. Earlier thinkers had of course faced the theological problem of reconciling human freedom and God's providential care and concern for humans. Rabbi Akiba, for example, is said to have taught that "everything is predestined, nevertheless will (or power) is given (*ha-qol tzafuy v̱e-ha-reshut netunah*)."[97] Recognizing the tension between these two concepts, Akiba however does not try to reconcile them.

The writings of Saadia Gaon and Judah Halevi clearly exemplify the Kalām preoccupation with the problem of reconciling providence with human freedom. In accordance with his Kalām predecessors, Saadia sees the problem of divine omniscience as intimately connected to issues of divine justice. Although not stated explicitly as an argument, Saadia does offer the following considerations to support his contention. He accepts as a given that God has granted to humans free will: "It accords with the justice of the Creator and his mercy towards man that he should have granted him the power (*koaḥ*) and ability (*yekholet*) to do what he commanded him to do, and to refrain from what he forbade him to do."[98] Saadia's discussion occurs within the Kalām context of trying to explicate these philosophical terms.

Kalām philosophers had fought long and hard over what constitutes a free action. At stake were two sets of questions: first, does freedom of will arise before the action itself, or simultaneously with it; and, second, is God's freedom subject to the same constraints? More specifically, Kalām philosophers debated whether the ability to act (*isti'dāt; yekholet*) must precede the act itself, or whether the ability to act could be said to exist simultaneously with the action in question.[99] On the latter view, supported primarily by the Mut'āzilite school, human freedom consisted in the mere act of consenting to an action. Hence the consent could accompany the action simultaneously without actually causing it. Saadia rejects this position, claiming that "The ability (to act) must necessarily exist before the act (*ha-yekholet re'uyah she-tihiyeh qodem ha-po'al*), so as to give man the (choice of doing the positive) act (*ha-po'al*) or abstaining from the act."[100] In line with this characterization, Saadia further claims that both acting and desisting to act count equally as free actions: "In the same way as when a man's doing a thing constitutes an action (*ka-'asher 'asot ha-'adam 'et ha-davar hu' po'el*), so too his desisting from it constitutes an action."[101] Saadia's point here is that the term "action" in its full sense implies free choice; even in not choosing, humans are exercising their free will. "I must further explain that man does not perform any action unless he chooses (*boḥer*) to do it, since it is impossible for one to act if he has no free will or fails to exercise his free will."[102]

Halevi, following Saadia, rejects as well the necessitarian position that possibility (*ṭeva' ha-'efshar*) does not exist: "Only a perverse, heretical person would deny the nature of what is possible. If he believed that what will be will inevitably be (*ve-'ilu hayah ma'amin ki yihiyeh 'al kol panim*) he would simply submit, and not equip himself with weapons against his enemy, or with food against his hunger."[103] But the acceptance of possibility is "not incompatible with a belief in Divine Providence" (*mi-din ha-'Elohim*).[104] In order to support the compatibility of possibility with providence, Halevi distinguishes between two chains of causation. The first chain is related directly as an immediate expression of the divine will, whereas the second represents an intermediary chain between God and natural events.[105] Based on these two causal chains, Halevi then distinguishes four types of actions: divine actions (*'elohiyim*) that issue forth directly from God; natural actions (*ṭiv'iyim*) that are derived from intermediate causes, and unfold naturally, barring any interference; accidental actions (*miqriyim*) that result not by nature or arrangement, or will power, but by accident (*be-miqreh*); and arbitrary actions (*mivḥariyim*) that have their "root in the free will of man (*sibatam ḥefetz ha-'adam be'et beḥirato, veha-beḥirah mi-klal ha-sibot ha-'emtza'iyot*)."[106] According to Halevi, human actions of will belong to the final category and possess causes which reduce them chainlike, to the Prime Cause. Reflecting Aristotle's concerns reflected in (1.6) and (1.7) above, Halevi claims that only actions in this latter class are susceptible to praise and blame.

Having established that humans have free will, both Saadia and Halevi are ready to turn to the issue of divine omniscience and whether God's knowledge interferes with this absolute freedom. Saadia's position will be that "the creator (be he exalted) does not allow his power to interfere in the least with the actions of men, nor does he compel them to be either obedient or disobedient."[107] But how can this be, given that God knows future events before they happen? Saadia, for the first time in medieval Jewish philosophy, crystallizes the dilemma of divine omniscience in the following terms:

> Perhaps someone will ask further: "If God knows that which is going to be before it comes into being, he knows already that a certain person will disobey him; and it is not possible that that person not disobey God, in order that God's foreknowledge prove to be correct."[108]

Saadia's solution is straightforward and introduces the position common in later scholastic thought, namely that God's knowledge is not causative. "He who makes this assertion has no proof that the knowledge of the Creator concerning things is the cause of their existence."[109] Importantly for our concerns, his support for this contention rests upon quasitemporal considerations. Saadia offers two reasons for rejecting the efficacy of God's foreknowledge. The first reason draws upon the relation between knowledge and causes; if God's knowledge were causative, then according to Saadia "they [future contingents] would have existed from eternity, since God's knowledge of them is eternal."[110] We shall return to the implications of this claim when we examine Crescas' adoption of the eternity thesis. Suffice it to say at this point that scholastic philosophers (Crescas included), parted company with Saadia on this point. Accepting the antecedent of this conditional, most philosophers were willing to accept the consequent as well. Thus, for example, Boethius and Aquinas were willing to argue that precisely because God's knowledge of events is eternal, it follows that God's knowledge is technically not foreknowledge at all but rather eternally existing knowledge. Boethius, for example, states explicitly that God's knowledge, technically speaking, is "the knowledge of a never fading instant rather than a foreknowledge of the 'future'."[111] Saadia clearly claims that God knows all things in advance, and in particular with respect to freely willed actions God "knew in advance that man was going to choose them."[112]

Halevi's support of compatibilism reflects that of Saadia. Halevi addresses several philosophical objections designed to deny God's knowledge. The first objection, attributed to the Mutakallimûn, suggests that inasmuch as human actions are possible, they are removed entirely from the sphere of divine knowledge. Halevi counters that these actions, although "completely outside the control of Providence," are indirectly linked to it.[113] The second objection concerns the ontological composition of what is possible, and suggests that "these matters are outside divine omniscience because the absolutely potential (*ha-'efshar ha-gamur*) is naturally an unknown quantity."[114] Halevi echoes Saadia's claim that "the knowledge of a thing is not the cause of its coming into existence."[115] It is here that Halevi states, along with his Kalām predecessors, that "the knowledge of events to come is not the cause of their coming to be (*sibah le-heyoto*), just as the knowledge of things which have been is not the cause of their having come to be."[116] Having established to his satisfaction that human freedom exists, Halevi concludes his discussion with his sixth principle, namely that "Man finds in himself (*mi-nafsho*) this power (*yekholet*) of doing evil or avoiding it in matters which are possible for him (*'efshar lo*)."[117]

To what extent does human action causally affect God's prior knowledge? Both Saadia and Halevi interestingly anticipate later scholastic discussions of this issue. Should one object, Saadia says, that if God knew that a certain person would speak, it is possible for that person to be silent, Saadia responds that "we answer quite simply that if that person was to keep silent instead of speaking we should have said in our original statement that God knew that this man would be silent, and we were not entitled to state that God knew that this person would speak."[118] Halevi offers a similar argument, claiming that speaking or being silent is in ones power so long as it is subject to reason and not subject to other accidents.[119] In this claim Saadia and Halevi both argue for freedom of choice on phenomenological grounds: our awareness of the power we have over our own speech attests to the existence of human freedom. The underlying issue raised by this argument, however, is whether Saadia and Halevi are willing to accept the implication that divine knowledge is ultimately affected by future contingent actions. In other words, while accepting the dictum that divine knowledge is not causative upon future actions, are they willing to grant that future human actions are to some extent at least causative upon divine knowledge? Neither Saadia nor Halevi addresses this critical issue, which becomes a focus for subsequent examinations. The full implications of this conclusion will be unpacked when we turn below to the indeterminist position.

Maimonides' Compatibilism

Compatibilism receives its fullest expression by Maimonides who deals with this topic in a number of contexts. By describing Maimonides as a compatibilist, I am placing myself in the midst of a controversy that has occupied much of recent Maimonidean scholarship. In recent years scholars have argued over whether an esoteric reading of the *Guide* reveals a determinist stance. Pines and Altmann have argued that with respect to the issue of absolute freedom, Maimonides holds a form of determinism that precludes the ontological uniqueness of human will and intention. They point out, for example, that nowhere in the *Guide* is the doctrine of absolute freedom espoused, nor does Maimonides definitively refute the necessitarianism that runs as a subtheme throughout the *Guide.*[120] Their position is based on two important passages: *Guide* 3.17 and *Guide* 2.48.

In *Guide* 3.17, Maimonides states that humans have an "absolute capacity" to act, meaning that they have the choice and will to do "everything that it is within the capacity of man to do." But in a subsequent statement in that chapter, Maimonides appears to equate human and animal volition, leading Pines to conclude that Maimonides leaves open the possibility that human choices are determined by outside or internal factors. Altmann reads this chapter more conservatively, suggesting that the equation of human and animal volition does not necessarily imply a deterministic reading. Both Pines and Altmann, however, find *Guide* 2.48 more clearly supportive of determinism. In this chapter, human and animal volition are explicitly compared:

> Inasmuch as the deity is, as has been established, he who arouses a particular volition in the irrational animal and who has necessitated this particular free choice in the rational animal and who has made natural things pursue their course. . . . [I]t follows necessarily from all this that it may be said with regard to what proceeds necessarily from these causes that God has commanded that something should be done in such and such a way.[121]

On the basis of this passage, Altmann concludes, as does Pines, that human choice is on a par with animal choice; that is, that "looking at this passage, one cannot avoid considering Maimonides a strict determinist."[122]

More recently, however, Gellman, Hyman and Stern have tried to controvert this determinist reading of Maimonides. In his interpretation of *Guide* 2.48 and 3.17, Gellman claims that according to Maimonides, the consequences of our actions are determined by God, although our actual choice is free.[123] Hyman, on the other hand, gives a more holistic reading of Maimonides, emphasizing the importance of the halakhic texts in understanding these two passages in the *Guide.* Noting the similarity between Maimonides' discussion and that of Aristotle in the *Nicomachean Ethics,* Hyman follows the suggestion of Sorabji that Aristotle's volitional actions can be either caused and determined, or caused without being determined. Hyman then applies this distinction to Maimonides and argues that on Maimonides' characterization of choice, some human actions are caused but not determined. On this view, then, Maimonides would be a libertarian, but "one who is well aware that the majority of human actions are not only caused, but determined as well."[124] Stern, finally, reiterates with Hyman the point that an Aristotelian cause is never a sufficient necessitating condition of an event. He takes the two passages as merely showing that events have prior causes, not that they are necessitated. Stern then points to other passages, for example, *Guide* 1.72, 2.4, 2.12, and 1.69 which strengthen the move from causation and necessitation of actions. Stern concludes that a full account of all these chapters may in fact "prove Maimonides a determinist even with respect to human action."[125]

On Pines' reading, and more conservatively that of Altmann, Maimonides' determinism upholds the causal necessitarianism implicit in divine providence and omniscience, at the expense of eliminating human freedom. Hyman and Stern, on the other hand, try to salvage a modicum of human freedom while at the same time recognizing the overarching causal structure of Maimonides' metaphysics. While I am sympathetic to this latter enterprise, nevertheless I shall argue that very little of human freedom remains once we recognize the force of God's activity in the world. I shall argue that Maimonides' compatibilism, as defined above, emerges most clearly in the context of his views on God's knowledge of future events and occurs against the backdrop of an examination of the objects of God's knowledge.

Already in Maimonides' early works we find discussion of the issue of free will and determinism. His earliest reference occurs in his *Mishnah* commentary upon *Berakhot* 9:7, but Maimonides does not expand upon the topic in great detail and does not enter into the philosophical complexity of the issue.[126] In *Shemoneh Peraqim,* Maimonides devotes chapter 8 to a discussion of human freedom in light of predestination, maintaining that despite God's general preordination of sublunar events, humans nevertheless have free will. As an example, he explains the biblical passage "I will harden the heart of Pharoah" by arguing that Pharoah and the Egyptians had already chosen to rebel against God "through the dictates of their own free will and the evil passions of their hearts, without any external constraint forcing them thereto."[127] The hardening of the heart, therefore, occurred after Pharoah's freely chosen transgression; by hardening Pharoah's heart, God deprived Pharoah of the opportunity to freely choose to repent of his sins and allow the Israelites to leave Egypt.[128] Maimonides then claims that God's knowledge is identical with his essence and hence incomprehensible to humans: "From what we have said, it has been demonstrated also that we cannot comprehend God's knowledge, that our minds cannot grasp it all, for he is his knowledge, and his knowledge is he."[129] On this basis, God's foreknowledge of human events can be said to be compatible with human freedom only in the context of recognizing the incomprehensibility of God's knowledge to our own intellect.

Maimonides' acceptance of human freedom in these legal texts is ostensibly the source of his profound antagonism to astrology. His treatment of the topic of astrology occurs against a textual tradition that, at least prima facie, did not eliminate either natural or judicial astrology from theoretical considerations.[130] In his *Letter on Astrology,* addressed to the Rabbis of southern France, the following sorts of considerations are adduced in opposition to astrology:

> *Argument from Error.* The Temple was destroyed because our fathers "erred and were drawn after them [books dealing with these themes of the stargazers, these things being the root of idolatry], imagining them to be glorious science and to be of great utility. They did not busy themselves with the art of war or

with the conquest of lands, but imagined that those studies would help them."[131]

Argument from Stupidity: "All these assertions [of judicial astrology] are far from being scientific; they are stupidity."[132]

Argument from Falsity: Astrology is identified with idolatry, star-worship of the Kasdim, Chaldeans, Egyptians and Caananites, which is seen by the wise men of Greece to be "falsity and a lie (*sheqer ve-khazav*)."[133]

Free will Argument: "There is no influence or constellation under which one is born that will draw him in any manner toward any of these ways."[134]

Argument from Utility: If astrology were true, "of what utility would the Torah and the commandments and the Talmud be to a particular individual?" [135]

Of these considerations, only the last two arguments have any philosophical bite. In the case of the "Argument from Stupidity," surely Maimonides has said nothing about the truth or falsity of astrology per se, but simply that our forefathers did a bad job of utilizing the discipline. Both the second and third are *ad hominem* arguments, addressed at the practitioners of astrology, not at the discipline itself. The most we can conclude is that in the early period, astrology was not esteemed by the Greeks, but this tells us nothing about whether astrology is per se estimable. What Maimonides has not offered us are the "lucid, faultless proofs" to which he himself alludes which supposedly refute astrology.[136] The real philosophical concern, then, and the one to which Maimonides devotes the most attention both in the *Letter* and in the *Guide,* concerns the relation between judicial and natural astrology. In other words, from the very real influence of the celestial beings upon sublunar reality, the question is whether judicial astrology can be postulated. It is this issue that directly affects theories of divine providence, retribution, and free will.[137]

In the *Guide* Maimonides suggests that what leads people to believe in astrology is their confusion of the difference between overflow and powers of the celestial bodies. Philosophers all agree that the governance of the lower world is "brought about through the forces overflowing from the spheres."[138] Maimonides extends this dictum to include the notion that sublunar entities have a *mazal* associated with them: "even individuals subject to generation have forces of the stars that are specially assigned to them."[139] Powers of the stars, however, obey the laws of physics and, being corporeal, should not be confused with overflow. Astral forces, in obeying the laws of physics, decrease over time and distance. Maimonides repeatedly states that stars exert their influence only as bodies acting upon other bodies.[140] This position reflects Aristotle's dictum that bodies alone can set other bodies in motion.[141] Furthermore, Maimonides follows the Aristotelian scheme of assigning intelligences to the orbs.[142] But what powers do these orbs have? While accepting the metaphysical importance of these celestial beings in his cosmology, nevertheless he limits their power to the physical realm, thus eschewing the inference to astrology. As Langermann states, Maimonides totally rejects the attribution to the stars of any capacity in the realm of overflow.[143]

At stake are two issues. The first has to do with the negative implications of astrology with respect to human freedom. Maimonides repeatedly points to the inverse relation between the two. Free will, he tells us, is bestowed upon all humans. If God had decreed human character, or "if there were some force inherent in his nature which irresistibly drew him to a particular course . . . as the foolish astrologers out of their own fancy pretend," what then would be the force of the prophetic exhortations to improve one's way; what would be the point of the Torah, or of reward and punishment?[144] Based on these passages, it follows that if Maimonides upholds human freedom, he will necessarily have to reject astrological determinism. If actions traceable to human will and intention differ from events resulting from natural causes alone, it might be that those causes that motivate rational choice cannot be traced back to the stars. But whether Maimonides is in fact a proponent of human freedom is an open question. This does not mean that Maimonides does not rule out astrology on other more pragmatic grounds. Let me say, however, that Maimonides' apparent rejection of astrology on the grounds that it rules out human freedom is only convincing if we are right in concluding that Maimonides is in fact a proponent of human freedom.

A second, more general issue has to do with general teleology, or more pointedly, the final cause of the heavenly bodies. Maimonides clearly asserts that "governance overflows from the deity, may he be exalted, to the intellects according to their rank."[145] From this general overflow, however, it should not be concluded that the more perfect entity exists for the sake of the less perfect entity. Freudenthal argues, for example, that Maimonides' theological cosmology precludes astrology, largely on the grounds that for Maimonides, celestial beings do not exist for the sake of humans, but rather for their own sake.[146] Astrology is much more embedded in a worldview in which the heavenly bodies exist for the sake of sublunar entities.[147] But as Freudenthal argues, Maimonides' worldview is unfavorable to a science predicting the effects upon humans of the celestial bodies.

Maimonides addresses the issue of God's knowledge more fully in *Guide* 3.15–21. In these chapters of the *Guide* the question of divine omniscience is posed in the context of whether God's knowledge extends to the infinite. Maimonides is concerned with two questions: whether God knows actually existing particulars, and whether God's knowledge extends as well to the domain of possibles. Possibles can refer to those things that are now existing and depend for their existence upon a necessary cause, or to those things which will be brought into existence at a future time.[148] Of these, the latter is of direct concern to us with respect to God's knowledge of future contingents, while the former pertains to God's knowledge of actualized particulars.

The first question for Maimonides is under what conditions God can be said to know concrete particular entities. Maimonides rejects the position that God's knowledge does not pertain to concrete particulars. According to Maimonides' formulation of this view, Aristotle and his followers have claimed that God is not omniscient on the grounds that since knowledge of particulars would imply change and increase on the part of God's essence, God's knowledge must be either limited to his own essence, or include in addition only unchanging universals. In either case God's knowledge could not extend to particulars. This conclusion is supported on the basis of the following three claims. First, they claim that particular things are apprehended by means of the senses only and not by the intellect; but God does not apprehend by means of a sense. Second, particular things are infinite, whereas knowledge consists in comprehending; but what is infinite cannot be comprehended through knowledge. And finally, the knowledge of things that are produced in time would necessitate some change in God; for such knowledge involves a renewal of knowledge.[149] In contradistinction to these arguments, Maimonides' own position (to be expounded more fully below) is a modification of the thesis that "He, may he be exalted, knows everything and nothing secret is at all hidden from him."[150] I take the term "everything" to apply literally to *everything;* the question, of course, is whether "everything" includes the domain of particulars.

How might God obtain knowledge of concrete particulars? For Maimonides, a critical issue is whether God's knowledge can be of infinite things. As we have seen in chapter 3, the term "infinite" has two senses: it can mean infinite in extent, or infinite in number. Maimonides is concerned primarily with the second sense. He in effect rejects the two philosophical positions adduced earlier, namely that since God's knowledge extends only to things immutable, and he cannot know infinite things that are changing, God's knowledge must extend only to universals or to his own essence.[151] He mentions the view that "knowledge has for its object the species, but in a certain sense, extends to all the individuals of the species."[152] Maimonides might very well have in mind here the position of Avicenna who argued that God knows the species directly but knows indirectly the individuals comprised in the species.[153] Although Maimonides does not explicitly reject this view, it is doubtful he would have espoused it, since in an earlier chapter he had upheld the quasinominalist thesis that "no species exists outside the mind, but (that) the species and the other universals are, as you know, mental notions and (that) every existent outside the mind is an individual or a group of individuals."[154] Friedländer, following Munk, has argued that although universals have no existence independently of the mind, nevertheless according to Maimonides they do exist in the intellect and hence form elements of knowledge. Ultimately, however, knowledge is of the individual, and not of the species. That God's knowledge extends directly to individuals is supported by Maimonides' statement that God knows "with one knowledge the many and numerous things."[155] Again, I emphasize the logical import of the "many" and "numerous" that reinforce the idea that God's knowledge comprises particulars. But how God can know an infinite number of multiples without this implying a multiplicity in his nature is not explained by Maimonides; he concludes simply that God's knowledge may have as its object something that is infinite.[156]

Let us turn, then, to the second issue adduced above, namely whether God can know unactualized particulars. Inasmuch as this issue introduces the problem of an immutable deity coming to know what is mutable, Maimonides pays particular attention to the difficulties attending to this issue. As we have seen above, Maimonides emphasizes that the term "knowledge" is predicated equivocally of God and humans, maintaining that God is in no way affected by what he knows. God remains one even though his objects constitute a plurality, and he remains unchanged even though the objects of his knowledge are mutable. These points are reflected in two brief assertions: first, that God's knowledge does not contain plurality, and second, that God cannot acquire at a certain time knowledge he did not possess previously. The first claim is supported by the statement that God knows "with one knowledge the many and numerous things,"[157] that is, God knows multiples without this implying a multiplicity in his nature. Reminiscent of Averroës' distinction between active and passive knowledge, Maimonides distinguishes between the knowledge that an artisan has of what he has made, and the knowledge somebody else has of the object in question. God's knowledge of objects, like that of an artisan's of his work, is not derived from the objects themselves; rather, "the things in question follow upon his knowledge, which preceded and established them as they are."[158] In other words, since divine knowledge is *a priori,* it is not affected by the ontological status of objects that results from this knowledge. Hence Maimonides argues that since the objects of God's knowledge do not causally act upon his knowledge, his essence is unaffected by their multiplicity. Although Maimonides' ultimate contention will be that the whole of Being comprises one individual, nevertheless his emphasis here is that God is not affected by the underlying plurality of the world.

The second claim, that God's knowledge is unaffected by any change in its objects, is supported in the context of a distinction between absolute and relative nonexistence. Absolute nonexistence is never an object of God's knowledge.[159] Relative nonexistents, or future contingents, are possible objects of God's knowledge. It is not impossible, Maimonides claims, that God's knowledge have as its object those "nonexistent things about whose being is brought into existence."[160] Maimonides illustrates this with the case of God's knowing that a certain man who is now nonexistent, will exist at a future time, will continue to exist for some time, and then cease to exist. "God's knowledge does not increase when this person comes into existence—it contains nothing that it did not contain before." Neither does his knowledge imply plurality or change: "something was produced of which it had been perpetually known that it would be produced in the way it came into existence."[161] However, Maimonides is quick to point out that God's knowledge does not bring about the necessary occurrence of the entity in question: the possible remains possible.

> . . . His knowledge, may he be exalted, that a certain possible thing will come into existence, does not in any way make that possible thing quit the nature of the possible. On the contrary, the nature of the possible remains with it; and knowledge concerning what possible things will be produced, does not entail one of the two possibilities becoming necessary . . . His knowledge concerning what will happen does not make this possible thing quit its nature.[162]

In other words, given two unactualized states of affairs *p* and *–p*, God's knowledge that *p* will become actualized does not affect the possible status of *–p*. Hence, like Saadia and Halevi, Maimonides asserts that God's knowledge of future possibles does not change their nature; nor is his nature altered by a change in the objects of his knowledge. Maimonides does not, however, attempt to flesh out the logical difficulties in holding such a view. As Benor has so aptly commented, it would have helped to know what Maimonides thought of Aristotle's sea-fight paradox in *De Interpretatione*, chapter 9.[163] This work is not mentioned in his *Treatise on Logic*, nor does Maimonides allude to it in any other text. Benor suggests that Maimonides might have been influenced by al-Fārābī's commentary on the *De Interpretatione*, in which al-Fārābī rejects the "definite" truth value of *p*, espousing a form of simple logical necessity that does not carry with it fatalistic assumptions.[164] But there is no textual evidence that Maimonides would in fact have followed this interpretation, other than that denying a definite truth-value to statements about future contingent events would accord nicely with his espousal of human freedom. Nor do I agree with Benor's claim that Maimonides adopts the eternity thesis that is later to be associated with Aquinas and others. Were he to do so, his theory of omniscience would be even more problematic, for reasons I shall lay out below when we turn to the eternity thesis in the context of Crescas' theory.

One way of highlighting the problems inherent in Maimonides' theory of omniscience is to turn to his theory of providence. I have already stated that God's knowledge comprises individuals. But to what extent does God's providence extend to individuals? It might be argued that if providence represents the theological ramification of knowledge, then God's knowing individuals *qua* individual can offer to individuals the guarantee of providential care. In *Guide* 3.17 Maimonides presents five different theories describing the relation between God's providence and human freedom.[165] Pointing to difficulties with each of them, he then turns to his own view, asserting that divine providence watches over humans in the sublunar world as a result of a divine overflow with which human intellect is united. That a ship goes down at sea is the result of chance, but "the fact that the people in the ship went on board . . . is, according to our opinion, not due to chance (*be-miqreh*), but to divine will in accordance with the deserts of those people as determined in his judgments."[166] In other words, providence is consequent upon the perfection of human intellect and reflects the causal and ontological grid whereby God orders reality: in order for God to permit a causal nexus according to which certain people board the ship and others not, he must know, as it were, whether or not they are deserving of providence. But Maimonides was not unaware of the difficulties inherent in this position, as evidenced by his attempt to explain why it is that often the righteous (who ostensibly should be most firmly united with the divine overflow) nevertheless suffer. Claiming that divine providence is constantly watching over those who have obtained the intellectual overflow from God, Maimonides argues that evil attends to those who withdraw their attentions from God: "Providence withdraws from him during the time when he is occupied with something else."[167] Prophets or excellent persons suffer evil only during times of distraction, the "greatness of the calamity being proportionate to the duration of the period of distraction or to the vileness of the matter with which he was occupied."[168]

I submit, however, that these passages are ambiguous with respect to what God actually knows, and hence are equally ambiguous with respect to the efficacy of divine providence. In other words, we need to say more about the relation obtaining between omniscience and providence. Benor has argued that while God's knowledge extends to particulars, his providence applies to particulars "in a universal way."[169] In support of the thesis that providence is limited to species and does not extend to particulars, it can be argued that inasmuch as I am distracted, I fail to benefit from God's providence. On this picture God does not know me other than as an instantiated member of the species in question, and so my actions are responsible for whether or not I am subject to providence. In this case providence is general rather than particular and is linked directly to a theory of divine emanation.[170] When we suffer evil, it is we who "are the cause of this hiding of the face (*hester panim*)."[171] It is not God who brings on the separation, but humans in their willful turning away from God. In the next section, I shall argue that this position is much closer to Gersonides than to Maimonides. Let me simply say here that given that God does ostensibly know particulars, it makes no sense for Maimonides to claim that God's knowledge has no correlation to providence. In support of the thesis that God's knowledge of particulars extends equally to providence, it can be argued that God withholds providence when God knows that I am distracted. On this reading, my suffering or benefiting from providence is the direct result of God's knowledge of me *qua* particular; this reading is supported by the statement that "outside the mind nothing exists except individuals . . . consequently providence watches only over these individuals."[172] I suggest that this reading is more consistent with Maimonides' own statements pertaining to the extent of divine omniscience. But, of course, on this reading we are left with two problems: First, that God is ultimately responsible for the suffering of the righteous, and second, that God's providence, coupled with his omniscience, do affect my ability (*yekholet*) to board or not board the ship.

Ultimately, therefore, Maimonides realizes that his compatibilism relies upon a leap of faith. By emphasizing the total otherness of the deity, the incomprehensibility of the divine attributes, and the inability of humans to understand God's knowledge, Maimonides attempts to reconcile the fact that on the one hand God has ordered our lives, and on the other hand human freedom is a prerequisite for moral and religious accountability. This attempted reconciliation reflects the similar efforts of Saadia and Halevi. All three thinkers share an attempt to argue that God is all-knowing (3.1), that *p* is a contingent state of affairs (3.2 and 3.3), and that God's omniscience does not preclude this contingency (3.4). But, as I have tried to demonstrate, compatibilism contains numerous metaphysical and theological tensions, not the least of which has been reflected in Maimonides' attempted theodicy.[173] Let us turn, then, to Jewish responses to compatibilism, which contain in their formulations a trenchant critique of the very position presented by Saadia, Halevi, and Maimonides.

Incompatibilist Response of Ibn Daud

Both Ibn Daud and Gersonides are deeply committed to the asymmetry of the past and the future. As Feldman has pointed out, this commitment entails that the future is open and indeterminate, whereas the past is closed and determinate.[174] That is, the truth-value of statements about future contingents are as yet undetermined, whereas those about past events are determinately true or false; there are no truth-value gaps in the past. For this reason many philosophers characterize the "truly possible" in terms of the future alone.[175]

Ibn Daud and Gersonides are the only Jewish thinkers to uphold a form of indeterminism as a solution to the paradox of divine omniscience. Ibn Daud tells his reader that he has undertaken to write his philosophical work

'Emunah Ramah in response to the problem of free will and omniscience. As he states in his preface, "[Someone] had asked the author, "Are the actions of man necessary (*mukhraḥ be-ma'asav*) or does he have choice (*beḥirah*) over them?" To answer this question the author stated [this treatise] in one Abstract and three books."[176] The problem is crystallized further in the introduction where Ibn Daud states the position of those who have difficulty accepting that God punishes those whom he necessitates to sin:

> If God, may he be exalted, necessitates him [to commit sins] (*makhriḥo 'aleihem*), then how can he punish him for them?[177]

Ibn Daud answers this quandary with a position not evidenced hitherto in Jewish thought, namely that God's omniscience does not extend to future contingents. His defense of free will is based on the very nature of the possible. In book 2.2 Ibn Daud develops this solution by upholding the domain of genuine possibility. Distinguishing between two types of possibility, he claims that only the first, epistemological possibility, leads to ignorance: "[One kind of] possibility is possibility with respect to ignorance (*'efshar mi-tzad ha-sikhlut*)."[178] Ibn Daud adduces two examples of epistemological possibility. The first example, whether the king of Babylonia died today or is alive, is a possibility

> because we, the men of Spain, do not know this [state of affairs]. Rather, both alternatives are equal to us. . . . But since [this] matter in itself is not possible (*ha-davar be-'atzmo 'eineno 'efshar*), of necessity one of the alternatives is correct, and God, may he be exalted, knows in [cases] like this that one of the alternatives is necessary as it is in itself.[179]

Clearly such cases are termed possible only because humans do not have sufficient knowledge of the matter at hand; in Ibn Daud's example, the men of Spain simply don't have prior knowledge of the determinate state of affairs that God does know. The second example is astronomical and concerns the future occurrence of an eclipse. To those ignorant of astronomy, eclipses appear to be indeterminate, but to a trained astronomer, and most certainly to God, astronomical events are not "a possibility" because they are known to be determinate, natural events.[180] Once again, the possibility adduced reflects an epistemic deficiency on the part of the knower and not a metaphysically indeterminate state of affairs.

Ibn Daud then distinguishes a second type of possibility; this genuine ontological possibility is one that God himself has created *qua* possible: God "created it as a thing that can bear one or the other of two contrary attributes (*bara' 'oto sovel ha-shnei te'arim ha-hafakhim im zeh ve-'im zeh*)."[181] The most God knows is that such a thing is possible; even God, however, does not know which of the two possibilities will be actualized. To the "sophist" who argues that this imputes ignorance to God, Ibn Daud replies that "this is not ignorance" because technically speaking, there is nothing determinate to be known.[182] Not to know *what is-not* does not constitute a deficiency in God's omniscience. And so Ibn-Daud feels that he has safeguarded future contingency without limiting God's power.

Omniscience and Human Freedom in Gersonides

Intimated but not fully developed by Ibn Daud, indeterminism finds its fullest expression in Gersonides' work *Milḥamot ha-Shem.*[183] Gersonides' concerns in treatise 3 are threefold: to determine whether God's knowledge extends to material singular objects, to future contingent objects, and to events. Although Gersonides himself does not always clearly separate these questions, I shall attempt to introduce a number of relevant distinctions into this discussion. The first question deals with knowledge of particular objects (*peratiyim*) and presupposes the following distinction between individuals and particulars: An individual is any object which is capable of unique reference, whereas a particular is a material individual.[184] Hence the first question in treatise 3 can be construed as whether God, an immaterial entity, can know materially constituted objects. The second question is a subset of the first and is concerned with whether God knows possible objects (*'efshariyim*). Although Gersonides does not distinguish in the text between knowing possible events and knowing possible objects, it is important to note that his discussion for the most part centers around the latter.

In addition, however, Gersonides is concerned with a further problem. Like the majority of his philosophical precursors, Gersonides does assert the existence of contingency (*ha-'efshar*) in the universe. However, as we have seen above, Gersonides is one of a minority of Jewish philosophers to uphold a theory of astrological determinism, according to which states of affairs are determined by the heavenly bodies. Within this deterministic scheme, Gersonides recognizes a domain of contingency that comprises chance events, possibilities, and human choices. Each of these represents an area that is not subject to the absolute necessity of the stars. Gersonides defines contingent events as those that are devoid of determining causes, as opposed to events that come to pass as a result of their own determinate causes. He goes on to say, however, that contingent events are nevertheless structured.[185] Gersonides' point is that inasmuch as the determination of contingent events occurs on the astral level, these events are ordered. However, their causes are incomplete. For as we shall see below, Gersonides will argue that the causes of human volition are not subject to the stars.

Contingent events are reflected most clearly in the domain of human agency. Following Aristotle, Gersonides argues that absolute necessity contravenes human deliberation and will.[186] A similar point is made in Gersonides' supercommentary on Aristotle's *De Interpretatione.* Arguing that arguments in favor of fatalism destroy the nature of the possible, Gersonides goes on to state that from fatalism many absurdities result, including the pointlessness of human deliberation. But, he claims, "we need no teacher from whom to learn that deliberation itself and the will is a cause from which many things arise." And furthermore, "this sort of possibility appears in the agent . . . because the agent has free choice and [the power of] decision, so that he can actualize each opposite."[187]

It should be noted that these distinctions are ontologically rooted in events and are not epistemically based. Gersonides is not suggesting that humans, acting out of ignorance of the determining causes of their behavior, believe (falsely) that their actions are based on human volition. Rather, he is claiming that human beings, by virtue of their intellect and will, can overcome the determining influences of their astrological signs. Granted, according to Gersonides, this ability is rare, and real instances of volition are uncommon: "Despite the fact that deliberate human choice (*beḥirah*) made out of rational consideration has the power to disrupt this determination, such an eventuality will occur very rarely."[188] As Manekin has argued, Gersonides has no equivalent for "free choice" or "free will;" the term "*beḥirah*" (choice) is more limited in scope, emphasizing the efficacy of human choice in the face of celestial causality.[189] Intellect and will, however, can move humans to do something other than what has been determined from the standpoint of the heavenly bodies. For God has placed within humans purposive reason "so as to move [humans] toward something other than that which has been determined from the aspect of the heavenly bodies, insofar as this is possible to make straight that which chance has convoluted."[190]

Gersonides claims as well that some states of affairs occur by chance (*be-miqreh*). Speaking in treatise 4 of *Milḥamot* of the origin of evil, Gersonides distinguishes between evils that arise from human choice and those which arise from chance.[191] In these examples, chance events occur without regard to human intention. However, not all chance events occur independently of human agency. In *Milḥamot* 2 Gersonides gives the following characterization of a chance event. In describing chance events that are viewed under the mode of indetermination, Gersonides says that characterizations of such events are "analogous to the claim that were we to dig a hole so deep as to make a pit, we would find a treasure—for this is infrequent, i.e., if we were to find a treasure each time that we dug so deeply as to form a pit."[192] The actual occurrence of contingency is fairly infrequent; in fact, human acts of will are the only example of contingency. Since only humans have free will, to ask whether God knows future contingent events is tantamount to asking whether God has foreknowledge of the free actions of humans.[193] I shall try whenever possible in the course of our examination to specify which of these questions is being discussed by Gersonides. In general, the term "particular" (*perat*) will refer to objects, whereas the phrase "future contingent" (*'efshari*) will refer to events.

Before stating his own view, Gersonides first examines the positions of his two philosophical predecessors, Aristotle and Maimonides, who represent diametrically opposed positions. According to Gersonides' understanding, Aristotle denied to God knowledge of particulars whereas Maimonides claimed that God in a single act knows all particulars.[194] Gersonides' own position will ultimately reflect his affinities with the Aristotelian position.

According to Gersonides' interpretation, Aristotle offered a number of reasons why God cannot know particulars. Of these, the most important for our purposes are the second and the sixth. The second reason states that particulars are temporal (*zemaniyim*), but since nothing of which neither motion nor rest is predicated can perceive temporal entities, God cannot perceive particulars. The sixth reason states that knowledge of changing particulars would itself be mutable; since God's essence is immutable, he cannot perceive mutable objects.[195] Both these arguments deny to God knowledge of particulars on the basis that particulars are qualitatively different from God's essence. Since a nontemporal being cannot know temporal entities, and an immutable being cannot know ever-changing entities, God could not know particulars without undergoing a radical change in essence. We shall return to both these points shortly.

Gersonides next describes the position of Maimonides. Reflecting the discussion in the *Guide*, Gersonides distinguishes five ways in which, according to Maimonides, God's knowledge differs from that of humans.

- God's knowledge, while being one, corresponds to many known things belonging to various species.
- God's knowledge may have as its object something that does not exist.
- God's knowledge may encompass an infinite number of things.
- God's knowledge "of future events does not entail that the predicted event will occur rather than its opposite is still possible."[196]
- God's knowledge undergoes no change in its apprehension of things produced in time.

Maimonides had claimed that despite the fact that in each of these cases a paradox would arise if the knower were a human being, nevertheless paradox is eschewed when the knower is God. Hence according to Maimonides, God knows multiples without relinquishing his unity, he knows nonexistents, he knows infinite objects, he is unchanging despite the change in the objects of his knowledge and he does not causally affect the objects of his knowledge.

Having examined the arguments in support of both Aristotle and Maimonides, Gersonides has enumerated a number of considerations that pertain to the problem of divine omniscience. These considerations can be specified in terms of three questions: Can God know *p*, when *p* stands for a future contingent event? Can God know future contingent objects without undergoing a change in his essence? And finally, does God's knowledge of a particular necessitate its actualization? In order to assess critically the arguments propounded by his predecessors, Gersonides must address each of these questions.

The first two questions concern God's knowledge of as yet unactualized particulars. According to Gersonides, Maimonides has argued that even though God knows entities before they come into existence, nevertheless his knowledge is unchanging. "His [God's] knowledge of this event that is nonexistent now is actually present, whereas the object of knowledge to which this knowledge refers is nonexistent."[197] Gersonides, however, finds two problems with this position. His first point is that Maimonides has violated both the identity theory of knowledge propounded by Aristotle and the correspondence theory of truth presupposed in any such theory of knowledge.[198] Knowledge is always of what is. However, in Maimonides' argument, the relation between the act of knowing and the nonexistent object of knowledge is necessarily vacuous, since there *is* nothing to which the former can correspond. Hence according to Gersonides, without an identity between knower and thing known, there can be no knowledge. For if the proposition known is true, it must refer to something that exists; without this referential relation, there can be no identity, and hence no knowledge.[199]

Gersonides then develops his main argument that an omniscient, immutable deity cannot know changing particulars. The underlying premise in this argument is that all future particular objects are in fact mutable: that is, they change from a state of nonexistence to one of existence. Gersonides claims that an immutable deity cannot be omniscient, if omniscience entails knowing objects that undergo change. In order to demonstrate that in ordinary-knowing situations omniscience and immutability are incompatible, Gersonides raises the following dilemma. Once it is granted that God knows future contingents, one of the two must obtain: Either (4.1) God knows that, from a set of possibilities, the alternatives which he knows would be generated have been generated. Or (4.2) God knows that the actualized alternative from a set of possibilities has been generated even though it is different from that alternative which he knew would be generated prior to the generation of this actualized alternative.[200] In other words, given a situation in which (p or $-p$) are both future contingents, one of which occurs at t, (4.1) states that God knows before t and at t that p would be generated, whereas (4.2) states that he knows at t that $-p$ has been generated even though he knew before t that p would be generated.

According to the first horn of the dilemma, God's knowledge is unchanging: He always knew, both before and after t that p would occur. Gersonides claims that this option allows for error on the part of God. The reason for this, he insists, is that "since the possibility whose occurrence he foreknew could not have occurred, it could happen in many of these things that what takes place is not the event that God foreknew would happen."[201] In other words, if both p and $-p$ are truly contingent, then $-p$ has as much of a chance of occurring as does p. Hence it is possible that even though before t God knew that p would occur, in fact $-p$ occurs. According to Gersonides, in such a case what God knows is false. Gersonides' argument can be reformulated as follows:

5.1 Either p or $-p$ will occur at t.
5.2 Assume that p occurs at t.
5.3 Before t God knows that p will occur at t.
5.4 But possibly $-p$ will occur at t.
5.5 Before t God knows that p will occur at t and possibly $-p$ will occur at t.
5.6 Therefore it is possible that (before t God knows that p will occur at t and $-p$ occurs at t).
5.7 If it is possible that (before t God knows that p will occur at t and $-p$ occurs at t), then God has erred.
5.8 It is possible that God has erred.
5.9 If it is possible that God errs, then God is fallible.
5.10 God is fallible.

Argument Summary: In this argument Gersonides shows that the assumption that God foreknows a future contingent event leads to the unacceptable conclusion that God is fallible. For if an event is truly contingent, it can either occur or not occur. But that means that when God knows that a contingent event will occur, the possibility exists that it will not occur. If it turns out not to occur, then the item that God knew (that the event would occur) turns out to be false. Hence God is fallible, that is God's knowledge is not absolutely certain.

A similar argument can be constructed assuming in (5.2) that at t $-p$ occurs. In both cases, (5.1) represents a future contingent event, while (5.3) represents God's infallible knowledge. Gersonides' argument is an attempt to demonstrate that (5.1), (5.2) and (5.3) are incompatible. His argument fails, however, in the transition from (5.5) to (5.6). (5.6) represents a scope fallacy from a *de re* to a *de dictu* modal form, that is, a move from $(p \& -q)$ to $-(p \& q)$. Gersonides' argument depends upon (5.6), but there seems to be no other way to generate (5.6). Hence he has not shown that God is fallible. The second horn of the dilemma suggests not only that God is fallible but that he is no longer immutable. For according to this option, before *t* God knows that $-p$ occurs and only at *t* does he know that *p* in fact occurs. Not only was God mistaken before *t* about the future occurrence of $-p$, but according to Gersonides, "this is certainly a novelty and change in God's knowledge."[202] Since God's knowledge changes from $-p$ before *t* to *p* at *t*, God is no longer immutable; rather he changes along with the change in the objects of his knowledge.

The arguments attributed to Aristotle further support Gersonides' thesis that divine omniscience and immutability are incompatible. In *Milḥamot* 3.4, Gersonides lays out the arguments of those Aristotelians who, following certain suggestions implicit in Aristotle, claimed that God does not know particulars. According to Gersonides, the Aristotelian claim that the thesis that God does know particulars is prone to two possibilities: Either (6.1), God knows objects before they come to be; or (6.2), God knows them only when they come to be and not before. (6.1) leads, however, to a number of untoward consequences, the most pressing of which is that God's knowledge is related to what does not exist. But this then leads to the following dilemma:

6.3 Either God knows objects according to their nature as contingent beings so that the contradictory of what he knows will be actualized remains a possibility.

6.4 Or God knows which one of these contradictory alternatives will be actualized and its contradictory does not remain possible.[203]

The Aristotelian rejects each of these alternatives as follows: (6.3) leads to the conclusion that God's knowledge of these entities, before they came to be, changes with their coming to be, since both were possibilities before being actualized, but afterwards both possibilities were eliminated. Hence on this reading God's nature is mutable. And the implication of (6.4) is that nothing in this world is a contingent that either may or may not be actualized. Since Gersonides, and the Aristotelians, are both committed to the existence of contingency, this option is rejected as absurd without further discussion. Having rejected all readings for (6.1), the only possible alternative is (6.2), namely that God knows future contingents only when they come to be. But this position entails that God's knowledge be subject to change; for every time that a new state of affairs occurred, God would acquire a new piece of knowledge. On the basis of these arguments, the Aristotelians, according to Gersonides, concluded that God has no knowledge of particulars. We shall see that Gersonides ultimately espouses a weaker formulation of this conclusion.

The third and final question concerns the epistemological force of God's foreknowledge. Does it follow from God's knowing a future contingent that it is necessary? In contradistinction to Saadia and Halevi, who claim that God's knowledge does not render the objects of his knowledge necessary, Gersonides will want to maintain that divine foreknowledge and contingency are incompatible. It should be noted that the main force of this controversy centers not upon causal necessity but rather upon epistemological and logical necessity. That is, Gersonides is not concerned in this context with whether God's knowledge itself has causal efficacy over the objects of his knowledge. This question, having to do with the relation between intellect and will, is a separate issue pertaining to the extent of God's creative will. Rather is he concerned with the apparent logical necessity that follows from the infallibility of God's knowledge. In other words, given that God knows *p*, it appears to follow from God's infallibility that *p* cannot fail to be the case, and hence that *p* is necessary.

Gersonides summarizes Maimonides' position, as stated in the *Guide*, as follows: "The divine knowledge of future events does not entail that the predicted event will occur; rather its opposite is still possible."[204] Only by denying the logical force of God's knowledge can Maimonides safeguard the contingency of events. Gersonides rejects Maimonides' arguments on the grounds that they attribute to God the *possibility* of fallibility. Arguing that divine omniscience severely compromises the contingency of the objects of God's knowledge, Gersonides dismisses Maimonides' form of compatibilism. Having rejected Maimonides' attempts to harmonize foreknowledge and contingency, and having upheld the existence of contingency in the universe, Gersonides adopts the one option left to him, namely that God does not know future contingents.

According to Gersonides, God knows that certain states of affairs may or may not be actualized. But insofar as they are contingent states, he does not know which of the two alternatives will in fact be actualized. For if God did know future contingents prior to their actualization, there could be no contingency in the world.[205] Echoing Ibn Daud, Gersonides claims that God's inability to foreknow future contingents is not a defect in his knowledge:

> The fact that God does not have the knowledge of which possible outcome will be realized does not imply any defect (*ḥisaron*) in God. For perfect knowledge of something is the knowledge of what that thing is in reality; when the thing is not apprehended as it is, this is error, not knowledge.[206]

In this fashion, Gersonides concludes, the problem of divine omniscience is resolved in favor of indeterminism. With respect to future contingents, God knows their ordered nature or essence, and he knows that they are contingent, but he does not know which alternative will become actualized. Hence contingency has been retained with no apparent sacrifice of omniscience. In an apparent attempt to mediate between the mutually contradictory views of Aristotle and Maimonides, Gersonides adopts the best of both sets of arguments, claiming that "there is no alternative but to say that God knows particulars in one respect but does not know them in another respect (*she-yed'am be-tzad velo' yed'am be-tzad*). But what these respects are, would that I knew!"[207]

God does not know particulars insofar as they are not ordered, that is, in their particularity, but rather insofar as they possess universal natures. Two interpretations of this statement suggest themselves. The first is that what God knows are the individual forms or concepts of particulars. That is, when a sufficient number of general properties is known, God can be said to know the particular in question. According to this interpretation, since every object represents a unique instantiation of an individual concept, knowledge of a particular consists in being able to read off the predicates obtained in the individual concept of that particular. So, for example, by knowing the descriptive predicates snub-nosed, Greek philosopher, living in Athens, drank hemlock, and so on, God can be said to know Socrates.[208]

It is unlikely for a number of reasons that Gersonides had this interpretation in mind, however. First it presupposes that it is possible to refer uniquely to a particular by this process of descriptive specifications without any reference to material specification. Gersonides, however, does feel that even though material composition of an object is accidental, it must enter into the description of the object in question.[209] Since God could not materially individuate the particular, he could not have referential knowledge of the particular in question. Furthermore, unable to individuate an object on the basis of its matter, God could not know whether the description in question referred to one object alone: in other words, God would be unable to ascertain whether the description in question were satisfied by only one member, or whether a number of objects shared the same description. Finally, this interpretation appears to rule out contingency. For if God can read off all the predicates contained in an individual concept, and these predicates are determinate, wherein lies the room for contingency?[210]

A second interpretation of the phrase is that what God knows are the universal causes of particulars. Insofar as particulars represent instances of these causes, God can be said to know particulars. This view was developed earlier by Avicenna, whose views Gersonides might have known.[211] In other words, according to Avicenna, God's knowledge extends only to the domain of genera and species; God knows individual persons, for example, only through knowing the species man. Although Gersonides does not explicitly mention Avicenna, various aspects of Avicenna's theory are present in his own theory of divine omniscience. For like Avicenna, Gersonides would like to argue that God knows universals and only indirectly does his knowledge pertain to particulars. What God knows are the constitutive properties of concepts; these properties are then applied indirectly to particulars.

Indeterminism and Prophecy

Gersonides is not unaware of the implications of this theory with respect to the status of temporality in general, and to prophetic statements about the future in particular. These implications can be summarized as follows. Many scriptural prophecies concern contingent events. Hence the existence of prophetic statements about future contingent events suggests that these events can be foreknown. If they are foreknown, how can they be contingent? And yet if they are not foreknown, how can the events in question be prophesied? In short, what sense can we make of prophetic statements that purport to relate information about future contingent events?[212]

Emphasizing the connection between prophecy and divine omniscience, Gersonides presents a theory of prophetic statements that is meant to complement his position with respect to divine foreknowledge. We have seen that according to Gersonides' theory, God does not have foreknowledge of future contingent events. Yet God is able to transmit to the prophet true statements about such events. In an attempt to reconcile the apparent conflict between God's lack of knowledge and the transmission of prophecies about future contingents, Gersonides construes these statements as disguised universal conditionals.

The logical implications of indeterminism affect prophecy in the following way. The indeterminist claims in effect that propositions about the future have no present truth-value. Gersonides, for example, states in his super-commentary to Aristotle's *De Interpretatione* that positing a truth-value for future contingents leads to absurdity.[213] According to Gersonides, therefore, future contingent statements "do not divide the true from the false completely when the statements are uttered."[214] That is, such statements are neither true nor false. But if statements about future contingent events have no determinate truth value when uttered, how do we account for the truth value of prophetic statements? Prophecies comprise a subclass of future contingent statements. If they are not true when uttered, they would seem to have no assertive value for their recipients. And yet if they are true when uttered, they suggest that the events prophesied are inevitable. Gersonides does not wish to eliminate the force of prophetic statements altogether. He is therefore hard-pressed to explain not only how prophecies are possible in a system that denies the possibility of foreknowledge of future contingents, but also how to construe their logical form. In order to extricate himself from possible theological embarrassment, Gersonides must demonstrate that his theory of divine omniscience does not preclude the possibility of prophetic statements.

Prophecy, Gersonides claims, is not the result of chance or guesswork but rather represents genuine knowledge on the part of the prophet. Inasmuch as prophecy concerns contingent human choices, Gersonides asks, how "[is it] possible that this communication received by us be concerned with future contingent events?"[215] According to Gersonides, two dilemmas arise from admitting that the prophet can foretell future contingents. The first is that if prophetic statements about future contingent events are true when uttered, then since the option prophesied has already been determined and structured, contingency will be abolished. That is, according to Gersonides, all those matters are then "by necessity" in the sense that "choice in accidents pertaining to in human affairs would be nullified (*yevuṭal ṭeva' ha-'efshar meha-miqrim*) . . . all things would exist by necessity and voluntary choice (*ha-beḥirah*) would be futile."[216] And since, according to Gersonides, contingency occurs only with respect to events involving human beings, once contingency (*davar 'efshari*) is removed from the human sphere, it follows that all things would exist by necessity.[217] The second problem concerns the relation of prophetic events to future contingency. For if these events cannot be foreknown, then how can statements about future contingent events be transmitted to a prophet?

Gersonides' theory must therefore answer two general questions: First, inasmuch as God does not know particulars *qua* particular, how is the prophet able to receive prophecies about particular human beings and events? And second, if contingent events cannot be foreknown, how can the prophet prophesy about future contingent events? The first question is stated by Gersonides as follows: "How is it possible for the [active] intellect to inform us about the particular thing *qua* particular, as is the case with this communication, since it can transmit only knowledge of what resembles itself, viz. the general structure?"[218] This question not only reflects the problem raised by Gersonides' own theory of divine omniscience, but echoes similar concerns of Averroës. In his *Epitome,* Averroës developed a theory of prophecy according to which the active intellect is devoid of matter and can comprehend only universal natures.[219] In his commentary on Averroës' *Epitome,* Gersonides did not take issue with this analysis. However, in *Milḥamot* he disagrees with Averroës, arguing that knowledge of a particular is transmitted by the active intellect not *qua* particular but rather "insofar as this individual is a contemporary of other human beings who were born in his zone; when the heavenly bodies were in the same position that they had when he was born."[220] That is, all those born under the same astrological sign have a similar general nature. The prophet then receives this general information and, with the aid of the hylic intellect that supplies him with "all those accidents through which he is individuated inasmuch as he is a particular human being of that description,"[221] the prophet's imaginative faculty then interprets the prophecy. The crucial difference between Averroës and Gersonides is that in Gersonides' analysis the prophet does not receive particular knowledge. At the point of origin, the emanation in question is general: the prophet, with the aid of the hylic intellect, then enters into the process and instantiates the message. Particularity is effected by what Gersonides calls the "available phantasms which constitute the instrumentalities for the production of that particularity."[222] In other words, the prophet, who receives an emanation of a general nature, specifies it in terms of his own phantasms.[223]

Gersonides has explained the particularity of prophetic statements by emphasizing that particular details are supplied not by the active intellect but by the prophet. But do these details include contingent events as well? In order to answer the second question posed above, Gersonides develops a theory of the conditional nature of prophetic statements. He maintains that although not all prophetic statements found in scripture are in conditional form, nevertheless they all contain implicit antecedent premises that must be satisfied prior to the actualization of the prophecy in question. These conditionals account for both the particularity and the contingency of the events prophesied.

A number of examples are adduced by Gersonides that reflect this conditional form. In Genesis 41: 4–48, Joseph prophesied to Pharaoh that a famine would occur and last seven years. Although the prophecy is presented in the form of an indicative statement, Gersonides suggests that the suppressed conditional implied by this prophecy is of the form

> 7.1 If Pharaoh does/does not do such and such, a famine will occur for seven years.[224]

Since it is up to Pharaoh either to realize the antecedent or not, the occurrence of the famine ultimately depends upon the satisfaction or nonsatisfaction of the antecedent.

In the next example, Gersonides' analysis is slightly more complex. In Exodus 32:10 God says to Moses

> 7.2 "Therefore let me alone, that my wrath may wax hot against them, and that I may consume them; and I will make of thee a great nation."

Interpreting this prediction in light of Gersonides' previous analysis, we see that the consequent expresses a benevolent situation which seems to depend upon a malevolent antecedent. For (7.2) can be construed in several ways. It might be read as an implied threat of the form

7.3 If you let me alone so that my wrath waxes hot, then I will consume them and make of thee a great nation.[225]

But (7.3) is unacceptable for Gersonides, for it suggests that God does not want Moses to commit an intrinsically good action. That is, (7.3) seems to imply both

7.4 If you let me alone (i.e., do not pray for the good of another), I will ultimately reward you (make of you a great nation).

and

7.5 If you do not let me alone (i.e., do pray for the good of another), I will punish you, not make of you a great nation).

Both (7.4) and (7.5), according to Gersonides, are unworthy of a benevolent God who supports the good behavior of his subjects. In order to avoid the untoward implications of (7.4) and (7.5), Gersonides interprets (7.2) as follows:

> the intent of this verse is to be taken as saying that even if Moses does not employ himself to save them, he should not be fearful that God will destroy Israel, because from Moses there will issue forth a great people. And thus, there was not removed from him the benevolent promise despite his praying for Israel.[226]

That is, Gersonides interprets (7.2) as

7.6 Even if you do not let me alone, I will consume them and make of thee a great nation.

In other words, Moses will be benefited, and the benevolent conditional will be actualized regardless of his (Moses') actions.

But doesn't (7.6) suggest that in some cases, at least, the antecedent and consequent are independent of one another? For (7.6) seems to suggest that God will reward Israel regardless of what Moses does. If that is so, what is the logical value of the conditional? In order to resolve this problem, Charles Touati has suggested that Gersonides' example marks a simple temporal succession from antecedent to consequent.[227] But Touati's solution is inadequate, for by Gersonides' own account, prophecies aim to accomplish more than simply state a temporal succession of events. Gersonides has wanted to claim that some benevolent conditionals at least reflect a nonaccidental connection between antecedent and consequent, but he has not explained how they differ from unconditional benevolent conditionals. What Gersonides has suggested is that not all benevolent conditionals state a connection between antecedent and consequent. But without any connection at all, the conditional in question no longer satisfies its original function, namely to capture the connection between antecedent and consequent.

In Gersonides' final example, this problem becomes even more apparent. In 1 Samuel 10:3–5, Samuel prophesies to Saul that Saul will encounter three men on the road; these men will give him two loaves of bread which he will then accept.[228] Recasting this prophecy as a conditional, we have

7.7 If Saul walks down the road, he will encounter three men who will offer him two loaves of bread which he will accept.

But what is the force of (7.7)? The whole point of such conditionals is that they do not follow laws, but rather reflect the free choices of a free agent. But if an event is truly contingent in Gersonides' sense, how can it function as the referent of a consequent in a universal conditional? It is difficult to see what information could be contained in the antecedent. For if the antecedent represents a necessary state of affairs, then the conditional is either trivial or false. And if it represents a contingent state of affairs, the conditional has not transmitted any information of a relevant nature. In short, Gersonides' analysis has not accounted for the conditional nature of prophetic statements.

One final prophecy is mentioned not by Gersonides but by Ibn Daud, namely 1 Samuel 23:12. In this passage, David is surrounded by Saul and asks God whether Saul will capture him and his men; David is told by God that if he remains in Keilah, his men will be surrounded and captured by Saul: "David continued, 'Will the citizens of Keilah deliver me and my men into Saul's hands?' And the Lord answered, 'They will.'" This prophecy became a paradigm in later scholastic sources and can be rendered as follows:

7.8 If David remains in Keilah, he and his men will be surrounded and captured by Saul.

This prophecy has been used by both scholastics such as Molina and Suarez, as well as recent philosophers, to utilize a form of middle knowledge which as Gaskin puts it, "shares with natural knowledge the fact that it is prevolitional, with foreknowledge the fact that its objects are contingencies."[229] The virtue of middle knowledge, according to its supporters, is that on this view God can plan for every eventuality without in any way compromising human volition. Although middle knowledge is not an explicit ingredient in

Jewish texts, nevertheless intimations of the conditional nature of prophecy, upon which middle knowledge is based, can be seen in Ibn Daud's discussion. In commenting upon 1 Samuel 23:12, Ibn Daud presents the following characterization of that passage:

> An example (of contingent statements) would be that of David, peace be unto him, "Your servant has indeed heard that Saul seeks to come to Keilah to destroy the city. Will they deliver me up?" He repeats the question and says, "Will the citizens of Keilah deliver up me and my men to Saul?" And the Lord said, "They will deliver you up." The meaning of "they will deliver up" in the response of God is that it is possible (*'efshar*) that they will do this. . . . If their capture were not contingent to God, may he be exalted, David would not have escaped from it.[230]

It is clear from his interpretation of the phrase "it is possible that they will do it" that according to Ibn Daud, whether or not David will be captured remains an indeterminate event. Ibn Daud's interpretation of (7.8) is similar to what Pollock has termed subjunctive or might-be conditionals. These conditionals are of the form

7.9 If p were true, q might be true.

Alternatively, (7.9) can be read as

7.10 It is not the case that q would be false if p were true.[231]

Applying this reading of might-be conditionals to our two prophetic examples, we can recast (7.10) as

7.11 If David were to remain in Keilah, he and his men might be surrounded and captured by Saul.

Similarly, we can recast (7.8) in terms of

7.12 If Saul were to walk down the road, he might encounter three men who would offer him two loaves of bread which he might accept.

(7.11) and (7.12) both retain the contingency of the consequent, a characteristic important for both Ibn Daud and Gersonides. However, they tell us no more than the fact that given a situation p, q may or may not occur. The occurrence of the antecedent in no way sheds light upon the future occurrence of the consequent. In these cases, the prophet knows no more than the casual observer. But if that is so, then Gersonides has not explained the important status of the prophet, for the prophet has not transmitted information of a relevant nature about the future. Once again, the logical status of prophetic statements remains deficient.

In short, the most anybody can know (God included) is that certain events might occur in light of certain antecedent conditions. But surely we don't need a prophet to tell us that! What I have tried to show is that ultimately Gersonides' indeterminism leaves little room for a theory of prophecy in which future contingent events are determinately prophesied. The most a prophet can tell us is that certain events are contingent; he cannot tell us which contingent possibility will be actualized. This inability on the part of the prophet to transmit the content of future contingent events is a severe limitation on the content of prophetic knowledge.[232]

The Challenge of Determinism: Crescas on Divine Knowledge and Possibility

These sorts of limitations on God's knowledge led Crescas to posit what Feldman has described as a "theological determinis[m] of the compatibility variety."[233] On Feldman's reading of Crescas, although human actions, including future acts, are determined by God's knowledge, these actions are not absolutely necessitated. In this section I shall argue that Crescas has abandoned compatibilism in favor of determinism, that is, that Crescas has so compromised the domain of possibility and of future contingency that he no longer permits a genuine sense of freedom. In other words, I claim that for Crescas the future is as fixed as is the past: there are no indeterminate events, and hence contingency has been eliminated.

In *'Or 'Adonai* Crescas lists three principles which are necessitated by tradition:

> *The infinite science of God:* "The first, that his knowledge, may he be exalted, encompasses the infinite (*be-mah she-'ein takhlit lo*)."[234]
>
> *God's foreknowledge:* "The second, that his knowledge, may he be exalted, extends over that which does not [now] exist (*be-mah she-lo' nimtza'*)."[235]
>
> *The non-causative power of omniscience:* "His knowledge, may he be exalted, extends over the (disjunctive) parts of the possible (*be-ḥelqei ha-'efshar*), without changing the nature of the possible . . . (*mi-bilti she-yishtaneh ṭeva' ha-'efshar*)."[236]

Crescas' stated goal in this work is to examine those arguments of the philosophers, and that of Gersonides in particular, which threaten these principles. In standard scholastic fashion, Crescas lists arguments both for and against the three principles, with the intention of supporting the former. The philosophical arguments against divine omniscience that concern Crescas include the following four. First, if God knows all events, then God is being perfected by this knowledge, but God cannot be affected by matters in the world. Second, if God's mind becomes identical with things known, God will become a multiplicity. Third, because particular things (*devarim*) are temporal (*zemaniyim*), and whatever relates to time is an accident of motion, it follows that whatever is not described in terms of motion will not know particulars. And finally, the whole notion of divine science is untenable given the existence of evil.[237] Other more specific objections are raised against God's prescience as well. For example—Can God's knowledge be infinite? Can the infinite be comprehended? If knowledge requires that its object exists, how can God know what doesn't exist? Does not God's knowledge of future contingents require a change in God—before they occur he knows them as future, and after that as past? These philosophical worries are summarized by Crescas as follows:

> . . . if his knowledge were of the attainment of one of the parts of the possible, and its contradictory remained possible, as would appear to be the case here, then when that one part has been attained, the possible will have vanished, and his knowledge will be of it [alone], and his knowledge will have undergone change. And since the intellect [according to the philosophers] becomes constituted as a substance out of what it knows, it would be necessarily implied that his essence undergoes change, and that is perfectly absurd.[238]

Because Gersonides exemplifies a paradigm of the philosophical stance, Crescas turns to Gersonides who claimed that God knows particulars only through their general order; that God knows only that certain things are possible; and God does not know of the happenings of one of the possible sides, even *a posteriori,* for knowing it would occasion a change in God. Setting out to refute this doctrine of Gersonides, Crescas offers several arguments against Gersonides' position. First, he argues that if God does not know particulars *qua* particular, then what he does not know is infinite, and what he knows is finite. Second, if God does not know which possibility will be actualized, he is ignorant of most of the happenings of the world. And finally, Crescas asks, can God be said to know the past on this theory? If not, then God is ignorant of most of human history.[239]

Crescas' own theory is presented as a response to what he sees as deficiencies in Gersonides position. He first claims that our knowledge is derivative, whereas God's knowledge is active and causal:

> from his knowledge and the concept of his will (*ve-tziyur retzono*) the objects known acquire existence, while our knowledge is emanated and acquired from the objects known by means of the senses and the imagination.[240]

That is, humans derive their knowledge from sense data, whereas God knows every event *ab initio:* "What God knows he in some sense makes."[241] This qualitative difference between human and divine knowledge, reminiscent of Maimonides' distinction, enables Crescas to resolve these problems with respect to the second argument. With respect to the second argument, knowledge of things adds nothing to God, for God is causal and creative, whereas things are dependent upon God:

> If that proposition is accepted, inasmuch as it imparts existence and essence to the things other than himself, he most evidently does not become constituted as a substance out of something other than himself, and his knowledge of (things other than himself) does not necessitate a plurality in his essence.[242]

God knows things not because he knows himself, but *eo ipso;* it is through his knowledge that they exist.[243] So too, knowledge does not impute to God multiplicity, because again God is the cause and knows them as many without being many himself. The third argument pertaining to time is similarly eliminated by Crescas. As we have seen in chapter three, Crescas has already shown that time is not an accident of motion (*she-'ein ha-zeman miqreh nimshakh bi-ṭenua'h*), and so he argues that the objection is not relevant.[244] Because time itself derives its existence from God, God's knowledge is not limited by temporality.

It is in the context of the first question, which implies a change from future to past, that Crescas' determinist stripe is introduced. Does this temporal change from future to past affect God's essence? Crescas responds that because God knows before the occurrence of an event that it will happen, God's essence does not change when the event actually occurs. But how can we call a thing possible if God knows before its occurrence how it will happen? It is here that Crescas upholds determinism in order not to minimize in any way God's knowledge. Crescas attempts to distinguish two senses of contingency, arguing that a thing may be necessary in one way and possible in another (*ve-hinei yitba'er lah be-mah she'-omar 'ein safeq she-heyot ha-davar meḥuyav be-tzad mah lo' yeḥayev ha-davar be-'atzmuto*).[245]

> when we reason analogously from our knowledge to his knowledge, we have no escape but to suppose that the (disjunctive) part which he knows is possible in some respect and necessary in some respect (*'efshar be-tzad mah u-meḥuyav be-tzad mah*), and that from the standpoint of its being necessary no change will occur in his knowledge and his essence, while from the standpoint of its being possible the nature of the possible will not be denied of possible things.[246]

This distinction is based on Avicenna's distinction between three types of possibility: the necessary in itself, the possible in itself, and the possible in itself but necessary from another. The second category comprises the "purely possible," what may or may not exist, and does not fully capture Crescas' notion of contingency.[247] Rather, Crescas argues that events known by God, although "possible in themselves," nevertheless are necessary with respect to their causal history. In other words, if God knows *p*, then the truth value of *p* is determinate and "is necessary in terms of its causes."[248] Giving the analogy of a man's knowledge that does not change the nature of the possibility of the thing known, so too does Crescas argue that the knowledge of God does not change the nature of the possibility in question. "Thus, the knowledge of God, may he be blessed, in the respect that [his] choice occurs in it, does not entail necessity in his essence (*lo' yeḥayev ḥiyuv be-ʿatzmuto*), and does not change the nature of the possible at all (*ve-lo' yeshaneh ṭevaʿ ha-'efshar klal*)."[249]

In book 2 principle 5, chapter 1 of *'Or 'Adonai* Crescas turns more specifically to the problem of free choice (*beḥirah*). What is the connection between choice and possibility? Crescas is unequivocal that free choice presupposes possibility: "If there did not exist the nature of the possible (*ṭevaʿ ha-'efshar*), man's actions would be necessary."[250] He offers five arguments in favor of the existence of possibility and seven arguments against the existence of possibility.[251] The last is most relevant to us and suggests that if God's knowledge extends to particulars, then if possibility exists, this would preclude God's prescience. Crescas' solution, adduced earlier but developed more fully now, is that the possible exists in some respects and in other respects not. The possible exists "only in regard to itself." That is, when regarded from the perspective of causality, it does not exist as possible. Crescas argues that natural phenomena are "possible in themselves and necessary with respect to their causes (*'efshariyim bi-vḥinat ʿatzmam u-meḥuyyavim bi-vḥinat sibotam*)."[252] What this means is that from the perspective of its causal history, every event is necessary. Only in light of human epistemological weakness (viz. our inability to know this causal history) can an event be said to be possible. As necessary, events can be foreknown; as possible per se, they are "*qua possibile.*" Crescas goes on to suggest that what is necessary before its existence is not possible (*'einenu 'efshari*). That is, "it is not possible from the perspective of its knowledge, but it is possible from its own perspective."[253]

Crescas' construal of divine omniscience can therefore be expressed as follows:

8.1 God knows at t_1 that I will do *p* at *t*.

8.2 That I will do *p* at *t* is a determinate event.

8.3 *p* is "possible only in regard to itself," not from the perspective of God's knowledge.

8.4 From the "perspective of God's knowledge" *p* is "necessary with respect to its causes."
8.5 From my perspective, it is possible for me to do *–p* at *t*.
8.6 However from "the perspective of God's knowledge," it is not possible for me to do *–p* at *t*.
8.7 Hence from "the perspective of God's knowledge," I am not free with respect to *p*.
8.8 Hence I am not free with respect to *p*.

Argument Summary: If God knows that I will do an action at a future time, then my doing that action is determined from the perspective of God's knowledge. Even though from my own perspective, I may still do the opposite, from God's perspective it is not possible for me to do the opposite. Hence from "the perspective of God's knowledge" I am not free with respect to performing future actions.

To the extent, then, that ontological differences between past and future have vanished, so too has contingency vanished. Human freedom exists no more in light of God's knowledge, and compatibilism has been replaced with a thorough-going determinism.[254] In order to counter the claim that he has eliminated human freedom altogether, Crescas in 2.5.3 distinguishes between acts that are compelled and those that are caused. Only in the former case can we say that the agent was not responsible for the action. In the latter, even though the action is caused, the agent according to Crescas nevertheless chooses to bring the action about, and so it is voluntary. A voluntary action is accompanied by a feeling of joy, or assent, which underscores its voluntary nature. Hence, Crescas argues, actions can be both causally determined and also voluntary.[255]

The temporal and ontological conflation of past and future is evidenced even more strikingly in Crescas' description of God's knowledge as timeless. By "timeless" Crescas means that God's knowledge is outside the domain of time altogether. Equating divine foreknowledge with the knowledge of present events, Crescas states that "the science of God is beyond time (*bilti nofelet bi-zeman*), his knowledge of the future is like his knowledge of things existing which does not necessitate compulsion (*hekhreḥ*) and necessity in the nature of things."[256]

That this position reflects analogous theories of eternity found in scholastic thought has not escaped notice. Feldman conflates the views of Boethius, Aquinas, and Crescas into what he terms the "BAC Theory." Boethius, for example, states that God's knowledge, "surpassing every temporal movement, remains in the simplicity of its own present and, embracing infinite lengths of past and future, views with its own simple comprehension all things as if they were taking place in the present."[257] And Aquinas echoes this position, responding in the *Summa Theologica* that "things reduced to act in time, are known by us successively in time, but by God [are known] in eternity, which is above time . . . [they are certain] to God alone, whose understanding is in eternity above time."[258] What these three theories share in common is that God, being an eternal rather than a temporal being, has a "timeless knowledge of temporal facts such that events apparently future are really present, at least to him; and that this cognition in no way annuls the inherent contingency in statements about future contingencies."[259]

Feldman and others have suggested numerous difficulties with the timelessness theory as it stands. Kretzmann, for example, has argued that a timeless God cannot be omniscient because he can never know what time it is now.[260] This conclusion is devastating to a theist who takes seriously God's intervention in history through providential acts and miracles. A second difficulty has to do with the notion of simultaneity implicit in the timelessness doctrine, that is, the notion that all events are occurring in a simultaneous "now" to God's eternal vision. Perhaps the most famous statement of this critique occurs in Kenny's suggestion that on the timelessness view associated with Aquinas (as well as Boethius and Crescas), "my typing of this paper is simultaneous with the whole of eternity. Again, on this view, the great fire of Rome is simultaneous with the whole of eternity. Therefore, while I type these very words, Nero fiddles heartlessly on."[261] Much ink has been spent in trying to respond to Kenny's challenge. Kretzmann and Stump, for example, define divine simultaneity as relative to reference-frames, and argue that given their notion of "ET-simultaneity," Kenny's sorts of problems simply cannot arise. Hasker attempts to further hone the notion of ET-simultaneity introduced by Kretzmann and Stump. Echoing Aquinas, Hasker argues that God timelessly knows that temporal entities are temporal, hence distinguishing the "mode" through which God knows temporal entities from the "mode" of entities themselves.[262] Let me say, however, that the most trenchant difficulty has to do with the elimination of possibility; it is this issue that is not resolved by these attempts. For, as Feldman has argued, the timelessness theory entails a world in which "all the episodes in the history of the universe have already been written."[263] These events cannot be other than what God already knows they are. More specifically, let us return to Crecas' statement that God's knowledge of future events is *like* that of present events. We know present events to be temporally necessary: to adopt Aristotle's language discussed above, while they are occurring they are necessarily occurring. By conflating present and future events, Crescas has eliminated the metaphysical openness of the future and reduced it to the necessity of the present. Any remaining possibility is "in the mind of the actor," as it were, and not in the event itself. And so it is the very doctrine of timelessness that removes any vestige of contingency from the world. Although Crescas is not able to accept the implications of this thorough-going determinism, his philosophical heir Spinoza does. In chapter 5 we shall examine Spinoza's contribution to the problem of divine omniscience.

Conclusion

We are now in a position to summarize and assess our findings. Jewish philosophers grappled with the tensions resulting from two traditions: on the one hand, a metaphysical system in which God was conceived as a unity, unchanging in essence, and all-knowing, and on the other hand a tradition that claimed that God is intimately involved with a mutable world of possible entities. Does God's knowledge extend to this world of possibles? If not, then it might be claimed that God's knowledge is deficient—that is, that God's unawareness of the realm of the possible represents a deficiency in his nature. If his knowledge does extend to the sublunar realm of possibility, then it might be argued that the possible must give way to the necessary—that is, that God's knowledge precludes the existence of the possible.

In answer to the questions raised at the beginning of this chapter, compatibilists attempt to uphold both omniscience and the genuineness of the future. Saadia and Halevi lay the groundwork for Maimonides' compatibilism. Maimonides upholds God's omniscience in the face of numerous objections, claiming that God knows both concrete and unactualized particulars. Concrete particulars are known even though they are individuated by corporeal matter, and unactualized particulars, or possibles, are known even though they do not yet exist. On this view God is not affected by what he knows. Maimonides attempts to forestall any objections with respect to the epistemological inconsistency of such a solution by emphasizing God's equivocal nature; because of the equivocal nature of the term knowledge when applied to God and man, the same consistency cannot be expected of both domains. An uneasy balance between God's omniscience and the logical and epistemological implications of such knowledge has therefore been achieved, but only at the expense of sacrificing the meaning of the term knowledge when it is applied to God.

On the other hand, Ibn Daud and Gersonides attempt to explain how God can know future contingents without this knowledge affecting the contingency of the objects known. Both argue that foreknowledge coupled with infallibility preclude the contingency of the objects of God's knowledge. Unable to adopt a compatibilist solution, Ibn Daud and Gersonides therefore uphold a form of incompatibilism in an attempt to preserve the existence of contingency in the world. Claiming that only a very small class of actions is truly contingent, Gersonides in particular argues that God's knowledge pertains to those actions only insofar as they are ordered and not *qua* contingent. That is, according to Gersonides, God knows particulars only as universals and not in their particularity. This knowledge is then transmitted to the prophet in the form of universal conditional statements. The prophet instantiates these statements and is able to apply them to future contingent events. I have argued, however, that this analysis does not adequately account for the conditional nature of prophecies.

More specifically, I have demonstrated that Gersonides employs two kinds of conditional prophecies: those in which the consequent expresses God's will, and those in which the consequent expresses human choices. Read in light of Gersonides' indeterminism, the first class of statements raises problems of its own, for it is not clear how God knows the particular events to which he is reacting. The second class of prophetic events seems to resemble law-like statements and is subject to the ambiguity inherent in all such statements. It would appear on this reading that the prophet is a good natural scientist. Two problems arise with this interpretation, however. The first is that it does not accord with Gersonides' own conception of prophecy. For as we have seen Gersonides wishes to distinguish the natural scientist from the prophet. Secondly, this analysis does not account for those prophecies, like (6.1), in which the consequent expresses a truly contingent event. Such events cannot be foretold by either a scientist or a prophet.[264]

For these and similar reasons, determinists such as Crescas have attempted to salvage what remains of divine omniscience. I have argued, however, that the end-result is a truncated sense of future contingency. On Crescas' reading, God knows in a timeless fashion those events that occur necessarily with respect to their causes, but that appear to be possible to one not familiar with the causes in question. By equating knowledge of future events with that of the past and present, Crescas has obliterated the difference between future and past. In so doing, Crescas returns full circle to Aristotle's sea-battle. Aristotle recognized the importance of positing an ontological difference between future events on the one hand, and past and present events on the other. Without this difference inherent in the very nature of reality, Aristotle felt that he could not give an adequate account of human action, deliberation and responsibility.

Crescas' own student Joseph Albo reflected this point when he criticized his teacher's theory of free will. Summarizing Crescas' position, Albo's reaction is that it is "very close to the view that all things are determined and that the possible does not exist. For since the things are necessary considered in relation to their causes, if God knows the causes, they are actually necessary."[265] Albo's point is that on Crescas' theory, future events cannot come into existence in any other way than in fact they *do* come into existence: "There is no thing that may equally be or not be when considered in relation to its causes."[266] In other words, having given up that p may "equally be or not be," Crescas has abandoned both (3.3) and (3.4) articulated at the start of this chapter, and has therefore eliminated the definition of a free action. Albo, therefore, finds Crescas' determinism unacceptable on the grounds that it has eliminated the openness of the future, and by implication human freedom. This critique will be developed even more coherently by Spinoza, to whom we now turn.

Chapter Five

Prelude to Modernity

Introduction

In this final chapter I turn to those philosophical developments in the late medieval period that contribute to the concept of time in early modern thought. In particular, I concentrate upon those conceptions of time that reappear in the works of Spinoza. Although Spinoza is not technically regarded as a Jewish philosopher, nevertheless he too is engaged with Jewish texts and themes, and he too is involved in the quest to reconcile Aristotelian and Neoplatonic metaphysics with broader cosmological and theological concerns.[1] This quest has been the subtheme of this book and culminates in Spinoza's conceptions of eternity, infinity, and temporality. I shall argue that Spinoza's formulations must be read and understood against the discussions of Jewish philosophy presented in the earlier chapters of this book. In a separate final chapter, I shall end with a theme that has not yet surfaced in this study but that is discussed by both medieval and modern philosophers, namely the concept of "the end of time." It is here, I shall argue, that the interplay between medieval Jewish philosophy and Spinoza is most clearly delineated.

Newton and His Philosophical Precursors

Notwithstanding the Condemnation of 1277 of the doctrine of the eternity of time, the Aristotelian emphasis upon eternity was nevertheless embraced and refined by the scholastic community of philosophers.[2] The Aristotelian definition of time reappears throughout the fifteenth and sixteenth centuries. But as we have already seen in previous chapters, Aristotle had his critics as well, stemming first from Plotinus and then Crescas, who emphasized that time is the product of the soul and is defined in accordance with duration rather than number. This critique is amplified by Copernicus whose heliocentrism threatened the Aristotelian cosmology adopted for centuries.[3] In the early Renaissance, many philosophers reintroduced Neoplatonic criticisms of Aristotle's theory of time. For example, in his commentary upon the *Enneads* of Plotinus, Ficino reiterates Plotinus' doctrine of the inseparability of time and soul.[4] Among others, Pico della Mirandola echoes Crescas in his rejection of Aristotle's theory of time, quoting Crescas' definition of time as "the measure of continuity whether of movement or of rest, between two moments [*Definit autem ipsum tempus ita (ut eius verbis agam) mensura continuitatis vel motus, vel quietis, quae inter duo momenta*]."[5]

Although the beginnings of modern science and philosophy were indebted to Aristotle and his medieval followers, the underlying intellectual structure of the medieval world was crumbling. This tendency is reflected in changing theories of time. Throughout this study I have emphasized the importance placed in the Aristotelian corpus upon time as the measure of change with respect to "before" and "after." As Van Frassen has argued, this theory of time is found precisely as part of both natural philosophy and natural science, and is concerned with successive continuous quantity.[6] But in the sixteenth and seventeenth centuries, natural science and philosophy become more clearly distinguished, and their subject matters become subject to different types of methodological investigation. "Time" as a philosophical subject differs markedly from "time" as a subject of natural science: metaphysics and physics employ different methods, and are asking new questions of their subject-matter. The heliocentrism of Copernicus threatens Aristotle's equation of time with motion in that, as Ariotti has argued, heliocentrism does not provide Copernicus and his successors with directly observable uniform motions or constant intervals of time so important to Aristotle's theory.[7] As we shall see below, this issue affects Newton's conception of absolute time.

Jewish philosophers in Renaissance Italy were influenced both by the Copernican revolution as well as by the Humanist revival of Platonism and Neoplatonism. The emphasis upon philosophy, theology and science found its way into Jewish philosophical texts in the late fifteenth century onward. Tracing the impact of the Copernican revolution upon Jewish thought in the fifteenth century, Levine suggests that European Jewry, although close to the Copernican debates, was "curiously unshaken" by the implications of the Copernican revolution upon metaphysical and epistemological speculation.[8] Ruderman and others, however, have surveyed the impact of astronomy upon sixteenth-century Eastern European Jewish philosophers. Citing the works of Moses Isserles of Cracow, the Maharal of Prague, and David Ganz, Isserles' most successful student in the sciences, Ruderman raises the tantalizing question of the extent to which developments in current astronomy affected their works.[9] David Ganz, for example, appears to be up to date on contemporary work in astronomy and science; in his work *Neḥmad ve-Na'im,* he traces recent developments in astronomy and mentions Copernicus as the greatest astronomer since Ptolemy. Nevertheless, as Neher and Ruderman have both pointed out, in his own astronomical writings Ganz adheres to the geocentric models of Brahe and Kepler.[10] Neher records a supposed conversation between David Ganz and Tycho Brahe in which Ganz valiantly upholds the rabbinic view over current astronomy.[11]

Joseph Solomon Delmedigo (1591–1655), on the other hand, recognized the challenge of the new Copernican astronomy. Ruderman has emphasized Delmedigo's tendency, along with that of his mentor Galileo, to understand the natural world outside the framework of Aristotelian physics; it is this tendency that is aligned with Delmedigo's interest in Kabbalah and Neoplatonic thought.[12] In his work *Sefer 'Elim* Delmedigo describes the "strange astronomy," as well as the dangers inherent in this new astronomy, which challenged the reigning metaphysics.[13] In *G'vurot Hashem,* a work appended to *Sefer 'Elim,* Delmedigo is more enthusiastic in his attitude toward Copernicus, demonstrating his knowledge of the new astronomy: "Happiness and joy were added to me when I heard that they (the researchers) have begun in our time (*be-zemanenu*) to think that the entire universe is like a lantern and is called 'lanterna'; and the candle burning within it is the solar body (*ha-guf ha-shemeshi*), which stands in the center and whose light spreads out until the sphere of Saturn (*ad galgal shabati*) which is at the outer limit of this universe."[14]

The work of Judah Abravanel (more commonly known as Leone Ebreo) represents an excellent example of the fusion of Hebraic thought with Ficino's revival of Greek philosophy. In his philosophical dialogue *Dialoghi di'Amore* (*The Dialogues on Love*), Ebreo constructs an allegory between two Jewish courtiers, Philo and Sophia, in order to illustrate the importance of the philosophical love of God.[15] Philo has explained to Sophia the origins of love, which raises the more general question of when love was born: was it produced from eternity or was it created in time? Philo immediately connects this question with the issue of the origin of the universe, which leads to a summary of the three regnant positions on the creation of the universe—that of Plato, that of Aristotle, and that of Moses. In the course of explaining why on the Aristotelian model the universe is eternal, Philo alludes to the view that time must likewise be eternal, "for any given instant is in reality the end of past time and the beginning of the future, and there can be no instant which is the first and the beginning of time. Time is therefore, eternal and without beginning."[16] Philo then attempts a reconciliation of Platonic theory with Mosaic law and Lurianic Qabbalah, applying this reconciliation to the doctrine of love.

Primarily as an alternative to Aristotle, numerous sixteenth- and seventeenth-century commentators of Aristotle incorporate the notion of duration into their discussion of time.[17] By the end of the sixteenth century, the term "*duratio*" has entered the terminology of the definition of time, among both Aristotelians and others.[18] Newton's teacher Isaac Barrow had already distinguished time entirely from motion, arguing that time existed before the world.[19] I shall not in this work enter the recent controversy over the extent to which Newton's metaphysics and physics are influenced by Jewish thought. That Newton was aware of and studied Christian Kabbalistic texts, most notably Christian Knorr von Ronsenroth's *Kabbala denudata* has been examined by Goldish and others.[20] And that Newton was aware of and read extensively in Maimonides has been argued suggestively by Popkin in a series of recent studies.[21] But I have not found evidence that these works actually influenced Newton's theory of time. What is clear, however, is that Newton sees his distinction between absolute and relative time as expressing the ancient distinction between permanent and successive duration.

Newton's most succinct statement of how space and time relate to existing things occurs in his short work *Tempus et Locus*.[22] The important point in this prelude to the third edition of the *Principia* is that something's duration is indistinguishable from its continued existence. This means that things are either in a state of resting or a state of moving, and both are related to space and time. To say that something endures is simply to say that it exists and is localized in place and in time.[23]

> Time and Place are common affections of all things without which nothing whatsoever can exist. All things are in time [*in tempore sunt omnia*] as regards duration of existence. . . . To exist in time and place does not argue imperfection, since this is the common nature of all things. For the duration of a thing [*Nam Duratio rei*] is not its flow, or any change, but permanence and immutability in flowing time. All things endure is so far as they remain the same at any time.[24]

This emphasis upon duration culminates in Newton's distinction in the *Principia* between absolute, common and astronomical time. Newton follows late Renaissance thinkers who try to distinguish time, duration, and motion. In his famous definition of absolute time as totally independent of motion, Newton claims that "Absolute, true and mathematical time of itself and from its own nature [*tempus absolutum, verum & mathematicum in se & natura sua*], flows equally without any relation to any thing external and by another name is called duration [*alioque nomine dicitur duratio*]."[25]

By this dissociation of absolute time from anything external, Newton regards moments or instants of absolute time as forming a continuous sequence much like points on a geometric line; the rate at which these instants succeed each other is independent of events external to time.

Absolute time is distinguished from relative or common time that reflects the measure of duration "by the means of motion . . . such as an hour, a day, a month, a year."[26] The "vulgar" tend to confuse the measures themselves (like days and years), with the things measured, namely time. In astronomy, absolute time is distinguished from relative time by the correction of what passes as apparent time; Newton points out that astronomers must correct the inequality of the measure of "natural days" in order to render celestial time more accurate. Arguing that the duration of the existence of things remains the same, Newton claims that it must be distinguished from the variable "sensible measures" of duration.[27] One implication of this conception of absolute time is that the order of the parts of time is immutable: "All things are placed in time as to order of succession."[28] Absolute space and time exist as independent entities in their own right, functioning as infinite containers: times and spaces are "the places as well of themselves as of all other things."[29]

The implication of this view with respect to our study is manifested in Newton's depiction of God found in the *General Scholium.* We have discussed in chapter four at least three ways in which God can be said to relate to time: We might say that God exists supratemporally, meaning that temporal concepts do not apply to divine existence. On this view God is above, beyond, or outside time. Or we might say that God exists in an unchanging present (a *nunc stans*), an ever-present eternity. Or, finally, we might say that God's existence is omnitemporal, meaning that God's nature endures through all possible times.[30] We have already examined in chapter four the implications of each of these views with respect to divine omniscience. McGuire suggests that while there is no direct historical connection between Spinoza and Newton, nevertheless it is curious that Newton's view of divine duration is similar to the very view that Spinoza criticizes. Rejecting the second view explicitly, Newton states that God's duration reaches "from eternity to eternity and is present from infinity to infinity [*durat ab aeterno in aeternum & adest ab infinito in infinitum*]."[31] His duration is not a *nunc stans* without duration, nor is his presence nowhere.[32] Newton emphasizes that existing always and everywhere, God "constitutes duration and space [*durationem & spatium constituit*]."[33] By this he means that space and time are not real attributes of God's nature, but rather they exist *because* God exists. As stated succinctly by McGuire, "if God's existence is therefore eternal and uncreated, and if all things including God must exist with respect to space and time to be actual existents, the fact of God's sheer existence 'constitutes' the existence of infinite space and time."[34]

As we shall see below, in holding that God's existence is timeless, Spinoza will reject the third view above, which is the very position that Newton holds. Spinoza present his view of "omnitemporality" in answer to those who believe that God does endure through all possible times.[35] In

Spinoza's view, to conceive divine eternity correctly is to understand that "temporal extension and temporal contrasts cannot apply to Divine existence."[36] Spinoza's work on time and duration not only reflects the culmination of the medieval tradition traced in this study, but is enormously influential upon the thought of Newton, Leibniz, and subsequent modern philosophy. Spinoza was not a medieval thinker, and perhaps was not a Jewish philosopher; nevertheless, to paraphrase Copenhaver, Spinoza honored the past while being astonishingly unconstrained by it.[37] Let us turn, then, to Spinoza's theory of time.

Spinoza's Metaphysical Monism

That Spinoza's substance monism entails that substance be without parts is not news. Neither is his claim that substance is infinite, and yet in some way representative of, or represented by, finite modes. And yet the relation between infinite, indivisible substance and finite, apparently divisible modes of substance, has been one of the major stumbling blocks in Spinoza scholarship. As noted by Wartofsky, Spinoza holds that substance is a unity, eternal and infinite; at the same time he holds that "in" substance there is infinite differentiation, that is, individuals. How, then, is individuation reconciled in the face of indivisible substance?[38] The apparent tension between infinity and divisibility is captured even more strikingly by Gueroult, who notes that "infinity and indivisibility being two sides of the same property, there results a radical antimony between the infinite and the divisible. If we affirm one we must deny the other."[39] Spinoza, affirming the infinite, denies divisibility. But, as Gueroult notes, the problem reappears on the level of modes, "where we must affirm infinite divisibility, that is, both the infinite and the divisible."[40] In order to resolve this tension between infinity, indivisibility, and individuation, we must understand what Spinoza intends by the cluster of issues surrounding these notions.

It is this cluster which forms the crux of Bennett's discussion. In his recent study of Spinoza's *Ethics,* Bennett suggests that Spinoza's theory of substance can be likened to contemporary quantum field theory. Dubbing Spinoza's theory a version of "the field metaphysic," Bennett claims that "the contemporary view in physics [of the fundamental particles of matter] is a version of Spinoza's own position."[41] In what follows I shall examine Spinoza's theory of infinite divisibility of substance in the context of his theory of time. I will end by pointing to the relevance of Bennett's characterization of the "field metaphysic" in understanding Spinoza. For if Bennett is right, his interpretation may help us to understand those murky passages in the *Ethics* that touch upon issues of indivisibility of substance, infinity, and individuation.

Bennett reminds us that on one level Spinoza shared with his contemporaries the view that ordinary physical objects have parts in the sense that they are infinitely divisible, or in Bennett's terms "splittable."[42] How, then, Bennett asks, would Spinoza react to recent theories according to which the ultimate particles of matter are unsplittable? These particles, or quarks, appear at least on a surface reading to negate Spinoza's presupposition that physical objects are infinitely splittable. On a second level, however, according to contemporary theory, quarks are both substance (in that they are the ultimate constituents of reality) and yet unsplittable. However, Spinoza has argued that there is only one substance that he regards as unsplittable. Furthermore, as Bennett points out correctly, Spinoza agreed with his Euclidean contemporaries that space was infinite in all directions; he lived too early to avail himself of the non-Euclidean notion of a finite but unbounded space.[43] So we have ahead of us two tasks: the first is to reconcile the metaphysical connection between ordinary objects and substance; the second is to give a defensible and metaphysically interesting reading of what substance actually is.

Bennett points to a possible way out of this dilemma. If, as Bennett argues, quarks are construed by contemporary physicists not as little lumps of matter but rather as spheres of influence, then the tension between Spinozistic monism and contemporary physics dissolves. Let us construe space to be the one extended Spinozistic substance. According to Bennett, space as postulated by Spinoza is analogous to fields as construed by quantum physicists: Just as the particles of matter represent the interactions of fields, so too does substance reflect the intercouplings of space. In other words, when Spinoza says that there is one extended substance, what he does is make particular extended things "adjectival on regions of space. . . . More specifically, it is space's containing a connected sequence of regions belonging to a string of place-times which satisfy certain conditions."[44] How well does this interpretation of Spinoza accord with other aspects of substance, most notably infinity and divisibility? In order to assess the cogency of Bennett's interpretation, we must examine in greater detail what Spinoza has to say about time and the infinite divisibility of substance.

Time, Duration, and Creation: Spinoza and Descartes Compared

In Spinoza's early work, time is characterized along with number and measure as a "mode of thinking," which serves to "explain a thing by determining it through comparison to another."[45] Time serves to explain duration and reflects Descartes' definition of duration "simply as a mode under which we conceive the thing in so far as it continues to exist."[46] Spinoza distinguishes between eternity and time as follows:

> Duration is an attribute under which we conceive the existence of created things insofar as they persevere in their actuality. . . . But to determine this duration we compare it with the duration of other things which have a certain and determinate motion. *This comparison* is called *time.* Time, therefore, is not an affection of things, but only a mere mode of thinking, *or* as we have already said, a being of reason.[47]

If, Spinoza argues, we say that duration is not attributed to God, certain questions simply do not arise. For example, whoever attributes duration to God in effect distinguishes God's essence from his existence, since duration pertains only to existence. To ask whether God has existed longer now than he did when he created Adam is to misunderstand the nature of God's essence. For to attribute "length" of time to God is to attribute duration to God, to "ascribe to God a greater duration each day and think of him as if he were created continuously by himself."[48] Furthermore, duration implies composition by parts, and God is not composed of parts. To ascribe to God eternity is to say that "there can be nothing which is before or after, we can never ascribe duration to him, without at the same time destroying the true concept which we have of God."[49]

In the *Ethics,* Spinoza uses the three temporal concepts duration (*duratio*), time (*tempus*), and eternity (*aeternitas*). Defining duration as the "indefinite continuation of existing,"[50] Spinoza claims that anything that endures must minimally exist. Time is the manner in which duration is made determinate; it is the duration of a thing less abstractly conceived. When temporality is added to an enduring entity, its existence is delimited temporally. Both duration and time apply only to *natura naturata,* the world mediated, whereas eternity pertains to the domain of *natura naturans.* But if duration cannot pertain to the domain of eternity, then Spinoza must be able to offer an alternative explanation of change, creation, and cause. For if, as Kneale argues, eternity entails sempiternity (and duration), then the notion of substance collapses.

In the Appendix to his commentary upon Descartes' *Principles,* Spinoza clarifies issues pertaining to creation that relate to duration. Spinoza rejects the locution creation *ex nihilo* on the grounds that the *nihilo* itself postulates a something that is real.[51] Further, he denies that time or duration pre-exist creation, but rather "these latter have begun with things." Because time is the measure of duration, or a mode of thinking, it presupposes the existence of particular thinking individuals; furthermore, Spinoza states that "duration ceases when created things cease to be, and begins when created things begin to exist."[52]

The starkness of Spinoza's position can be better appreciated when we compare his conception of time and duration to that of his immediate philosophical predecessor. For, as Mason has recently argued, "Descartes' questions were Spinoza's questions,"[53] and as we shall see in the next section, Spinoza countenances his theory of infinity in contradistinction to that of Descartes. Recognizing the importance of Descartes' questions, Spinoza knew that he had to deal with them in order to construct his own metaphysics. For Descartes the topic of time arises already in the first of his six *Meditations on First Philosophy*. Descartes doubts the veracity of not only past and future, but of the present as well. It is not just that Descartes feels that he cannot distinguish between waking and sleeping, but that he cannot verify the present perceptions that supposedly he is experiencing. Even the one thing that Descartes knows with certainty, namely that he exists, he knows only for *as long as* he is thinking: "I am, I exist—that is certain. But for how long? For as long as I am thinking."[54] The duration of temporal thought is thus introduced in the very context of the *cogito*. In *Meditation* 3 this line of argument is continued. "I perceive that I now exist, and remember that I have existed for some time."[55] Descartes thus introduces not only the veracity of the present instant, but past instants as well. But the recognition of past instants of the self does not necessarily imply that the self has always existed as it does now. Descartes suggests that

> [1] a lifespan can be divided into countless parts, each completely independent of the others, so that it does not follow from the fact that I existed a little while ago that I must exist now, unless there is some cause which as it were creates me afresh at this moment—that is, which preserves me. For it is quite clear to anyone who attentively considers the nature of time (*ad temporis naturam*) that the same power and action are needed to preserve anything at each individual moment of its duration as would be required to create that thing anew if it were not yet in existence. Hence the distinction between preservation and creation is only a conceptual one, and this is one of the things that are evident by the natural light.[56]

This point is reinforced in a slightly different way in the *Principles*, where Descartes suggests that the duration of our existence is sufficient to demonstrate the existence of God: "For the nature of time is such that its parts are not mutually dependent, and never coexist. Thus, from the fact that we now exist, it does not follow that we shall exist a moment from now, unless there is some cause . . . which continually reproduces us, as it were, that is to say, which keeps us in existence."[57] For Descartes the question of the continuity of time is relevant to intuition: is intuition itself instantaneous, or does it take time. But how can intuition, and by inference thinking, take time if time is made up of discontinuous instants? Is the present of consciousness specious? In short, what do we make of Descartes' statement that intuition is given "*total simul et non succesive?*"[58]

The relevance of these questions to the issue of creation is demonstrated when we turn to the recent scholarly controversy over whether

Descartes holds that the world and time are continuous or discontinuous. With respect to the issue of creation, the standard reading of Descartes has been that God sustains the world through the successive creation of discrete, independent, and atemporal instants. On this model, termed the "Continuous-Recreation Theory," duration is composed of a collection of indivisible temporal atoms, and motion emerges as the successive recreation of a body in different places. The Continuous-Recreation Theory can be stated as follows:

> *C-R Thesis*: The world, created by God at an initial instant, requires the same creative act at each subsequent instant to conserve it (and every item within it) in existence.

The real question, then, is whether Descartes' Continuous-Recreation Theory stated in the third meditation entails the discontinuity of matter and time. Following Garber, I shall distinguish between two main strands of the debate as follows: The Classical Discontinuity Thesis, associated with Gueroult, asserts the following:

> *D-Thesis*: The world sustained by God is a world of durationless instants; each temporal atom is created by God in succession.

This view is contrasted with the Continuity Thesis of Beyssade and Arthur and asserts the following:

> *C-Thesis*: The created world sustained by God is one in which any portion of time is further divisible into smaller portions; durationless instants mark the boundaries between one part and another.

The question for these commentators is whether the *D-Thesis* is required by the *C-R Thesis*. If not, an argument may be made for the compatibility of the *C-R Thesis* and the *C-Thesis*. That duration and motion are conceived as discontinuous, is evidenced in [1] as well as in the following passage.[59]

> Now some attributes or modes are in the very things of which they are said to be attributes or modes, while others are only in our thought. For example, when time is distinguished from duration taken in the general sense and called the measure of movement, it is simply a mode of thought. For the duration which we understand to be involved in movement is certainly no different from the duration involved in things which do not move. This is clear from the fact that if there are two bodies moving for one hour, one slowly and the other quickly, we do not reckon the amount of time to be greater in the

latter case than the former, even though the amount of movement may be much greater. But in order to measure the duration of all things, we compare their duration with the duration of the greatest and most regular motions which give rise to years and days, and we call this duration 'time'. Yet nothing is thereby added to duration, taken in its general sense, except for a mode of thought.[60]

Beyssade and more recently Arthur, however, have argued for a reading of this passage according to which inasmuch as every part of duration is, itself, a finite portion of time, duration itself is not a succession of durationless events. On the *D-Thesis,* Descartes emphasizes the internal consciousness we have of duration: thought always takes place in time and so our existence cannot be a succession of timeless instants. What God sustains is body extended in time. More specifically, Arthur argues that inasmuch as there is no convincing evidence for the usual presentation of Descartes' doctrine of continuous creation as asserting a discontinuous succession of discrete acts of divine creation, there is no convincing evidence that Descartes denied the continuity of time. On this reading God sustains bodies extended in time; both thinking and physical events always take place in time, and not in a succession of timeless instants.

Let us return to the main points of [1], which can be broken down as follows:

1.1 A lifespan can be divided into countless parts, each completely independent of the others,

1.2 so that it does not follow from the fact that I existed a little while ago that I must exist now,

1.3 unless there is some cause which as it were creates me afresh at this moment—that is, which preserves me.

1.4 For it is quite clear to anyone who attentively considers the nature of time (*ad temporis naturam*) that the same power and action are needed to preserve anything at each individual moment of its duration

1.5 as would be required to create that thing anew if it were not yet in existence.

1.6 Hence the distinction between preservation and creation is only a conceptual one, and this is one of the things that are evident by the natural light.

On the basis of this passage, proponents of the *D-Thesis* argue that time is composed of innumerable mutually independent parts or moments (1.1); that I myself am composed of mutually independent parts (1.2); and that it is necessary that God recreate every part in each separate moment at which it endures (1.4). Thus thc world is constantly recreated and preserved in a discontinuous succession of discrete acts (1.5) and its duration is a "discontinuous sequence of discrete moments."[61] On this reading, then, the *C-R Thesis* requires the *D-Thesis.* But as Arthur acutely points out, from the fact that parts of time are separable, independent, and contingent, it does not necessarily follow that they are actually separated and hence discrete.[62] In his reply to Gassendi, Descartes distinguishes between abstract time and the concrete duration of an enduring substance.

> This can plainly be demonstrated from what I explained about the independence of the parts of time, which you in vain attempt to elude by propounding the necessary character of the connection between the parts of time considered in the abstract. Here it is not a question of abstract time, but of the time or duration of something which endures; and you will not deny that the single moments of this time can be separated from their neighbors, i.e. that a thing which endures through individual moments may cease to exist.[63]

In this passage Descartes does not assert that every moment of concrete duration is *actually* separated from the next, but rather he emphasizes that they *could* be separated, that the enduring thing could at any moment cease to exist. What matters here is the emphasis upon cause: since the existence of an enduring thing is dependent upon God's continuous action, and a thing has duration only as long as it exists, there is no moment that is not contingent; without God's action there would be no link between one moment and those that follow it.[64] This would be an argument in favor of the claim that the *D-Thesis* is not necessary condition for the *C-R Thesis.*

Garber, however, rejects both these readings and offers an alternative to Descartes' view of temporal atomism. Secada has argued, based on several passages, that Descartes refers to parts of time or moments of duration, but not to instants or temporal atoms: "Any lapse of time, or indeed all of time, can be divided into segments which are causally independent."[65] Garber agrees, arguing that all the statements Descartes makes about the Reliance of the world upon God are consistent with the continuity thesis. Take for example (1.1) above which states the doctrine of the independence of portions of time. In this passage it is the parts of time that are independent of one another; but as Garber points out, Descartes does not tell us what counts as a part of time, or whether or not atemporal instants are such parts.[66] Even the *C-R Thesis,* on this reading, is not definitive, for from the fact that every instant is created and is the beginning of a newly created world, it does not follow that what is created is a bare instant or that "God could create an atemporal instant without creating a finite duration for that instant to bound."[67] Passages such as *Principles* 1.121 and 3.59 are indecisive and can be interpreted in a way so as to uphold "finite moments of time, finite stretches of time."[68] With respect to the *C-Thesis,* Garber concludes that although these passages are consistent with the continuity thesis, they do not require it, and so the issue of the continuity or discontinuity of time was not an issue which worried Descartes greatly.[69] That this is not the case, however, with Spinoza, for whom issues of continuity and divisibility are critical to his metaphysics, is the subject of the next section.

Substance, Infinity, and Divisibility in Spinoza

Spinoza examines the issues of infinity and divisibility in three central texts: in his early commentary on Descartes' *Principles of Philosophy,* his note to Proposition 15 of the *Ethics* (1p15s) and in his *Letter to Isaac Meyer on Infinity (Letter 12)*. Inasmuch as the historical background to 1p15s has been examined in detail by Wolfson,[70] we need only highlight several important points. Remember the context of *Ethics* 1p15. In propositions 1–12 Spinoza has demonstrated that God or substance exists (E1p7), is necessarily infinite (E1p8), and consists of infinite attributes (E1p11). Furthermore, he has wanted to maintain that God, or substance, is both intellect and extension. In order to uphold this contention, he must defuse those objections that rest upon the premise that extension, falling under the category of quantity, is divisible. Spinoza will certainly not want to claim that substance is divisible; hence he must show that extension is not per se divisible. How then is extension to be construed? This question really reflects the issue introduced earlier in this volume, namely whether a magnitude is divisible everywhere, that is, perpetually divisible into smaller units, or divisible only down to some atomic magnitude, beyond which subdivision is no longer possible.

So, when Spinoza turns to the issue of infinite divisibility, he does so against a battery of Aristotelian arguments that have been mediated by the scholastics and reformulated by Descartes. Aristotle's discussion, we will remember, was stimulated in part by an attempt to counteract Zeno's paradoxes. In order to avoid the untoward implications of Zeno's paradoxes, we saw that Aristotle distinguishes two senses of infinity: actual infinity refers to a complete divided state, whereas potential infinity refers to the process itself of dividing. Aristotle's discussions, as we have seen, functioned as a springboard for Jewish and scholastic discussions of infinity. This tradition is reflected in Descartes and Spinoza as well. Descartes, while he accepts Aristotle's conception in general, reserves the term "infinite" for God alone: "we must take the precaution of always bearing in mind as carefully as possible both that God, the creator of all things, is infinite, and that we are altogether finite."[71] He characterizes as "indefinite" things that appear to be without limit, or that may be divided indefinitely: "For our part, in the case of anything in which, from some point of view, we are unable to discover a limit, we shall avoid asserting that it is infinite, and instead regard it as indefinite."[72] In Descartes the "indefinite," in contrast to the "infinite," is an epistemic notion, stemming from a limitation of our understanding. The infinite is incomprehensible to our finite intellect.

> Our reason for using the term 'indefinite' rather than 'infinite' in these cases is, in the first place, so as to reserve the term infinite for God alone. For in the case of God alone, not only do we fail to recognize any limits in any respect, but our understanding positively tells us that there are none. Secondly, in the case of other things, our understanding does not in the same way positively tell us that they lack limits in some respect; we merely acknowledge in a negative way that any limits which they may have cannot be discovered by us.[73]

In this passage two reasons are given for denying the term "infinite" to things and applying it to God alone: First, there are no limits in God, whereas with respect to other things we only "negatively admit their limits . . . cannot be found by us." Second, God is unlimited in all respects, whereas things are unlimited only in some respects. These are what Wilson considers the epistemological and metaphysical criteria for the distinction. Wilson further suggests that Descartes' metaphysical criterion is but a short step to Spinoza's conception of *res extensa* as an infinite attribute of God, a step that Descartes attempts to reject on the grounds that extension implies divisibility that is inapplicable to God.[74] Although Descartes' point in this passage is primarily epistemological in that he emphasizes that "any limits which they may have cannot be discovered by us,"[75] Descartes does allow for the divisibility of extended substance. In fact, it is not even clear whether Descartes' treatment of Zeno's paradoxes is consistent with other Cartesian doctrines. In his *Letter to Clerselier*, Descartes summarizes Zeno's paradoxes. His solution to the Achilles paradox is based on a mathematical twist: demonstrating his finitism about the world, Descartes argues that the series $\frac{1}{10} + \frac{1}{10^2} + \frac{1}{10^3} + \ldots + \frac{1}{10^n}$ is equal to $\frac{1}{9}$, rather than tending toward 1/9. According to Descartes, Zeno's paradox arises because people imagine that this $\frac{1}{9}$ is an infinite quantity because it is divided by the imagination into infinite parts.[76]

Against both Aristotle and Descartes, however, Spinoza will want to argue for both the infinity and indivisibility of substance. Although he follows Descartes in identifying matter and extension, he disagrees with Descartes' claim that extended substance must be divisible. Spinoza knows Descartes' doctrine, expounds it, and eventually rejects it. More pointedly, in the *Principles*, Spinoza refers to the traditional puzzles against actual infinity and according to Ariew, he "answers such problems as would Descartes."[77] But Ariew then suggests that Spinoza is already separating himself from Descartes on the question of the infinite. In the *Principles of*

Philosophy, Spinoza examines the issue of infinite divisibility against the backdrop of atomism. Principle 5 is a demonstration of Descartes' continuity thesis: Atoms are defined as indivisible parts of matter; but inasmuch as the nature of matter consists of extension and hence is divisible, there is no part of matter which is not divisible, and hence there are no atoms.[78] In the *scholium* to this principle Spinoza connects the postulating of atoms as an attempt to avoid the possibility that one infinite might be greater than another, and quotes Descartes' dictum that with respect to the question of infinity "we therefore perceive only quite inadequately."[79]

But Spinoza goes about Zeno's paradox somewhat differently than did Descartes. Although he would agree with Descartes that we imagine infinite quantities of ⅙, he disagrees with Descartes that the infinite is incomprehensible. For Spinoza the infinite is not imagined or numbered, but is comprehended. He therefore must try to defuse the paradoxes by refuting their underlying conception of infinity.[80] It is in principle 6, in the context of examining the motion of matter, that Spinoza analyzes Zeno's paradoxes. Interestingly enough, Spinoza is primarily concerned to refute Zeno because his paradoxes suggest that "the senses show us something (viz. motion) which the intellect finds absolutely contradictory, so that the mind would be deceived even about those things that it perceives clearly and distinctly with the aid of the intellect."[81] The example Spinoza chooses to discuss is that of a wheel, ABC (see fig. 5.1).[82] If this wheel moves around its center X with a certain speed, point A will complete the circle around X more quickly than if it moved more slowly. By increasing the speed of the motion of the wheel, we increase the speed at which A moves through B and C around X. If we increase the speed to infinity and the time diminished to a

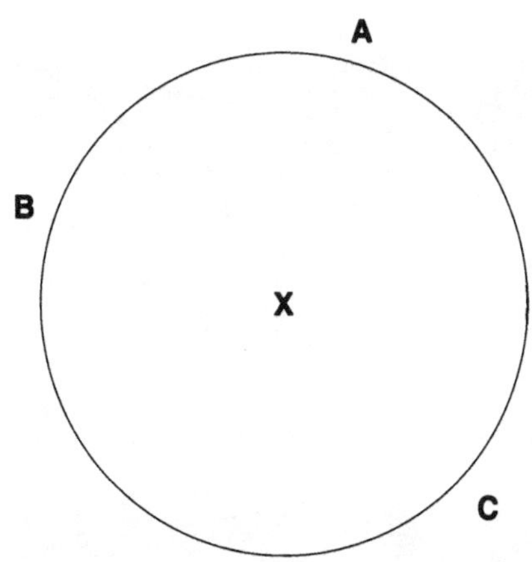

Figure 5.1. Spinoza's Understanding of Zeno

moment, then when point A is at that greatest speed, it "will be at every moment, or continuously, in the place from which it began to be moved; so it will always remain in the same place."[83]

Spinoza refutes Zeno's argument by rejecting its two working assumptions: first, that there is a greatest speed of motion for bodies, and second, that time is composed of moments. Both assumptions are false since we can also conceive a motion faster than the fastest motion, and a time shorter than the shortest one. In order to demonstrate his point, Spinoza envisions a smaller wheel DEF (fig. 5.2) connected to ABC such that when ABC moves around its center X, DEF is made to move around its center as well. If DEF is half the size of ABC, then DEF moves twice as fast as ABC. And so if ABC were to move infinitely fast, D would still be moving "faster" than A, which supposedly is at every point continuously.

After mentioning several other of Zeno's paradoxes, as well as Descartes' discussion of them in his *Letter to Clerselier* (1646), Spinoza concludes his discussion by reiterating his main point, namely that by refuting these paradoxes he has upheld the supremacy of reason: the senses "can never persuade him that something is false that the intellect has clearly and distinctly found to be true."[84] This analysis of the indivisibility of substance is resumed in E1p13: "A substance which is absolutely infinite is indivisible (*substantia absolute infinita est indivisibilis*)."[85] His proof runs as follows: If substance were divisible, then the parts into which it would be divided would either retain the nature of the original substance or not. If they do, then there would be a plurality of infinite substance, which is absurd. If not, then infinite substance would cease to be, which is also absurd. Hence it follows (as a corollary) that no substance of any metaphysical kind is divisible.

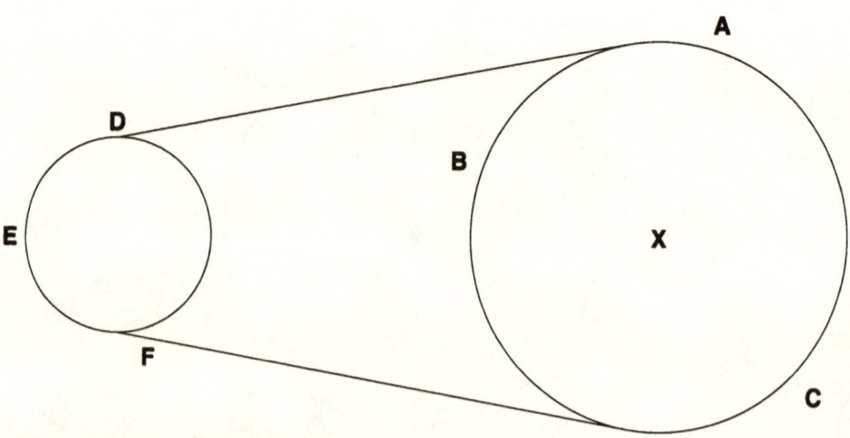

Figure 5.2. Spinoza's Refutation of Zeno

But how does the indivisibility of substance accord with its extension? In the scholium to proposition 15, Spinoza offers a set of arguments supporting the indivisibility of extended substance. Curley is right to note that "the main theme of the scholium is the defense of the doctrine that extended substance is an attribute of God."[86] I disagree, however, with Curley's further remark to the effect "that extended substance is infinite is a subordinate theme."[87] For, as we shall see, it is precisely the *infinity* of extended substance that is at issue and that distinguishes him from his historical predecessors. Spinoza first addresses those arguments of his opponents that claim that inasmuch as corporeal substance consists of parts, it cannot be infinite and hence cannot pertain to God. (We have already seen Aristotle and Descartes expounding versions of this view.) Spinoza examines three versions of this argument, which we can label the argument from division, the argument from measure, and the argument from determinate distance. The argument from division can be restated as follows: Assume that corporeal substance is infinite. This corporeal substance is divisible into two parts, each part of which must be either finite or infinite. If each part is finite, then the infinite is composed of two finite parts, which is absurd. But if each part is infinite, then the infinite is composed of two infinite parts, which would result in one infinity's being twice as great as another. This too is absurd.[88] In this argument Spinoza apparently assumes that both parts of substance must be homogeneous, and thus does not consider the possibility of one part being finite and the other infinite. Interestingly enough, he also rules out one infinity's being larger than another, a possibility he flirts with in his *Letter to Meyer.*

The argument from measure relies on the relation of parts to the whole, and can be stated as follows: Assume that an infinite substance is measured by equal parts, all of which are finite in extent. Such an infinite substance can be measured by different parts, say an inch and a foot. But then the sum of inch-parts will be smaller than the sum of foot-parts. Hence one infinite number will be twelve times greater than another, which is absurd.[89] As Wolfson has noted, Spinoza reflects here a similar point made by Descartes in his *Principles.* Both Spinoza and Descartes agree that the infinite (or indefinite) cannot be expressed by a precise number; in other words, it cannot be measured. That being so, surely one infinite cannot have a greater quantity than another, a point already amply recognized by medieval Jewish and Arabic philosophers.[90] That it cannot is further demonstrated in the argument from determinate distance. Let us conceive, Spinoza suggests, from one point of an infinite quantity, two lines AB and AC that extend to infinity. At the beginning point A they are a determinate distance apart. However the distance between B and C becomes indeterminable. This, however, is absurd.[91]

Interestingly enough, Spinoza does not offer a response to this third argument, although an allusion is made in the *Letter to Meyer*.[92] Spinoza's general reaction is that the three arguments are all based on the erroneous assumption that they suppose an infinite quantity to be measurable. He first refers back to propositions 12 and 13, both of which have shown that bodily substance is not composed of parts. More importantly, however, is his subsequent claim that "such absurdities stem from supposing infinite quantity to be measurable and hence made up of finite parts."[93] As we have seen, Spinoza's own contention is that infinite quantity, and hence substance, is not measurable. In support of this contention Spinoza draws upon several additional analogies geometrical in nature. For example:

> So others, after they feign that a line is composed of points, know how to invent many arguments, by which they show that a line cannot be divided to infinity. And indeed it is no less absurd to assert that corporeal substance is composed of bodies, or parts, than that a body is composed of surfaces, the surfaces of lines, and the lines, finally, of points.[94]

Spinoza finally brings in the example of a vacuum to show that substance must be indivisible. Spinoza clearly believes, although he does not argue the case here, that "there is no vacuum in nature."[95] If there were a vacuum in nature, then one piece of the continuum could be annihilated without affecting the remainder. But according to Spinoza the parts of substance are indistinguishable, for its parts are "all so fitted together" that one part cannot remain in its condition without the other.[96]

Almost as an aside, Spinoza mentions what I consider to be one of his most interesting metaphysical points. Even if, he says, his reply to the original objectors were insufficient,

> I do not know why [divisibility] would be unworthy of the divine nature . . . so it cannot be said in any way that God is acted on by another, or that extended substance is unworthy of the divine nature, even if it is supposed to be divisible, so long as it is granted to be eternal and infinite.[97]

This apparently throw-away comment can only be meant as an entree to the topic of modes, which follows in proposition 16. Interestingly enough, though, most commentators have not noted the import of this comment, which opens the door for Spinoza to argue that substance can be either divisible or indivisible so long as it is infinite.

The Role Played by Imagination

The infinity of substance is characterized further in *Letter 12* to Meyer, what has come to be known as Spinoza's *Letter on the Infinite*. In this letter Spinoza attempts to clarify several points concerning his characterization of substance, infinity, and divisibility. Spinoza distinguishes six cases of the infinite, which are contained in three pairs:[98]

2.1 What is infinite as a consequence of its own nature, or by the force of its definition [*Id quod sua natura, sive vi suae definitionis sequitur esse infinitum*]	2.2 And what has no bounds, not indeed by the force of its essence, but by the force of its cause [*Id quod nullos fines habet, non quidem vi suae essentiae, sed vi suae causae*]
3.1 What is called infinite because it has no limits [*Quod infinitum dicitur, quia nullos fines habet*]	3.2 And that whose parts we cannot explain or equate with any number, though we know its maximum and minimum [*Id cuius partes, quamvis eius maximum et minimum habeamus, nullo tamen adaequare et explicare possumus*]
4.1 What we can only understand, but not imagine [*Id quod solummodo intelligere, non vero imaginari*]	4.2 And what we can also imagine [*Id quod etiam imaginari possumus*]

According to Spinoza, if we had attended to these three sets of distinctions, we would understand both that some kinds of infinite are divisible while others are not, as well as that some infinites are greater than others.[99] (4.1) and (4.2) pertain to epistemological distinctions. If substance is truly indivisible, as Spinoza wishes to maintain, how then do we account for the epistemological inclination to divide quantity, which, of course, is one way that substance manifests itself? In *Letter 12* Spinoza argues that

> So if we attend to quantity (*quantitas*) as it is in the imagination (*in imaginatione*), which is what we do most often and most easily, we find it to be divisible, finite, composed of parts, and one of many. But if we attend to it as it is in the intellect, and perceive the thing as it is in itself, which is very difficult, then we find it to be infinite, indivisible and unique.[100]

Spinoza's answer is that we tend to conceive substance either "abstractly or superficially as we imagine it, or as substance, which is done by the intellect alone."[101] It is from the perspective of the imagination that substance appears to be finite, divisible and composed of parts.[102]

The imagination is brought in to talk about perception of time and number as well. Time and measure, for example, arise from the way we imagine quantity abstracted from substance: "Measure, Time and Number are nothing but Modes of thinking, or rather, of imagining."[103] Construed as aids to the imagination (*entia imaginationis, auxilia imaginationis*), measure, time, and number, which are discrete measurable quantities, have no status as real things (*entia realia*) existing outside the intellect (*extra intellectum*).[104] Further, many of the mathematical notions are not considered by Spinoza as *entia realia* but as *entia rationis,* or even as *entia imaginationis.* The implications of such a construal are revealed in the example of somebody who, conceiving duration abstractly and, conflating it with time, gets confused and does not understand how an hour can pass.

> When someone has conceived duration abstractly, and by confusing it with Time begun to divide it into parts, he will never be able to understand, for example, how an hour can pass. For if an hour is to pass, it will be necessary for half of it to pass first, and then half of the remainder, and then half of the remainder of this. So if you subtract half from the remainder in this way, to infinity, you will never reach the end of the hour.[105]

This example, Spinoza says, is "the same as composing Number merely by adding noughts."[106] On these grounds Spinoza concludes that "neither Number, nor Measure, nor Time (since they are only aids of the imagination) can be infinite."[107] On this basis, modes turn out to be infinitely divisible and hence continuous. Spinoza in short comes out in support of Zeno, "defending the continuity of the Parmenidean sphere against the numerical discontinuity of the Pythagoreans."[108] In other words, Spinoza rejects atomism in favor of the continuity of substance. Interestingly enough, Spinoza uses this last point to recall an argument of Crescas' that involves the denial of an actual infinite.[109] Giving a mathematical example to demonstrate that there can be an actual infinite, Spinoza stands in marked contrast to the Aristotelians who denied such a case.[110]

Summarizing his discussion now, Spinoza offers three meanings to the term infinite: Some things are "infinite by their nature and cannot in any way be conceived to be finite." Others are infinite " by the force of the cause in which they inhere, though when they are conceived abstractly they can be divided into parts and regarded as finite." Others, finally, are called infinite or indefinite because "they cannot be equated with any number, though they can be conceived to be greater or lesser."[111] Referring back to our table, Spinoza has highlighted the importance of (2.1), (2.2), and (3.2). Of the three, (2.1) most clearly captures the per se infinity of things, whereas (2.2) and (3.2) are used in an ancillary sense. Interestingly enough, (4.1) and (4.2), which represent epistemological distinctions, do not at all appear in this final characterization.

Spinoza on Divine Omniscience and Contingency

The notion that substance *is all* that exists finds its most rigorous expression in Spinoza's theory of divine omniscience. Because whatever is, is in God, it follows that by knowing himself God knows all finite modes, and conversely by knowing all modes *sub specie aeternitatis,* God knows himself. In chapter 4 we have already examined the philosophical and theological complexities inherent in this difficult issue. In that chapter I distinguished the main positions as being compatibilism, incompatibilism, and determinism, and suggested that determinism was the position most difficult to reconcile with traditional Jewish thought. Of all the philosophers examined in this work, only Spinoza accepts the implications of a thorough-going determinism. In this work I cannot begin to explore the implications of Spinoza's theory of omniscience upon his theory of human action and will; the very possibility of human action within the context of rigid determinism remains the focus of much recent scholarly discussion.[112] I shall, however, focus upon Spinoza's analysis of divine omniscience as it pertains to our overall theme of temporality.

Spinoza's theory of omniscience is developed within the context of his rejection of creationism. As I have argued in chapter 3, creationism holds the following theses to be true: God created the world *ex nihilo;* this act involves the exercise of divine will and intellect; this will occurred freely and with certain ends in mind; and unless the universe exists contingently, God is not omnipotent. Spinoza attacks this position on at least four levels: he claims that God does not have separate intellect/will; he tries to reconcile omnipotence and the absence of contingency; he denies free will to God; and he denies teleology in nature.

With respect to the status of intellect and will in God, Spinoza presents the following argument in favor of his contention that they do not pertain to God:[113]

5.1 Assume that intellect and will belong to God in some sense.
5.2 Divine intellect, insofar as it is conceived as the essence of God, is the cause both of the essence and existence of all things.
5.3 So Divine Intellect is the cause of our intellect.
5.4 But an effect differs from its cause in that which it has from its cause.
5.5 So Divine Intellect differs from human intellect in both essence and existence.
5.6 And so Divine Intellect has nothing in common with human intellect except the name.
5.7 God, therefore, does not have intellect/will in any traditional sense.

Argument Summary: If intellect and will belong to God and are the cause of the essence and existence of all things, then insofar as Divine Intellect is the essence of God, Divine Intellect is the cause of our intellect. But the essence and existence of Divine Intellect differ from that of its effects; and so God's intellect and will are totally different from our own.

The crucial points in this argument, are of course (5.4) and the move from (5.4) to (5.5). From this argument Spinoza concludes that predicates such as intellect and will, when applied to God, mean entirely different things when applied to humans: "They would not agree with one another any more than do the dog that is a heavenly constellation and the dog that is a barking animal."[114] The problem, however, is that in this argument Spinoza has not demonstrated that God and human intellect really do differ in essence: the most he has shown is that *if* God's intellect has the essence that it causes ours to have, then one can cease to have that essence without the other's ceasing to have it.

Spinoza rejects as well the Cartesian view that the world could have been otherwise. He must therefore be able to argue for divine omnipotence in light of the noncontingency of the world. First he offers the position of his adversaries, as represented by Descartes and his followers, who themselves offered the following type of argument to show that in fact God does not actualize all possibilities:[115]

6.1 Assume there is no state of affairs *p* such that both it was in God's power to bring it about that *p*, and God did not bring it about that *p*.
6.2 For any *p*, if it is logically possible that *p*, then it was in God's power to bring it about that *p*.
6.3 Let *p* be 'Adam sinned'.
6.4 For any *p*, if *p*, then it is because God brought it about that *p*, that *p*.
6.5 God brought it about that Adam sinned.
6.6. For any *p*, if God brought it about that *p*, then God did not have the power not to bring it about that *p*.
6.7. That is, God necessarily brought it about that *p*.
6.8. Hence God necessarily brought it about that Adam sinned.
6.9. But Adam's sinning is a contingent state (i.e., his not sinning is possible).
6.10. So, it was in God's power to bring it about that Adam did not sin.
6.11. But (6.8) and (6.10) are contradictory.
6.12. Hence (6.1) cannot be true.
6.13. Thus there are certain states of affairs that God does not actualize.
6.14. Therefore contingency exists in the universe.

Argument Summary: Let us assume that Adam's sinning is a contingent state of affairs. Let us also assume that God can actualize whatever is logically possible. If it is the case that Adam sinned, then God necessarily brought about Adam's sinning. But if Adam's sinning is contingent, then it is in God's power to bring it about that Adam not sin. But it cannot be the case both that Adam's sinning is necessary and contingent. Hence it is not case that God actualizes all events.

What Spinoza's adversaries want to show is that "if he [God] had created all the things in his intellect [6.1] then he would have been able to create nothing more, which they believe to be incompatible with God's omnipotence."[116] Spinoza, however, maintains that all possible states of affairs are actualized. On the basis of E1p16 he argues that all things necessarily flow from God's power, and God's omnipotence is maintained by the necessary actualization of all possibilities. Spinoza's refutation is aimed at (6.9), namely at the alleged contingency of Adam's sinning. This thrust is apparent when we turn to the freedom, contingency and necessity of modes and substance. The important texts here are E1p17, 18, 29, 32, 33, and the appendix. In order to determine whether or not humans (finite modes) and God (infinite substance) have free will, we must first ascertain what the terms "freedom," "necessity," and "contingency" mean for Spinoza. Spinoza's primary concern is to ascertain whether things or events can be uncaused. To this end he defines free and necessary as follows:

> *free* = what "exists and acts by necessity of its nature alone." (E1d7)
> *necessary* = what is "determined by another to exist and to produce an effect in a certain and determinate manner." (E1d7)

A thing is called necessary either by reason of its essence or by reason of its cause (E1p33s). Based on this twofold distinction, Spinoza claims that there is no contingency; all things are necessary in one of these two ways. Spinoza's main claim is that nothing is uncaused, although at least one thing is self-caused. God alone is a free cause (E1p17c2) in that there is no cause external to God, and God acts from the laws of his own nature alone. But the fact that God is a free cause does not imply that God has free will. Spinoza will want to refute the position of those who claim that God could have created a world different from the one he in fact created. The main argument goes as follows:[117] Assume that for any *p* such that it is the case that *p*, then *p* solely because God wills that *p*. Assume also that for any *p* such that God wills that *p*, God wills from eternity that *p*. The term eternity can mean either timelessness or omnitemporality. If by eternity we mean timelessness, then if from eternity *p*, then timelessly *p*. For any *p* such that it is the case that *p*, then timelessly God wills that *p*. If timelessly *p*, then it cannot at any time be (or have been) that –*p*. If by eternity we mean omnitemporality, then if from eternity *p*, then at all times *p*. For any *p* such that *p*, then at all times God wills that *p*. If at all times *p*, then it cannot at any time be (or have been) that –*p*. In both cases, for any *p* such that it is the case that *p*, it cannot at any time be, and could not at any time have been, that God does not will that *p*.

Both timelessness and omnitemporality rest on the notion that eternity entails the occurrence of *p*. The first two premises are accepted by his opponents. Spinoza then shows that on either understanding of eternity, the conclusion is that God could not have willed any other state of affairs, and so God does not have free will. All other things/events are not free in the sense defined in E1d7: According to E1p26 and E1p27, modes are determined to act by God, and cannot *not* act in those ways; and according to E1p29, all modes are determined from the necessity of God to exist and act in a certain manner. "In nature there is nothing contingent."[118] Spinoza concludes that what we call "contingency" is the result of ignorance: "a thing is called contingent only because of a defect of our knowledge."[119] This point is reinforced when Spinoza states that what we view as contingent is the result of ignorance: "It depends only on the imagination that we regard things as contingent, both in respect to the past and in respect to the future."[120] In E2p48d Spinoza reiterates the point that the human mind has no absolute or free will, that is, it is determined to action by causes. The entire dilemma of divine omniscience, therefore, is rendered otiose. The truly free individual is the one who is led by reason, who has adequate ideas of his actions (E4p68), and who follows the necessity of his own nature (E4p69) by recognizing the causal nexus governing his actions. No human finite intellect, however, can ever have fully adequate knowledge. This freedom represents an ideal limiting case; only God can be fully free, for only in God's self-knowledge, as reflected in his knowledge of *natura naturata,* can the total knowledge required for this freedom be achieved.

Conclusion

I started my analysis of Spinoza with a thesis of Bennett, to wit, that Spinoza's characterization of infinite substance stands in striking synchrony with theories within contemporary physics. I then examined selected texts within the vast Spinozistic corpus dealing with issues pertaining to infinity, divisibility, and the relation between substance and modes. I think that Gueroult is correct when he claims that:

> To prove the continuity (or the infinite divisibility) of modes is to prove the absolute indivisibility of substance, and vice versa . . . the endless divisibility of the continuous, which is that of modes, is conceivable only through the indestructible subsistence in them of an indivisible absolute, which necessitates that no truly separate part can ever be reached and that the division, since it can never be completed, be absolutely infinite. This absolute indivisibility, which is immanent in the modes, is that of their substance.[121]

That is, in rejecting (in accord with Zeno) atomism, discontinuity, and number, Spinoza erases the radical gulf between substance, which is both infinite and indivisible, and modes, which are infinitely divisible and thus continuous.[122] Is Bennett, right, then, in his contention that Spinoza's conception of infinite substance approaches contemporary theories? One way of determining this is to examine the *implications* of the field metaphysics in explaining reality. For example, how does Spinoza answer the metaphysical question "how many extended substances *are* there?" To his credit, Spinoza does not resort to the answer "none"; neither does he give the answer "many." Nor, in fact, according to Bennett, does he give the answer "one," if by "one" we mean picking out a particular referent to which the term "substance" refers. Like contemporary field theorists, he would claim that we are simply asking the wrong question. On this reading, the infinite substance characterized by Spinoza just is "*space plus what it contains at a time.*"[123] Both Spinoza and field physicists have rejected a view according to which substance is "lumps of matter" and have replaced it with a view in which what exists are regions of space/time. On this picture, substance *is all* that exists, but is not any one thing. The implications of this position become apparent when we turn, in the conclusion, to the final topic of this study, namely the end of time.

Conclusion: Eternity *a parte post*, Individuation, and Immortality

In this book I have addressed the interrelationship obtaining between time, cosmology, and creation in medieval Jewish philosophy. I have argued, through a detailed analysis of notions of infinity, creation, knowledge, and individuation, that there exists a coherent ontology of temporality within the frameworks of Jewish thinkers writing on these issues. I have tried throughout this work to emphasize the importance of reading this ontology against the backdrop of theological constraints. Whereas in some cases the theological constraints are challenged, as reflected in the works of Maimonides and Gersonides, in other cases, as in the case of Spinoza, these constraints are rejected altogether. It is the interplay between metaphysics, cosmology, and theology that has formed the subtext of this study.

What more fitting way to end this work than to address a concomitant issue, namely, the end of time. I have in mind two issues: the incorruptibility of the universe as a whole, and the incorruptibility of human life as we know it, as reflected in death. Earlier philosophers had grappled with the first question primarily from the Aristotelian perspective. Whether or not the universe as a whole had a *terminus ad quem* was discussed in the context of Aristotle's conception of an eternal universe. Not surprisingly, Aristotle's arguments for the eternity of time and motion pertain as well to the impossibility of a last instant of time, as to a first instant of time. For if any instant must have both a previous and successive instant, there can be no last instant of time, just as there can be no first instant of time. So for Aristotle the issue of whether there is an "end to time" is inextricably linked to arguments pertaining to a "beginning to time." If the universe is finite *a parte ante,* it will be finite *a parte post;* but given that there is no first instant, so too should there be no last instant.[1]

Medieval philosophers confronted the issue of the end of the universe against this Aristotelian backdrop. But, of course, the issue is complicated by the fact that if the universe is eternal *a parte post* as well as *a parte ante,* then it is eternally co-existent with God. And so the question becomes whether one can be logically committed to both dimensions of eternity. Saadia Gaon, for example, argues backwards, as it were, from the ultimate destructibility of the universe to its necessary creation.[2] Maimonides, however, disagrees and devotes several chapters in the *Guide* to this issue, arguing that the universe is eternal *a parte post.* Rosyln Weiss has suggested that the world's destruction represents for Maimonides "so complete and irreparable a rupture in the nature of things that Maimonides is compelled to reject it." For, as she rightly continues, such a rupture or permanent change threatens the very rationality of the universe: "Only a perpetually ongoing never-changing universe can be a rational one."[3] For this reason, Maimonides claims that although the doctrine of the world's indestructibility is not a foundation of the Torah, nevertheless many rabbinic dicta support the eternity *a parte ante* thesis: "the passing away of this world . . . [is] not affirmed in any prophetic text or in any statement of the Sages."[4] The philosophical question for Maimonides is whether a materially created entity is necessarily subject to destruction. The view of the vulgar according to which "all this world will pass away" is rejected by Maimonides.[5] Terms such as *le-ʿolam, ʿolam va-ed,* and *nitzḥi* attest to the imperishability and everlastingness of the world.

While for Maimonides the doctrine of incorruptibility is not logically necessary, Feldman has pointed out that for Gersonides the doctrine reflects a necessary feature of the universe.[6] In *Milhamot* 6 Gersonides defends what Feldman terms a thesis of "strong indestructibility" according to which the universe is inherently indestructible. Given that the object of God's creation is perfect, there is no sufficient reason for God to destroy his creation.[7] Although the universe is contingent, it is indestructible, as evidenced by the celestial bodies that are themselves indestructible.

But the question of the end of time has a metaphysical dimension as well, which is reflected in the ultimate finitude of human time represented by death. In chapter 1 we already alluded to the preoccupation with death reflected in Ecclesiastes. That preoccupation served to determine a theory of time that carried with it moral consequences. In general, problems associated with death and immortality of the human soul occurred for Jewish philosophers within the context of cosmological and psychological concerns.[8] I agree with Gracia and Thiel that we can distinguish a number of issues pertaining to the problem of individuation of persons:[9]

- The metaphysical question of what makes an individual the individual it is and distinguishes it from all other individuals of the same kind;

- The epistemological question of how we know individuals and their distinctiveness from one another, that is, our abililty to pick them out;
- The specific problem of identity through time—the conditions of an individual's remaining the same over time despite its having changed.

These issues appear in medieval Jewish texts but primarily within the context of problems associated with the immortality of the soul. More specifically, starting with Ibn Gabirol's characterization of a universal hylomorphism and continuing throughout the Jewish Neoplatonic tradition, Jewish philosophers have been concerned with what is the ultimate composition of particular entities and whether matter is sufficient to particularize these entities. The Neoplatonic emanation scheme forced philosophers to offer an explanation for the very existence of material entities: how, within an ontology that emphasizes the unity of the whole of substance, does one account for the proliferation of entities within a predominantly nonphysical hierarchy?[10] Further, Jewish philosophers were clearly influenced by the Islamic controversy over the unity/plurality of forms.[11] Jewish discussions took several dimensions. First, in light of the fact that the soul is the form of the body, does it retain its individuality upon separation from the body? And second, philosophers were concerned with the implications of their views with respect to the process of knowledge; that is, the connection obtaining between human intellect and the Active Intellect, as characterized by Avicenna and Averroës.[12]

The issue of immortality had been broached already by Maimonides in the context of examining the problem of particularization. In *Guide* 2.19, the problem of particularization is raised in terms of the difference among things. Given that all things in the sublunar universe have one common substance, why, then, are the species, as well as the individuals within each species, different from one another? Maimonides first summarizes the Aristotelian response according to which prime matter is transformed into the four elements each with different qualities; these in turn form the basis for the myriad compounds found in the sublunar spheres. Since matter has great latitude with respect to its specific forms, "the individuals of the species differ in a way corresponding to this latitude, as has been elucidated in the natural science."[13] Maimonides agrees with Aristotle that differences in the spherical motions imply that they have different forms. However, unlike Aristotle, he is concerned as well with where these differentiating forms originate. "Since every substratum in [the heavenly spheres] has been particularized so as to receive a certain form other than the forms received by the others, who is it that has particularized these substrata and has predisposed them to receive various forms?"[14]

Kalām philosophers had maintained that the very fact that a thing has been determined in terms of one particular size, place, location, and so forth is proof that there exists a being that freely chooses these determinations. Inasmuch as objects in the atomist ontology do not have specific natures, the fact that they exhibit one set of characteristics rather than another must be explained. Hence the Mutakallimūn concluded that a Being must be responsible for the characterization of each entity in the universe: "Thus the world as a whole requires someone to particularize it as a whole and each of its parts by means of one of the various admissible possibilities."[15] Although as we have seen in chapter 3, Maimonides rejects atomist occasionalism as a metaphysical doctrine, he agrees with its conclusion, namely that the fact that there is particularization is evidence of a particularizer.

This doctrine of particularization is developed further in *Guide* 1.72, where Maimonides draws a comparison between the universe as a whole and the human entity. According to this analogy, the universe is one individual being, whose parts constitute a single entity. Just as the members of the human body cannot exist by themselves, so too the parts of the universe cannot exist independent of one another.[16] However, with respect to the heavenly spheres, Maimonides has several important stipulations. Although the heavens are material, "its matter is not like that which is in us."[17] And, further, the terms "matter" and "body," "when used of us and of the heavenly bodies, are equivocal; that is, they have totally different referents."[18] This is not to say that the spheres are individuated on the basis of their matter; rather, Maimonides follows the Aristotelian doctrine that each heavenly sphere is a separate intelligence.[19] Inasmuch as the motions of the spheres differ from one another, it stands to reason that the explanation of these differences must rest in their form. Following Aristotle, Maimonides claims that "there exist separate intellects whose number is equal to that of the spheres, that every sphere desires the intellect that is its principle and is the mover causing it to move according to the movement proper to it, and that that intellect is the mover of that sphere."[20] Inasmuch as these intellects are separate from matter, "no multiplicity due to a difference between their essences is at all possible with regard to them because they are not bodies."[21] Maimonides concludes that every intellect, and hence every sphere, is distinguished from the others with respect to its motive powers.

Can this characterization of the spheres be used to understand the issue of immortality of the soul? Interestingly enough, Maimonides devotes very little space in the *Guide* to the issue of immortality. In *Guide* 1.41, Maimonides distinguishes three meanings to the term *nefesh,* or soul, and suggests that only in the third sense, that is, as a rational intellect, can the soul be considered immortal. However, the soul that survives after death is not like the one a person has when alive, for the former enjoys a separate existence after death and is a distinct ontological reality: "For that which comes into being at the time a man is generated is merely the faculty consisting in preparedness, whereas the thing that after death is separate from matter is the thing that has become actual and not the soul that also comes into being; the latter is identical with the spirit that comes into being."[22] In this passage Maimonides seems to be suggesting that when separated from the body, the immortal soul enjoys a special sort of existence akin to the Active Intellect. Like the Active Intellect, this soul has no personal features. But given that there are no such personal features, it follows that immortal, incorporeal souls cannot be individuated. Maimonides rejects the attempt of Avicenna who claims that there must be a distinction of cause and effect among souls; he favors the doctrine of the unity of intellect as propounded by Ibn Bājja. For inasmuch as what remains of one soul is neither the cause nor the effect of another, Maimonides suggests that "all are one in number," as Ibn Bājja was wont to suggest.[23] This interpretation is supported by *Guide* 2.16, in which Maimonides states that multiplicity is ultimately founded upon materiality.

> In whatsoever is not a body, multiplicity cannot be cognized by the intellect, unless the thing in question is a force in a body, for then the multiplicity of the individual forces would subsist in virtue of the multiplicity of the matters or substances in which these forces are to be found. Hence no multiplicity at all can be cognized by the intellect in the separate things, which are neither a body nor a force in a body, except when they are causes and effects.[24]

Regardless of whether this proposition is read as a metaphysical or an epistemological claim, it upholds the contention that, inasmuch as immortal souls have no causal or motive features, they (unlike the heavenly spheres) are not individuated after death.

The main features of Maimonides's views on immortality can now be summarized. Having distinguished between material and purely formal things, Maimonides posits different criteria for each. Material things are individuated on the basis of their matter, as evidenced by the proliferation of entities in the sublunar world. The heavenly spheres are individuated on the basis of formal considerations, that is, the motive powers of the intellect. However, inasmuch as souls have neither a material element nor a motive power, they are not individuated after the death of the body. In accord with Averroës, Maimonides concludes that all immaterial souls form a united whole.

In the remaining pages, I shall compare the views of Gersonides and Spinoza, whose theories of individuation in time and immortality of soul represent a philosophical response to the Averroism represented by Maimonides. I have already mentioned in chapter 4 that for Gersonides, considerations of an epistemological nature prompted his interest in individuation. Gersonides' discussion of divine knowledge raises the general question concerning the role individuals and particulars assume in his ontology. For Spinoza as well, the formal isomorphism between *natura naturans* and *natura naturata* raises specific concerns having to do with the individuation and immortality of the human being. To assess this role, we must look more closely at formal considerations in both authors concerning immortality and individuality.

Spinoza's famous discussion of eternity of the mind, which is couched against the backdrop of what the human mind can know, is summed up in *Ethics* 5p23, "The human Mind cannot be absolutely destroyed with the Body, but something of it remains which is eternal."[25] Spinoza's demonstration of this proposition rests on E2p8c, namely, that we do not attribute to the human mind any duration that can be defined by time except insofar as it expresses the actual existence of the body, which is defined by duration and time. After the separation from or cessation of the body, however, mind no longer pertains to duration; because it is constituted by God's essence, mind will necessarily be eternal. Spinoza continues that "we feel and know by experience that we are eternal . . . this existence that it [our mind] has cannot be defined by time or explained through duration."[26] It would appear, then, that when the body dies, we are left with the *idea* of extended essence that eschews personality, thus eliminating individual immortality. But, as Curley asks, how can Spinoza maintain on the one hand that we are eternal, and on the other hand that the eternal has no relation to time, knowing that as individuals we do exist in and through time?

Commentators on *Ethics* 5p23 have run the gamut from devotion to despair. Hampshire, for example, suggests that

> It cannot be claimed that we can easily understand what exactly Spinoza meant when he wrote [E5p23]. . . . It seems, but this must be conjectural, that we sometimes have experiences of complete and intuitive understanding, and that on such occasions we feel and know ourselves to be mentally united or identified with the eternal order of Nature. . . . But everyone must be left further to interpret these propositions as he can, or perhaps to confess that at this point he finds himself beyond the limits of literal understanding.[27]

Bennet is more openly critical of these passages and writes that

> I don't think that the final three doctrines (of Book 5) can be rescued. The only attempts at complete salvage that I have encountered have been unintelligible to me and poorly related to what Spinoza actually wrote. . . . After three centuries of failure to profit from it, the time has come to admit that this part of the *Ethics* has nothing to teach us and is pretty certainly worthless . . . this material is valueless. Worse, it is dangerous: it is rubbish which causes others to write rubbish.[28]

With these sober caveats in mind, let me rush in where others have dared to tread, and suggest a coherent way of reading this portion of the *Ethics*. Using Gersonides as a foil, I shall argue that Spinoza's theory of human immortality can in fact be rescued in a way that preserves individuality. If there is a subtheme to this chapter, it is that only by recognizing the implications of Spinoza's thorough-going hylomorphic monism, and understanding its place in the long line of Jewish Neoplatoic hylomorphisms, can we fully understand Spinoza's views on individuation and eternality.

We saw in chapter 5 that Gersonides' discussion of the question of whether God knows future contingents was examined extensively in the context of issues having to do with the nature of particulars, the distinction between individuals and particulars, and problems associated with the individuation of forms. Gersonides' discussion of divine knowledge thus raises the general question concerning the role individuals and particulars assume in his ontology. Gersonides' discussion of the immortality of the soul, contained primarily in *Milḥamot* 1, raises similar concerns. His discussion must be understood against the backdrop of Greek and Arabic commentators on Aristotle's notoriously difficult passage in *De Anima* 3.5:

> And in fact mind as we have described it is what it is by virtue of becoming all things, while there is another which is what it is by virtue of making all things: this is a sort of positive state like light; for in a sense light makes potential colors into actual colors. Mind in this sense of it is separable, impassible, unmixed, since it is in its essential nature activity. . . . When mind is set free from its present conditions it appears as just what it is and nothing more: this alone is immortal and eternal . . . and without it nothing thinks.[29]

In this passage Aristotle appears to postulate the existence of an Active Intellect that is separable from the passive intellect and primarily responsible for the intellectual activities of the human mind. But what is the relation between the Active and passive intellects, and which, if either, of the two is immortal?[30] In answer to these questions, Gersonides first summarizes four representative positions as follows:

1.1 *The view of Alexander of Aphrodisias,* according to whom the material intellect is in the individual soul in its entirety; this intellect is "nothing but a disposition . . . that is, a passive receptor. In contrast, the "other intellect" is identified with the divine Agent Intellect, which is eternal and totally separate from humans. This latter intellect represents the total intellectual cognitions acquired by an individual and is to some extent immortal.[31]

1.2 *The view of Themistius,* according to whom the material intellect itself is a "separable intellect that is neither generated nor corruptible."[32]

In other words, the material intellect is incorporeal and eternal, and ultimately identifiable with the Agent Intellect.[33]

1.3 *The view of Averroës,* which is a conflation of (1.1) and (1.2) in that the material intellect is a separate substance and identified with the Agent Intellect, but "insofar as it attaches itself to the human soul, it is a disposition and has a potentiality for knowledge."[34] That is, both the Agent Intellect and the material intellect are ultimately one and the same in all men.[35]

1.4 *The view of the "moderns,"* according to whom the material intellect is "a separable intellect that is generated essentially but not from something else."[36]

Inasmuch as Gersonides' own position reflects that of Alexander (1.1), he raises a number of objections to the other three accounts. Against, for example, Themistius' position (1.2) that the Agent and material intellects are equivalent, Gersonides first argues that Themistius' position is unable to account for the individuation of persons on the grounds that separate forms cannot be numerically individuated:

> [Material] form is individuated in so far as it is manifested by different subjects, so that, for example, this form of Reuben is numerically different from the form of Simon. If this were not so, one and the same form would be both knowledgeable and ignorant at the same time on the same point. . . . But this is utterly absurd. And so individuation of this form can be accounted for on Alexander's hypothesis, but this is not the case on Themistius' hypothesis, for separate substances cannot be numerically individuated, except if they differ as species, as has been demonstrated by arguments proper to this subject.[37]

Gersonides' point can be restated as follows. If both Reuben and Simon share the same material form, and this form is reflected in cognitive abilities, then when Reuben knows *p* and Simon is ignorant of *p*, it follows that one and the same form both knows and fails to know *p*. Hence, Gersonides concludes that there must be another criterion for individuating these cognitive processes. Themistius attempted to minimize these difficulties by claiming that Alexander is unable to account for the identity of the act of knowing and that which is known. Gersonides, however, claims that Themistius' gambit is insufficient on the grounds that epistemological identity does not necessarily presuppose that the intellect be a separate form, as Themistius insists. Rather, Gersonides supports Alexander's attempt to account for this identity on the grounds that cognition need not be of particulars, but can be of general natures that are identified with the material intellect.

The import of Gersonides' critique of his predecessors can be reduced to three main issues. From a theological perspective it is clear that the doctrine of unity of intellect threatens the notion of personal immortality. For if all humans share the same intellect, then upon physical death, all that remains of the person is an unindividualized intellect. Epistemologically, the doctrine is unable to account for how it is that two (or more) knowers can entertain contrary items of knowledge; or more stringently, how one person can be in error of an item that another person knows. And from a metaphysical perspective, the main problem is how to individuate this separate intellect when it is manifested in many individuals: for if it is individuated materially on the basis that individual bodies differ, then the substance is no longer incorporeal or separate. As Feldman has pointed out, on this theory an incorporeal substance is either a unique member of a species or is not a member of a species at all.[38]

Gersonides avoids these untoward consequences by adopting the position of Alexander of Aphrodisias. Gersonides agrees with Alexander of Aphrodisias that immortality consists in the intellectual perfection of the material intellect. He disagrees with Alexander, however, over the precise nature of this intellectual attainment. For Alexander (according to Gersonides) had claimed that immortality is achieved when the intellect acquires knowledge of the Agent Intellect (hence the term "acquired intellect" is introduced). Immortality is thus understood by Alexander as a form of conjunction between the Agent and acquired intellects. "They [the followers of Alexander] maintain that the material intellect is capable of immortality and subsistence when it reaches that level of perfection where the objects of knowledge that it apprehends are themselves intellects, in particular the Agent Intellect . . . [material intellect] is immortal when it is united with the Agent Intellect."[39]

Gersonides rejects this notion of conjunction, however, and replaces it with a model of immortality according to which it is the content of knowledge of the acquired intellect that matters. When the content of the acquired intellect mirrors the rational ordering of the Agent Intellect, immortality is achieved. What is the content of this knowledge? The Agent Intellect must possess complete knowledge of the sublunary world; that is, it "contains a conception of the rational order obtaining in all individuals."[40] The anti-Platonic tenor of this position is emphasized when Gersonides describes in more detail what it is that the Agent Intellect knows. For according to Gersonides, the knowledge of the Agent Intellect must be grounded in the domain of particulars.

> Universality accrues to [this order] by virtue of its grounding in perceived particulars existing outside the mind. . . . The definition is the very order that is in the mind of Agent Intellect according to which the genus is generated. This order is exhibited in some sense in each and every individual instance of that genus, as we have seen. It does not follow from this, however, that all these

> individuals are numerically one, as would be the case for those who believe in [Platonic] universals. In this manner knowledge of accidental properties also is established, not just of essential properties.[41]

From this important passage it is clear that Gersonides' position avoids the epistemological difficulties apparent in a realist ontology. Inasmuch as the material intellect reflects the knowledge inherent in the Agent Intellect, and inasmuch as this knowledge is grounded in particulars, it follows that humans can have knowledge of particulars; in this acquisition of knowledge lies immortality.

It is precisely this aspect of Gersonides' theory that is so important when we turn our attention to Spinoza. When Spinoza's theory of immortality is read against the backdrop of Gersonides' hylomorphic ontology, coupled with its concomitant epistemology based upon knowledge of particulars, the various strands of *Ethics* 5 form a coherent whole.[42] The final propositions of *Ethics* 5 can be reduced to three main doctrines:

2.1 A part of the mind is eternal and survives decay of the body.
2.2 Intuitive science (third kind of knowledge) yields blessedness.
2.3 Blessedness consists in intellectual love of God.

These doctrines represent the culmination of several strands: the notion of the free human, the overcoming of the passions in order to achieve this notion of freedom, and the incorporation of knowledge as a criterion for freedom. I have already described the "free man" as an ideal limiting condition for humans. As Hampshire suggests, "humans beings as finite modes, cannot in principle be completely free and unaffected by external causes; human freedom must be a matter of degree. . . . [Spinoza contrasts] the actual and normal conditions of servitude with the humanly unattainable ideal of permanent and perfect freedom."[43] The free person lives according to the dictates of reason alone (E4p67). Since he lives by reason, he has no fear, certainly not of death. Since good and evil are the result of inadequate ideas, the free person is not ruled by this conventional type of morality (E4p68). In the appendix to book four, Spinoza begins to extol the virtue of reason, and connects rational life with "blessedness." Blessedness is connected with rational perfection and intuitive knowledge of God. Hence the more we know (understand the cause of) a passion, the less we suffer from it. The main idea is that by understanding the causes of the emotions, we dissolve the power they have over us; by having adequate knowledge (true knowledge, clear and distinct ideas), the passions become neutralized.[44]

We can now move to the triad of doctrines in *Ethics* 5. Two charges have been levelled against these propositions. Against (2.1) commentators have argued that, given *Ethics* 2p7 according to which the human mind is the idea of the human body, it follows that there should be an eternally existing idea of body as well. That is, Spinoza should espouse an eternal existence for body parallel to that of mind. Given what Spinoza has said about isomorphism of mind and body, as well as the indivisibility of substance, I would contend that positing a parallel existence to body is in fact what Spinoza proposes. This point is reflected in Spinoza's theory of individuation and reflects the same distinctions made by Gersonides.[45] On Spinoza's theory of individuation, part of what makes me who I am is that I am affected by other individuals; individuation on this model turns out to be relational, incorporating both material and formal elements.[46] Mason reinforces the point that it is bodies that are the source of identification of persons: The ideas that make up an individual mind acquire their identity by being ideas of a particular body.[47] This identification with body remains embedded in the mind after the "death of the body." The crucial point which many commentators lose sight of is that for Spinoza there is a difference of degree, not of kind, between persons and other objects. To admit that the eternity of mind incorporates the idea of the body, as expressed in E2p7, and that therefore the corresponding body is eternal as well as the mind, simply reinforces Spinoza's rigorous monist ontology. In fact, I would maintain that this is precisely what the isomorphism of God/nature amounts to, and should not be regarded as an inconsistency in Spinoza's ontology. That Spinoza emphasizes the eternity of part of the mind, which incorporates the body *qua* idea of the mind, is undeniable in the sense that mind has an idea of the "essence of body under the form of eternity."[48] As Margaret Wilson has correctly argued, the eternity of mind includes the knowledge of the essence of body and so is not as separate from body.[49] The knowledge I have of my body differs in kind from the knowledge I have of anybody else's body, or that anybody else has of my or another body, in large part because of the isomorphism between *my* mind and body. Hence the worry that if two minds have eternal and adequate ideas of another body (not their own), there is no way to distinguish their token ideas, does not apply for Spinoza.[50]

Doctrines (2.2) and (2.3) suggest that eternity is tied to the third kind of knowledge, or intuition. This third level, which is constituted as knowledge of individuals, is different from the other two degrees (E5p36s). It is this level of knowledge that leads to intellectual love of God, or salvation. At this level, the individual achieves a level of intuitive union/knowledge, and is characterized by a "sort of repose" (E5p36s) or "peace of soul" (E5p62s) that results from recognizing the necessity inherent in himself, in God, and in things. But this repose raises yet another charge, which is aimed at the relation between God and mind. Spinoza has said in E1d8 that eternity pertains to God alone inasmuch as God is the only entity in which essence is identical to existence; but in E5p23 he ascribes eternity to mind. If we are to make sense of the isomorphism between God and mind, then Spinoza's theory of eternity of mind must eschew individual immortality, on the grounds that in mind essence is not identical to existence.[51] This point is made in Nadler's recent biography of Spinoza. In this work Nadler claims that it was Spinoza's supposed denial of personal immortality that led to his excommunication (*ḥerem*); while I agree with this assessment of thecauses for Spinoza's *ḥerem,* my point is that the Jewish community, along with recent commentators, misunderstood this doctrine in their attribution to Spinoza the denial of personal immortality.[52]

It is on this point, I argue, that Spinoza comes closest to reflecting Gersonides' depiction of immortality in terms of knowledge of particulars. Just as Gersonides argued in (1.1) above that the Agent Intellect represents the total intellectual cognitions acquired by an individual, so too Spinoza argues that the more singular things the mind understands, the more it understands God. On the basis of this characterization, I maintain that what we call immortality of soul, characterized as eternity of mind, for Spinoza must be personal. Within this unity of mind with God/Substance, there is still something of "me" that remains. Another way to make this point is to claim that we are already what we will become, and so what we become in the future comprises the duration that we have lived through. We have seen that individuals are separate one from the other both before death, and even after death; individuals retain their individuality with respect to one another and God. Of course, there is an important difference between claiming that eternal minds differ from one another based on levels of knowledge achieved, and claiming that they contain individuating features that define their personal, individual nature. The first claim has to do with criteria of differentiation among minds, whereas the second has to do with criteria of individuation within God. This was the point made by Gersonides in his claim that "it does not follow from this, however, that all these individuals are numerically one." Just because the knower becomes one with God or Active Intellect does not entail that the individuating features of the knower are dissolved into a general unity.

It might be argued that in an ontology that privileges identity of mind and body, a mind that is in any way "separated" from body cannot by definition be the "same" entity it was before the separation, and hence the eternal mind that unites with God in eternal blessedness cannot retain its personality in this union. Harris, for example, tries to argue that it is not that in death something "remains" or persists after the death of the body, but rather that the mind transcends temporality altogether in the mode of knowing: human mind becomes one with God, and God/Substance become one with human mind.[53] But such readings overlook the importance played by matter and temporality, as exemplified by body and matter, in both Gersonides and Spinoza. On Harris' picture, by transcending temporality, and entering into union with the Active Intellect or with God, the individual does leave its identity behind. But this is not Spinoza's view.

But how can something that is finite and part of duration/time enter the domain of eternity?[54] This question acknowledges that eternity cannot simply mean for Spinoza a "merging of part of the mind into a stream of divine consciousness."[55] Spinoza's answer can only be that the eternity of substance, inasmuch as it comprises "everything that is," comprises as well the realm of duration in general. The domain of "time in general" as opposed to particular instantiations of "space-time" is part of what we take to be eternity. That is precisely what Spinoza means when he states that "this something that pertains to the essence of the mind will necessarily be eternal."[56] If my analysis is correct, then I have shown that just as matter is integral to the individuation of the person and to God, so too temporality is not always antithetical to the divine. By incorporating both temporality and the concomitant materiality expressed in a hylomorphic ontology, the individual retains its individuality while becoming one with God. Once again, we see that issues of temporality play a crucial role in determining the relation between the human and God. By understanding time, and the cluster of issues surrounding temporality, we are better able to penetrate the veil that separates the domain of eternity from that of temporality.

NOTES

Introduction

1. Stephen Jay Gould, *Time's Arrow, Time's Cycle: Myth and Metaphor in the Discovery of Geological Time* (Cambridge, Mass.: Harvard University Press, 1987), p. 11.

2. A. Néher, "The View of Time and History in Jewish Culture," in *Cultures and Time,* ed. L. Gardet et al. (Paris: The Unesco Press, 1976).

3. See Aviezer Ravitsky, *History and Faith: Studies in Jewish Philosophy* (Amsterdam: J.C.Gieben, 1996), for a penetrating discussion of the status of Jewish philosophers vis-à-vis their intellectual environment.

4. Stephen Hawking, *A Brief History of Time* (New York: Bantam Books, 1988), p. 21.

5. Hawking, *A Brief History,* p. 33.

6. Steven Jay Gould, *Time's Arrow, Time's Cycle,* p. 10.

7. Ibid., p. 11.

8. Ibid., p. 11.

9. Ibid., p. 13. It is worth mentioning Nietzsche's theory of eternal recurrence in this context, as a modified example of the cyclicity of time.

10. Ibid., p. 16.

11. Hawking, among others, tries to answer such questions by characterizing three arrows of time: the thermodynamic arrow of time, the psychological arrow of time, and the cosmological arrow of time. His main thesis is that, notwithstanding the fact that "the laws of science do not distinguish between the forward and backward directions of time," nevertheless all three arrows necessarily point in the same direction and do in effect distinguish between past and future. See Hawking, *A Brief History,* p. 152.

12. M. Eliade, *The Sacred and the Profane* (New York: Harcourt, Brace & Co., 1959).

13. See Friedrich Kümmel, "Time as Succession and the Problem of Duration," in *The Voices of Time: A Cultural Survey of Man's Views of Time as Expressed by the Sciences and the Humanities,* ed. J. T. Fraser (New York: George Braziller, 1966), p. 31.

14. For an example of the difference between relativistic and absolute perceptions, see A. Cornelius Benjamin, "Ideas of Time in the History of Philosophy," in *The*

Voices of Time: A Cultural Survey of Man's Views of Time as Expressed by the Sciences and the Humanities, ed. J. T. Fraser (New York: George Braziller, 1966), p. 8ff.

15. This question is the focus of Bas van Frassen in his work *An Introduction to the Philosophy of Time and Space* (New York: Random House, 1970).

16. Gould, *Time's Arrow, Time's Cycle,* p. 80.

17. Ibid., p. 49.

18. Ibid., p. 59.

19. Ibid., p. 194.

20. See H. A. Wolfson, *The Philosophy of Spinoza* (New York: Schocken Books, 1969), for a detailed history of the concept of eternity.

21. This project has had enormous repercussions with respect to Christian scholasticism as well. The influence of Maimonides upon subsequent scholastic thought has not gone unnoticed by scholars. Bonaventure, for example, bases his discussion of the eternity of the world upon Aristotelian, Arabic, and Maimonidean arguments. See Stephan Baldner, "St. Bonaventure on the Beginning of the Temporal World," *The New Scholasticism* 63 (1989), pp. 206–28 for a discussion of Maimonides' influence upon St. Bonaventure. See also Richard C. Dales, "Early Latin Discussions of the Eternity of the World in the Thirteenth Century," *Traditio* 43 (1987), pp. 171–97 for an extensive discussion of Maimonides' influence.

22. Norman Kretzmann, "Ockham and the Creation of the Beginningless World," *Franciscan Studies* 45 (1985), pp. 1–33.

23. Kretzmann, "Ockham and the Creation of the Beginningless World," p. 3.

24. Wolfson, *Philosophy of Spinoza,* p. 133.

25. In his unpublished M.A. thesis (Heb.) "Remarks on the Concept of Infinity in Medieval Jewish Philosophy" (Hebrew University, 1969), Rosenberg reflects Wolfson's distinction when he claims that according to Aristotle the term infinite can refer both to quantitative and to essential notions (pp. 1–3).

26. Rosenberg, "Remarks on the Concept of Infinity," p. 29.

27. In *Sefer Bahir,* for example, the compound *'ein-sof* appears in a descriptive form in section 48: "the thought with which one can think up to infinity and the endless" *(le-'ein-sof ve-takhlit)*, but there is no sense of *'ein -sof* as a noun. See Gershom Scholem, *Origins of the Kabbalah,* p. 130.

28. Scholem, *Origins,* p. 265.

29. Ibid., p. 268.

30. I. Efros, *The Concept of Space in Jewish Mediaeval Philosophy* (New York: Columbia University Press, 1917).

31. Norman Kretzmann, ed., *Infinity and Continuity in Ancient and Medieval Philosophy* (Ithaca, N.Y: Cornell University Press, 1982). See also the seminal work of Richard Sorabji, *Time, Creation and the Continuum* (Ithaca, N.Y.: Cornell University Press, 1983) for discussion of similar issues.

32. Herbert Davidson, *Proofs for Eternity, Creation and the Existence of God in Medieval Islamic and Jewish Philosophy* (New York: Oxford University Press, 1987).

Chapter 1. Time and Cosmology in Athens and Jerusalem

1. E. Jenni, "Time," in *The Interpreter's Dictionary of the Bible*, vol. 4 (Nashville, Tenn: Abingdon, 1962), p. 643.

2. J. Barr, *Biblical Words for Time* (Naperville, Ill.: Alec R. Allenson, 1962), pp. 117–18.

3. See L. Stadelman, *The Hebrew Conception of the World: A Philological and Literary Study* (Rome: Pontifical Biblical Institute, 1970), note 5; pp. 1–3, 39.

4. Neher for example, argues that time was never perceived as a reality by Greek philosophers, for whom the world remained essentially changeless and orderly: "Of all ancient peoples, the migratory and conquering Semites, who had a very turbulent history, were the first to become aware of the divergent functions of time and space." See André Neher, "The View of Time and History in Jewish Culture," in *Cultures and Time*, ed. Louis Gardet et al. (Paris: The Unesco Press, 1976), p. 150.

5. A. J. Gurevich, "Time as a Problem of Cultural History," in *Cultures and Time*, ed. L. Gardet et al. (Paris: The Unesco Press, 1976), p. 234. According to Gurevich, Christianity took from the Old Testament the notion of time experienced as an eschatological process, a fervent waiting for the great event in which history is fulfilled—the coming of the messiah.

6. B. J. Malina, "Christ and Time: Swiss or Mediterranean?" *Catholic Biblical Quarterly* 51 (1989): 4.

7. Ibid., p. 5.

8. Ibid., p. 19.

9. According to Mumford, the mechanical conception of time, arising from the regular measure of time, arose out of the routine of the monastery; for the clock is "not merely a means of keeping track of the hours, but of synchronizing the actions of men." The clock, by its nature, "dissociated time from human events and helped create the belief in an independent world of mathematically measurable sequences: the special world of science." Mathematical time is "strung out in a succession of mathematically isolated instants," whereas organic time, or duration, is cumulative. In the mid-1340s the division of hours into minutes and minutes into seconds became common. The abstraction of time took over. See Lewis Mumford, *Technics and Civilization* (New York: Harcourt Brace & Co., 1934), p. 15ff.

10. Barr, *Biblical Words for Time*, p. 137. What do the Greeks and Hebrews have in mind by "cyclic"? Quoting Boman, Barr claims that when theologians say that the Hebrews did not have a cyclic view of time, what they tend to mean is that "in Hebrew thought the sequence of historical events, or of some historical events, is a purposive movement towards a goal; it is certainly not cyclic in the sense of something recurrent, but is non-recurrent, non-reversible and unique." But when the Greeks say of time that it is cyclic, they often (e.g., see Aristotle) have in mind the perfection of the circle, and not historical events in time. "In so far then as Greek cyclic theories of time are based on the circular movements of the heavenly bodies, they are based on

something which was quite naturally understood as cyclic by the Hebrews also." Barr concludes, then, that the appearance of something that could perhaps be called a cyclic view of time on the Greek side, or of something that could perhaps be called a linear view on the Hebrew side, does not in itself justify postulating a fundamental difference between them.

11. The ostensible connections, if any, between linguistic usage and ontological conceptions of time have been examined by many. Whitrow notes that in some ancient forms of language the dominant temporal characteristic was duration rather than tense and argues that it is only in Indo-European languages that distinctions between past, present, and future have been fully developed. Arguing that in Hebrew, for example, the verb treats action as either incomplete or perfected, future events in Hebrew are always expressed as coming after us (G. J. Whitrow, *Time in History* [Oxford: Oxford University Press, 1988], p. 13). In a similar vein, Boman, asking how it is that the Hebrews, who have no verb expression of time, have become the people of history, while the Greeks whose time-words are complex, have no ostensible interest in historical development, suggests that "it is the Hebrews and Semites who have the adequate understanding of time, not the Greeks and we Europeans. We must then ask whether the tenses of Hebrew verbs do not express time more clearly than do our tenses" (Boman, *Hebrew Thought*, p. 144). For a critical rejection of these positions, see Barr, *Biblical Words for Time*, p. 131.

12. Barr, *Biblical Words for Time*, p. 132. Barr spends a fair bit of time talking about Orelli's book and showing how his view of time is "now entirely antiquated." (Ibid., p. 82).

13. Ibid., p. 90.

14. See P. Steensgaard, "Time in Judaism," in *Religion and Time*, ed. A. N. Balslav and J. N. Mohanty (Leiden: Brill, 1993), p. 77ff., for a discussion of the importance of religious eschatology.

15. Steensgard as well wants to argue that in the Old Testament we find both historical, linear time as well as mythical time that has a cyclical character. For example, he points to early cultic, cyclical conceptions and then asks "when did the consciousness of historical time become so strong that it repressed the idea of a cultic-cyclical repetition?" In the prophets, for example, time is understood in a linear fashion. See Steensgard, "Time in Judaism," p. 65. Agreeing with Eliade that both Judaism and Christianity view time in a linear/historical fashion, Rubinstein nevertheless reiterates with Mowenckel that history is not the antithesis of myth, but is integrated into the rabbinic cult: "While such events took place at a certain point in history, they were re-enacted and repeated, and thereby experienced, on the festival." Rubenstein correctly argues that rabbinic sources portray the festivals in mythic terms, emphasizing both history and myth, as a "recovery of a sacred time." The festivals express what occurs in the present instant, but mark stages of an extended, ongoing drama, between God and Israel. The important point here is the "otherness" of biblical time which is superimposed onto the historical flow. See

Jeffrey Rubenstein, "Mythic Time and the Festival Cycle," *The Journal of Jewish Thought and Philosophy* 6 (1997), pp. 157–83.

16. *Mishnah* 1.1.9b.

17. See N. Kretzman's collection *Infinity and the Continuum in Ancient and Medieval Thought* (Ithaca, N.Y.: Cornell University Press, 1982) for examples and discussion of this scholastic literature. See also *The Cambridge History of Later Medieval Philosophy,* ed. N. Kretzmann, A. Kenny, and J. Pinborg (Cambridge: Cambridge University Press, 1982).

18. For a discussion of the ways in which these rabbinic discussions influence subsequent Jewish philosophers, see Norbert Samuelson, *Judaism and the Doctrine of Creation* (Cambridge: Cambridge University Press, 1994), p. 140ff.

19. For a survey of early Jewish views on this topic, see Alexander Altmann, "A Note on the Rabbinic Doctrine of Creation," *Journal of Jewish Studies* 7 (1956), pp. 195–206; Jonathan A. Goldstein, "The Origins of the Doctrine of Creation *Ex Nihilo,*" *Journal of Jewish Studies* 35 (1984), pp.127–35. In this paper Goldstein tries to argue, against recent scholars, that early Jews and Christians in fact had a theory of creation *ex nihilo* that was tied to conceptions of bodily resurrection.

20. See Efraim Urbach, *The Sages: Their Concepts and Beliefs,* trans. I. Abrahams (Jerusalem: Magnes Press, 1975), p. 211, for discussion of these passages. See also Menachem Kasher, "The Concept of Time in Scripture and the Early Rabbis" (Heb.), *Talpiot* 5 (1952), pp.799–827.

21. *Nedarim* 39b.

22. *Genesis Rabbah* 1.4.

23. *Shabbat* 88b.

24. *Ḥagigah* 14a.

25. *Rashi.*

26. Ibn Ezra, *Commentary on the Pentateuch (Genesis),* trans. and annotated by H. Norman Strickman and Arthur M. Silver (New York: Menorah, 1988), p. 24. Ibn Ezra, *Perushe Ha-Torah* (Jerusalem: Mossad Ha-Rav Kook, 1976), p. 12.

27. See Altmann, "A Note on the Rabbinical Doctrine," p. 197ff.; Goldstein, "Origins."

28. See Winston, "The *Book of Wisdom*'s Theory of Cosmogony," *History of Religions* 11 (1971), 186.

29. The Greek "'*ek tou me ontos*" was translated into Latin as *ex nihilo,* into Arabic as *lā min shay*', and into Hebrew as *yesh me-'ayin.* For a history of this formulation, see H. A. Wolfson, "The Meaning of *Ex Nihilo* in the Church Fathers, Arabic and Hebrew Philosophy, and St. Thomas," in *Studies in the History of Philosophy and Religion,* vol. 1, ed. I. Twersky and G. Williams (Cambridge, Mass.: Harvard University Press, 1973), pp. 207–11. See also Winston, "The *Book of Wisdom*'s Theory of Cosmogony," pp. 185–202. Samuelson explores the many Hebrew and Latin ways to say "out of nothing" in Samuelson, *Judaism and the Doctrine of Creation,* p. 101ff.

30. *Ḥagigah* 12a.

31. See Altmann, "A Note on the Rabbinical Doctrine," p. 206: "the light is an emanation from the Divine *Logos* or Wisdom. . . . It was not by pure chance, therefore, that the thirteenth-century mystics of Spain found it possible to connect their doctrine with the *midrash* of the third century teacher of Palestine."

32. *Genesis Rabbah* 9.2.

33. *Ḥagigah* 12a.

34. Naḥmanides, *The Commentary of Naḥmanides on Genesis, Chapters 1–6.8,* trans. and notes by Jacob Newman (Leiden: E. J. Brill, 1960), p. 39.

35. Ibid., p. 39.

36. See Cassuto, *The Sages,* p. 28 for a discussion of this point. With respect to the first question, Cassuto, a modern bible interpreter, argues that whenever in Scripture a clear reference is made to the relationship between a given day and the next, it is precisely sunrise that is counted as the beginning of the second day. But with respect to festivals and appointed times, they are observed on the night of the preceding day, which is acknowledged as a separate day. Hence "throughout the Bible there obtains only one system of computing time."

37. Naḥmanides, *Commentary,* p. 37.

38. Ibid., p. 40.

39. Ibid.

40. Sforno, *Commentary on the Torah,* in *Mikraot Gedolot: Sefer B'reishit* (New York: Shulsinger Bros., 1950–78), p. 3.

41. Ibid.

42. Ibid., *Commentary,* p. 10–11.

43. Ibid., *Commentary,* p. 32

44. Samuelson, *Judaism and the Doctrine of Creation,* p. 148.

45. Anne Higgins, "Medieval Notions," *Journal of Medieval and Renaissance Studies* 19 (1989), pp. 232ff. See also the excellent discussions in J. LeGoff, *Time, Work and Culture in the Middle Ages,* trans. Arthur Goldhamer (Chicago: University of Chicago Press).

46. Yosef Yerushalmi, *Zakhor: Jewish History and Jewish Memory* (Seattle: University of Washington Press, 1982), p. 17.

47. As stated so eloquently by Neusner: "What seems to me Mishnah's system expresses is a totally ahistorical, possibly even anti-historical, conception of sanctification. In this conception, curiously, historical time plays no role, beginning, middle or end. Before us is a theory of sanctification focused upon timeless ontology, and set wholly apart from, conceived as entirely other than, sanctification provoked by the advent of an event, whether in history, supernature or nature." See Jacob Neusner, *History of the Mishnaic Law of Appointed Times,* vol. 5 (Leiden: Brill, 1981–83), p. xv.

48. Ellen Robbins, "Time Telling in Ritual and Myth," *The Journal of Jewish Thought and Philosophy* 6 (1997), p. 73.

49. Sacha Stern, "Fictitious Calendars: Early Rabbinic Notions of Time, Astronomy, and Reality," *The Jewish Quarterly Review* 88 (1996), p. 104.

50. For a succinct overview of theories of time in Greek philosophy, see Richard Sorabji, *Time, Creation and the Continuum* (Ithaca, N.Y.: Cornell University Press, 1983).

51. Heraclitus, Fragm. 10.77(30) in *Philosophy Before Socrates,* ed. Richard D. McKirahan Jr. (Indianapolis: Hackett, 1994), p. 124. He insists that the world was not created: "The cosmos, the same for all, none of the gods nor of humans has made, but it was always and is and shall be: an ever-living fire being kindled in measures and being extinguished in measures."

52. For a discussion of the import of Parmenides' language, and different understandings of his expression *esti,* see McKirahan, *Philosophy Before Socrates,* p. 160. I follow McKirahan's translation of *esti* as *"it"* or *"it is."* In his poem fragment "The Way of Truth" he characterizes "What is" or "*it*" (*esti*) as ungenerated, imperishable, whole and continuous, complete, timeless, motionless, and one. Some of the phrases used to describe *esti* include—"whole and of a single kind"; "nor was *it* ever nor will *it* be, since *it* is now, all together, one continuous"; "nor is *it* divided, since *it* is all alike"; "*it* is all continuous."

53. G. E. R. Lloyd, "Views on Time in Greek Thought," in *Cultures and Time,* ed. L. Gardet et al (Paris: The Unesco Press, 1976), p. 138.

54. Plato, *Timaeus* 29d.

55. Ibid., 31a–b.

56. Ibid., 37c–d.

57. Ibid., 38a.

58. Ibid.

59. For a detailed discussion of the ramifications of this issue, see Sorabji, *Time, Creation and the Continuum,* p. 108ff.

60. Plato, *Timaeus* 38b. Although in general I follow Cornford's translation in *Plato's Cosmology,* I have replaced his term "ever-existent" with the term "everlasting" for the Greek word *aidios.*

61. See for example: "Concerning the whole Heaven or World . . . it has come to be" (28b); "at the same time that he ordered the Heaven, he made . . . Time" (37d); "For there were no days and nights, months and years, before the Heaven came into being; but he planned that they should now come to be at the same time that the Heaven was framed" (37e); "In virtue, then, of this plan and intent of the god for the birth of Time, in order that Time might be brought into being" (38c); "Now so far, up to the birth of Time" (39e).

62. This latter interpretation takes into account Plato's introduction of a third ingredient in the process of becoming, which he terms the receptacle, the "nurse of all Becoming." Various examples are given to explain this third entity, which Plato designates as a "this." Plato suggests that when we try to speak, for example, of fire or water, these things "slip away and do not wait to be described as 'that' or 'this' or by any phrase that exhibits them as having permanent being." The receptacle is the only element that can be called "this" for its nature is permanent and unchanging, having no character or qualities of its own. The receptacle is then identified with

space (*chora*) in the sense in which it always is (*aei on*). But whether *chora* pre-exists time is unclear in the *Timaeus*. Plato argues that being, *chora*, and becoming all existed "even before the Heaven came into being." Using the famous example of a winnowing basket to describe the disordered motion out of which the elements were separated, Plato states that the Demiurge "began by giving them a distinct configuration by means of shape and numbers." And so on this reading it appears that chaos does in fact have existence prior to the creative act of the Demiurge. That this suggestion is not contradictory to the original creation account is evident if we understand the Demiurge to be creating order out of the disordered chaos represented by *chora;* as Plato himself points out, the Demiurge creates the world not out of nothing, but out of a pre-existent stuff: "When He took over all that was visible, seeing that it was not in a state of rest but in a state of discordant and disorderly motion, He brought it into order out of disorder."

63. See for example Plato, *Timaeus* 39e3, 38e3–5, 34b–c, 53a, 69b.

64. Sorabji, *Time, Creation and the Continuum,* p. 272ff.

65. See Shmuel Sambursky, "Place and Space in Late Neoplatonism," *Studies in the History of Philosophy of Science* 8 (1977), pp. 173–87.

66. See Plotinus, *Enneads* 2.5.3 for this important introduction.

67. See G. J. Whitrow, *Time in History* (Oxford: Oxford University Press, 1988), chapter 1 for further details.

68. His argument can be summarized as follows: An object y can have the capacity simultaneously for both x and $-x$. While actually possessing x, y has the capacity for $-x$. But y cannot actualize both x and $-x$ simultaneously. Let y be something that exists forever. In this case y does not have the capacity for $-x$; for if y actualized this power, it would be in the simultaneous state of x and $-x$. Therefore, "the same thing which has the power of always being cannot also have the power of always not being." Aristotle, *De Caelo,* 281b33.

69. Aristotle, *De Caelo* 283a30.

70. Aristotle, *Physics* 8.1.25lb ff.

71. Aristotle, *Metaphysics* 12.6.1071b ff.

72. See Sorabji, *Time, Creation and the Continuum,* p. 210ff. for further discussion of this argument in Aristotle.

73. See Aristotle, *De Caelo* 1.9.279a15 ff.

74. Aristotle, *De Caelo* 279a20.

75. Many commentators have pointed out the tensions inhering in these chapters. Annas, for example, writes that (pp. 112–3) "it is not altogether simple to combine Aristotle's two ways of looking at time: that deriving from the arguments of *Physics Z* and relating time as a continuum to the other continua of motion and magnitude, and that deriving from the ideas in *Metaphysics* I and finding expression in the thesis that time is a number." Julia Annas, "Aristotle, Number and Time," *The Philosophical Quarterly* 99 (1975), pp. 97–113. In this chapter I shall not be concerned with issues of the continuum, but will return to them and to their implications with respect to time, in chapter 3 below.

76. Aristotle, *Physics* 4.10.217b32.

77. Aristotle, *Physics* 4.11.218b22–30.

78. In chapter 3 we shall discuss Crescas' interpolation of this example, which is based on Arabic examples. Abu-l-Barakāt, for example, has an example taken from the Quran of sleepers in a cave who, unconscious, are unable to perceive the flow of time. See Shlomo Pines, *Studies in Abu'l Barakāt al-Baghdādi: Physics and Metaphysics* (Jerusalem: The Magnes Press, 1979), p.112.

79. Cf. David Bostock, "Aristotle's Account of Time," *Phronesis* 25 (1980), pp. 148–69.

80. Aristotle, *Physics* 4.11 219a 4–9. How and whether Aristotle moves from the epistemological to ontological dimension has been the subject of commentators; cf. Sorabji, *Time, Creation and the Continuum,* p. 75.

81. Aristotle, *Physics* 4.11.219b1–2. Annas points to the anti-Platonist thrust of this definition.

82. Aristotle, *Physics* 4.11.219b5–8.

83. See Sorabji, *Time, Creation and the Continuum,* pp. 84–89, for further discussion.

84. Aristotle, *Physics* 4.12.220b15; *Physics* 4.12.220b15.

85. *Physics* 4.12.221b8.

86. *Physics* 4.14.223a21–28.

87. See Annas, "Aristotle, Time and Number," p. 101, for further discussion of this point.

88. For further discussion of the extent of commentaries on this issue, see Sorabji, *Time, Creation and the Continuum,* pp. 93–97; Wolfson, *Crescas' Critique,* pp. 661–2.

89. Quoted in Grant, *Planets, Stars, and Orbs,* pp. 36–7.

90. Edward Grant, "Cosmology," in *Science in the Middle Ages,* ed. David Lindberg (Chicago: University of Chicago Press, 1978), p. 266.

91. Aristotle, *De Caelo* 2.7–12.

92. Aristotle, *De Caelo* 1.2–4; 2.4. For a discussion of this doctrine of *aether,* see G. E. R. Lloyd, *Early Greek Science: Thales to Aristotle* (New York: W.W. Norton, 1970), pp. 109–11.

93. See Aristotle, *Metaphysics* 7.7 ; *Physics* 8.6.

94. See Grant, "Cosmology," p. 285.

95. See Maimonides, *Guide* 2.6.

96. Olaf Pederson, "Astronomy," in *Science in the Middle Ages,* ed. David Lindberg (Chicago and London: University of Chicago Press, 1978), p. 321.

97. For a brief introduction to these works, cf. G. E. R. Lloyd, *Greek Science after Aristotle* (New York: W.W. Norton, 1973), pp. 53–74; Pierre Duhem, *Le Système du Monde* (Paris, Hermann); M. Clagett, *Greek Science in Antiquity* (New York: Abelard-Schuman, 1955), pp. 83–99.

98. See Ptolemy, *Almagest* 13.2. For a discussion of the importance of Ptolemy's description of theory, see P. Duhem's classic work *To Save the Phenomena:*

An Essay on the Idea of Physical Theory from Plato to Galileo (Chicago: University of Chicago Press, 1969).

99. See Pederson, "Astronomy," who suggests that Ptolemy's material reached Western Europe through the Arabic works of Thabît ibn Qûrra.

100. Edward Grant, "Eccentrics and Epicycles in Medieval Cosmology," in *Mathematics and Its Applications to Science and Natural Philosophy in the Middle Ages,* ed. Edward Grant and John E. Murdoch (Cambridge: Cambridge University Press, 1987), p. 195.

101. Ibid.

102. See Grant, "Cosmology," p. 281. Grant goes on to describe a third alternative as well, namely one in which additional orbs are introduced according to which the variation in planetary distances was incorporated into a system of concentric planetary spheres. In this way both Aristotelian and Ptolemaic systems are salvaged.

103. R. B. Harris, "Preface" in *Neoplatonism and Jewish Thought,* ed. L.E. Goodman (Albany: State University of New York Press, 1992), p. xi.

104. Underlying the longer version of the Theology of Aristotle is an additional pseudographical work discovered by S. M. Stern he calls *Ibn Ḥasday's Neoplatonist.* In an important article Stern has argued that the independent treatise *Ibn Ḥasday's Neoplatonist* was incorporated into the long version of the *Theology of Aristotle,* that it strongly influenced Isaac Israeli's philosophy, and that it was preserved almost in its entirety in a Hebrew translation incorporated into Ibn Ḥasday's work *Ben ha-Melech veha-Nazir* (*The Prince and the Ascetic*). For further discussion of the relation between these two versions, see the discussions by Sh. Pines, "La Longue Recension de la Théologie d'Aristote dans ses Rapports avec la Doctrine Ismaelienne," *Revue Études Islamiques,* 1954. Altmann and Stern argue along with Pines that the shorter version is the original and the longer is the result of editorial refashioning of the text. Cf. A. Altmann and S. Stern, *Isaac Israeli: A Neoplatonic Philosopher of the Early Tenth Century* (Oxford: Clarendon Press, 1958). Detailed discussion of recent editions and translations of the *Theology of Aristotle* can be found in R. C. Taylor "A Critical Analysis of the Structure of the *Kalām fī māhd al-Khaīr* (*Liber de Causis*)," in *Neoplatonism and Islamic Thought,* ed. Parviz Morewedge, *Studies in Neoplatonism: Ancient and Modern 5* (Albany: State University of Albany Press, 1992), p. 26, note 5. Paul Fenton has recently discovered that Shem Tov Ibn Falaquera translated quotations directly from the original "vulgate" Arabic version of the *Theology,* into his own work *Sefer ha-Ma'alot,* making Ibn Falaquera the only medieval Jewish author to have done so. See Paul Fenton, "Shem Tov Ibn Falaquera and the Theology of Aristotle," *Da'at* 29 (1992), pp. 27 - 39. For a detailed study of the importance and transmission of the *Theology of Aristotle,* cf. Paul Fenton, "The Arabic and Hebrew Versions of the Theology of Aristotle," in *Pseudo-Aristotle in the Middle Ages: The Theology of Aristotle and Other Texts,* ed. Jill Kraye, (London: Warburg Institute, 1986), pp. 241–64.

105. For the extensive history of this work, see R. C. Taylor, "A Critical Analysis," p. 11ff. In his article "Gleanings from Moseh Ibn Ezra's *Maqalat al-Hadiqa,*"

Sefarad 36 (1976), pp. 285–98, P. Fenton traces the many Neoplatonic influences upon Jewish philosophers.

106. For a critical examination of this work, see David Kaufmann's important article "The Pseudo-Empedocles as a Source of Salomon ibn Gabirol," in his *Meḥqarim be-Sifrut ha-'Ivrit shel Yemei ha-Binayim,* ed. David Kaufmann (Jerusalem: Mossad Harav Kook, 1962), pp. 78–165.

107. Lloyd Gerson, *Plotinus* (London & New York: Routledge, 1994), p. 116.

108. Plotinus, *Enneads* 3.7.11, p. 263.

109. Ibid., 3.7.2 p. 254.

110. Ibid., 3.7.2.

111. Ibid., 3.7.3, p. 255.

112. Ibid., 3.7.3, p. 255.

113. Ibid., 3.7.6, p. 257.

114. Gerson, *Plotinus,* p. 121. For an extended analysis of Plotinus' critique, see Gordon H. Clark, "The Theory of Time in Plotinus," *The Philosophical Review* 53 (1944), pp. 337–358.

115. See Lloyd, "Views on Time in Greek Thought," p. 143.

116. Plotinus, *Enneads* 3.7.11 p. 262.

117. Ibid.

118. In his article "Matter and Time in Plotinus," in *Dionysius* 9 (1985), pp. 53–74, John Simons suggests that this passage reinforces his reading that time is the principle of nonbeing and hence of evil for the disembodied soul. My interpretation obviously differs from this reading.

119. Plotinus, *Enneads* 3.7.12, p. 236.

120. Augustine, *Confessions* 11.26.

121. Plato, *Timaeus* 37d.

122. For a discussion of the issues and texts involved in this debate, see Sorabji, *Time, Creation and the Continuum,* pp. 272–75.

123. Plato, *Timaeus,* 38c.

124. For a discussion of additional similarities between Genesis and Plato's *Timaeus,* see Samuelson, *Judaism and the Doctrine of Creation,* pp. 194–97.

Chapter 2. Time, Creation, and Cosmology

1. Cf. André Neher, "The View of Time and History in Jewish Culture," in *Cultures and Time,* ed. Louis Gardet et al. (Paris: The Unesco Press, 1976), p. 50: "The primordial element is 'time' itself. Creation was manifested in the appearance of time. This time is entirely new. That is the significance of the verb *bara'*."

2. Aristotle's discussion of the eternity of the universe is contained in several places, most notably *De Caelo* 1, *Physics* 8.1, and *Metaphysics* 12.6. For a recent discussion of these and other relevant passages, see R. Sorabji, *Time, Creation and the Continuum* (Ithaca, N.Y.: Cornell University Press, 1983), p. 276ff.

3. See, for example, Idit Dobbs-Weinstein's discussion of Maimonides' theory of creation in *Maimonides and St. Thomas on the Limits of Reason* (Albany: State of New York Press, 1995), p. 61ff.

4. Note the following passages in the introduction to the *Guide:* "For my purpose is that the truths be glimpsed and then again be concealed, so as not to oppose that divine purpose which one cannot possibly oppose and which has concealed from the vulgar among the people those truths especially requisite for his apprehension." Also, "God, may he be exalted, knows that I have never ceased to be exceedingly apprehensive about setting down those things that I wish to set down in this treatise. For they are concealed things; none of them has been set down in any book." Maimonides further describes seven sorts of contradictions commonly found in philosophical works and suggests that two of these (5 and 7) may be used specifically to conceal potentially controversial or even heretical doctrines from the masses. He then states that any contradictions found in the *Guide* itself are intentional and are of type 5 or 7. Finally, Maimonides characterizes the doctrine of creation as a potentially volatile, and certainly an extremely challenging, topic. For a recent summary and analysis of scholarly controversy surrounding the esoteric reading of the *Guide,* see Herbert Davidson, "Maimonides on Metaphysical Knowledge," *Maimonidean Studies* 3 (1992–3), pp. 49–103. For a detailed and rich discussion of the development of this esoteric understanding of Maimonides, starting in the thirteenth century with his translator Samuel ibn Tibbon, and continuing down to the twentieth century with Shlomo Pines and Leo Strauss and others, see chapters 7 and 8 in Aviezer Ravitsky's recent work *History and Faith: Studies in Jewish Philosophy* (Amsterdam: J. C. Gieben, 1996)

5. Maimonides, *Mishneh Torah: The Book of Knowledge,* trans. Moses Hyamson (Jerusalem, 1962), 37a.

6. Ibid.

7. Maimonides, *Mishneh Torah* 37a-37b.

8. Maimonides, *Mishneh Torah* 37b.

9. Ibid.

10. Maimonides, *Mishneh Torah* 38a.

11. Maimonides, *Guide* 1.72, p. 184.

12. Ibid.

13. Ibid.

14. Ibid., p. 185.

15. Maimonides, *Guide* 2.19, p. 308. See also the comment in 2.24 to the effect that in Aristotle's time "mathematics had not been brought to perfection."

16. Maimonides, *Guide* 2.24, p. 322.

17. Ibid.

18. Ibid., pp. 322–33.

19. See Joel L. Kraemer, "Maimonides on Aristotle and Scientific Method," in *Moses Maimonides and His Time,* ed. Eric L. Ormsby (Washington, D.C.: Catholic University of America Press, 1989), pp. 53–88. Langermann points out that it is not

at all clear whether Maimonides himself subscribes to all these arguments. See Y. Tzvi Langermann, "The True Perplexity: *The Guide of the Perplexed,* Part II, Chapter 24," in *Perspectives on Maimonides,* ed. Joel L. Kraemer (Oxford: Oxford University Press, 1991). Maimonides does, however, lay claim to an original point in *Guide* 2.24, p. 323.

20. Maimonides, *Guide* 2.24, pp. 326–27.

21. Surely that is one of the bases of Langermann's point in his article "The True Perplexity." The intriguing question, of course, is whether Maimonides thought that all the issues in the *Guide of the Perplexed* could ultimately be traced back to this basic perplexity. As Kraemer points out, p. 80, "perplexity is not a permanent condition."

22. The original text is found in *Guide* 2.24, pp. 326–27.

23. Ibid. Compare this characterization with *Guide* 2.11: "Now the master of astronomy does not mind this [that there has been no demonstration whether the sun has an eccentric sphere or an epicycle], for the object of that science is to suppose as a hypothesis an arrangement that renders it possible for the motion of the star to be uniform and circular." On the basis of this and similar passages, Goldstein has concluded that "Maimonides has compartmentalized physics and astronomy, such that the astronomer is seemingly free to base his calculations on any mathematical model he invents that can produce agreement with the observations." See Bernard R. Goldstein, *The Astronomy of Levi ben Gerson (1288–1344): A Critical Edition of Chapters 1–20* (New York: Springer Verlag, 1985), p. 6.

24. This sentence is ambiguous in that it is not clear whether this is Maimonides' own perspective, or his understanding of what Aristotle thought.

25. Menachem Kellner, "On the Status of the Astronomy and Physics in Maimonides' *Mishneh Torah* and *Guide of the Perplexed:* A Chapter in the History of Science," *British Journal for the History of Science* 24 (1991), p. 454; and also idem, "Maimonides on the Science of the *Mishneh Torah:* Provisional or Permanent?" *American Jewish Studies Review* (1993), pp. 169–94.

26. Kellner, "Status," p. 461.

27. Langermann, "True Perplexity," p. 161.

28. Ibid., p. 165.

29. On Langermann's reading, Maimonides is alluding to our inability to understand the nature of the fifth element of which the heavens are formed, and not to the actual physical configuration of the heavens as a whole.

30. My student Sarah Pessin has distinguished twelve such modes of knowing. These distinctions, as well as her subsequent discussion, upon which I draw heavily in my own interpretation, can be found in her unpublished paper "Maimonides' View on Astrology: Inconclusive Evidence, Unconvincing Arguments."

31. Maimonides, *Guide* 2.31, p. 360.

32. Representative interpretations can be found in Sara-Klein Braslavy, "The Creation of the World and Maimonides' Interpretation of Gen. I–V," in *Maimonides and Philosophy,* ed. S. Pines and Y. Yovel (Dordrecht: Martinus Nijhoff, 1986), pp. 65–

78; Herbert Davidson, "Maimonides' Secret Position on Creation," *Studies in Medieval Jewish History and Literature*, ed. I. Twersky (Cambridge, Mass.: Harvard University Press, 1979), pp. 16–40; Warren Zev Harvey, "A Third Approach to Maimonides' Cosmogony-Prophetology Puzzle," *Harvard Theological Review* 74 (1981), pp. 287–301; Arthur Hyman, "Maimonides on Creation and Emanation," in *Studies in Medieval Philosophy*, ed. J. F. Wippel (Washington, D.C.: Catholic University of America Press, 1987), pp. 45–61; Alfred L. Ivry, "Maimonides on Possibility," in *Mystics, Philosophers, and Politicians: Essays in Jewish Intellectual History in Honor of Alexander Altmann*, ed. J. Reinhartz et al. (Durham, N.C.: Duke University Press, 1982), pp. 67–84; Lawrence Kaplan, "Maimonides on the Miraculous Element in Prophecy," *Harvard Theological Review* 70 (1977), pp. 233–56.

33. It might be fruitful to compare these attempts with the work of Ian Barbour, who similarly distinguishes in the modern period four ways of relating religion and science: conflict, independence, accommodation, and conversation. See Ian Barbour, *Religion and Science: Historical and Contemporary Issues* (San Francisco: Harper Books, 1997).

34. For a penetrating critique of the personalities and texts involved in the ongoing interpretation of Maimonides' esotericism, see Ravitsky, "History and Faith," p. 247ff.

35. Maimonides, *Guide* 2.13, p. 281.

36. Ibid., 2.13, p. 283.

37. Ibid., 2.13, p. 284.

38. See Harvey, "A Third Approach," p. 289, note 9, for further discussion of this point. See also the extensive discussion by H. A. Wolfson, "The Meaning of *Ex Nihilo* in the Church Fathers," pp. 207–21.

39. See the discussion of these four propositions in Hyman, "Maimonides on Creation," p. 49ff.

40. See, for example, the recent discussion by Roslyn Weiss, "Maimonides on the End of the World," *Maimonidean Studies* 3 (1992–93), pp. 195–218.

41. Maimonides, *Guide* 2.13, p. 284.

42. Ibid., 2.13, p. 285.

43. Ibid., 2.13, p. 285.

44. Ibid., 2.25, p. 328.

45. Ibid., 2.25, p. 328.

46. Ibid., 2.25, p. 329.

47. See Ravitsky, *History and Faith*, p. 248ff.

48. Sara Klein-Braslavy makes her case for a skeptical interpretation of Maimonides in her article "Interpretation of Maimonides of the Term 'Create' and the Question of the Creation of the Universe" (Heb.), *Da'at* 16 (1986), pp. 39–55. For additional discussions of epistemological and metaphysical skepticism in Maimonides, see the essay by S. Pines, "The Limits of Human Knowledge according to Al-Fārābī, Ibn Bajja and Maimonides," in *Studies in Medieval Jewish History and Literature*, vol. 1, ed. I. Twersky (Cambridge, Mass.: Harvard University Press, 1979), pp. 82–

109; see also the unpublished work by Josef Stern, "Remarks on a Skeptical Theme in Maimonides' *Guide of the Perplexed*." B. Kogan summarizes some of the other interpretations regarding Maimonides' own view of creation in his essay "Creation and Time in Maimonides and Gersonides: A Response to Tamar Rudavsky," note 1, in *God and Creation: An Ecumenical Symposium*, ed. David Burrell and Bernard McGinn (Indianapolis: University of Notre Dame Press, 1990), pp. 147–53. A response to Pines can be found in Herbert Davidson, "Maimonides on Metaphysical Knowledge," *Maimonidean Studies* 3 (1992–93), pp. 49–103.

49. See the discussion of this point in Joel Kraemer, "Maimonides on Aristotle and Scientific Method," in *Moses Maimonides and His Time*, ed. Eric Ormsby (Washington, D.C.: The Catholic University Press of America, 1989), pp. 53–88.

50. Maimonides, *Guide* 2.15, p. 291.

51. Ibid., 2 Intro, p. 241.

52. Ibid., 2 Intro, p. 240.

53. Ibid., 2.15, p. 290.

54. Ibid., 2.15, p. 293. For a careful study of the importance of demonstration to Maimonides' discussion of creation, see Kraemer, "Maimonides on Aristotle," pp. 64–76.

55. For a similar interpretation of Maimonides' position on creation, but one drawn from different evidence, see Alfred Ivry's recent article "Maimonides on Creation," in *Creation and the End of Days*, ed. David Novak and Norbert Samuelson (Lanham, Md.: University Press of America, 1986), pp. 185–214. See, for example, Ivry's comment on p. 198: "He [Maimonides] has, accordingly been viewed as either a closet Aristotelian on creation, or a more revealing Platonist. The view which I have adduced . . . puts him in neither camp completely, though brings him closer to that of the Platonists."

56. Hyman, "Maimonides on Creation," p. 52.

57. Maimonides, *Guide* 2, Intro prop. 15, p. 237. This definition is reflected in two other contexts as well. In *Guide* 1.52, Maimonides defines time as (*Guide* 1.52) "an accident attached to motion when the motion of priority and posteriority is considered in the latter and when motion becomes numbered." And in a letter to Ibn Tibbon he defines time as "the measure of motion according to prior and posterior in motion."

58. Sara Klein-Braslavy, *Maimonides' Interpretation of the Story of Creation* (Heb.) (Jerusalem: Israel Society for Biblical Research, 1987), pp. 230–31.

59. Maimonides, *Guide* 2.17, p. 296.

60. See Ivry, "Maimonides on Possibility," p. 69ff.; Jonathan Malino, *"Aristotle on Eternity: Does Maimonides Have a Reply?"* in *Maimonides and Philosophy*, ed. S. Pines and Y. Yovel (Dordrecht: Martinus Nijhoff, 1986), pp. 52–64.

61. Maimonides, *Guide* 2.18, p. 299.

62. Ibid., 2.18, p. 300.

63. See Sorabji, *Time, Creation and the Continuum*, p. 269ff.

64. Maimonides, *Guide* 2.18, pp. 300–301.

65. Ibid., 2.13, p. 281.

66. Ibid.

67. For the importance of this passage for subsequent Jewish philosophers, see Warren Zev Harvey's discussion in "Albo's Discussion of Time," *Jewish Quarterly Review* (1981), p. 220ff.; see also Jonathan Malino's discussion in his unpublished rabbinic dissertation "Maimonides' Guide to the Perplexities of Creation" (Hebrew Union College, 1979), pp. 61–62.

68. Maimonides, *Guide* 2.13, p. 282.

69. Ibid.

70. The questions raised by Levi ben Gerson (Gersonides, 1288–1344) are contained in his major work *Milḥamot ha-Shem.* Reference to book 5.2–3 and book 6 will be made primarily to the Hebrew edition, which was reprinted in Leipzig in 1866 (*Milḥamot*). References will be to treatise, chapter, and page number. Unless otherwise specified, all translations from the Hebrew are my own. In addition, the following recent English translations of portions of Gersonides' works will be used (*Wars*): S. Feldman, trans. and ed., *The Wars of the Lord (Books 1–4)* (Philadelphia: Jewish Publication Society of America, 1984). For an extensive bibliography of scholarly works on Gersonides, see the recent bibliography compiled by Menachem Kellner, *Bibliographia Gersonideana: An Annotated List of Writings by and about R. Levi ben Gershom,* in *Studies on Gersonides,* ed. Gad Freudenthal (Leiden: E. J. Brill 1992), pp. 368–410. For an analysis of the chronology of Gersonides' work, see Ruth Glasner, "The Early Stages of Gersonides' *The Wars of the Lord,*" *Jewish Quarterly Review* 87 (1996), pp. 1–46.

71. Gersonides, *Milḥamot* 5.1.1, p. 23 (text found in Bernard R. Goldstein, *The Astronomical Tables of Levi ben Gerson* (New Haven, Conn.: Connecticut Academy of Arts and Sciences, 1974), henceforth referred to as "Goldstein, *Astronomy.*"

72. See Gad Freudenthal, "Épistemologie, astronomie et astrologie chez Gersonide," *Révue des études juives* 146 (1987), p. 360, for a discussion of the implications of this realism.

73. This passage is quoted in Goldstein, *Astronomy,* p. 24. Goldstein claims that Gersonides is almost unique among medieval philosophers for his effort to base a new system of astronomy on his own observations rather than on those of his predecessors. The point is made even more succinctly by Glasner, "The Early Stages," p. 9.

74. Gersonides, *Milḥamot* 5.1.3, p. 27 (in Goldstein, *Astronomy*).

75. See examples of these inaccuracies in *Milḥamot* 5.1.14, p. 93ff. (in Goldstein, *Astronomy*).

76. Gersonides, *Milḥamot* 5.1.46 in Bernard R. Goldstein, "A New Set of Fourteenth Century Planetary Observations," *Proceedings of the American Philosophical Society* 132 (1988), p. 386.

77. Norbert Samuelson, "Elements and Matter in Gersonides' Cosmogony," in *Gérsonide en Son Temps,* ed. Gilbert Dahan (Louvain and Paris: E. Peeters, 1991), p. 213.

78. Gersonides, *Milḥamot* 5.1.1, 23 (in Goldstein, *Astronomy*).

79. See Gersonides, *Milḥamot* 5.1.2, p. 24 (in Goldstein, *Astronomy*). Langermann emphasizes this point in his appendix "Gersonides on Astrology," in Gersonides, *The Wars of the Lord (Books V–VI)* ed. and trans. Seymour Feldman, (Philadelphia: Jewish Publication Society of America, forthcoming).

80. In *De Caelo* 1.10 Aristotle associates this view with Empedocles and Heraclitus. It is also found in *Genesis Rabbah* 3.7 and 9.2.

81. Gersonides, *Milḥamot* 6.1.2, p. 294.

82. For an examination of the underlying logical moves implicit in Gersonides' attack, see Seymour Feldman, "Gersonides' Proofs for the Creation of the Universe," *Proceedings of the American Academy for Jewish Research* (1967), pp. 113–37.

83. This reconstruction of Gersonides' understanding of Aristotle's argument is based on the text in *Milḥamot* 6.1.3, p. 298.

84. A similar argument is propounded in *Physics* 4.12. In *Physics* 4.12 Aristotle demonstrates that inasmuch as time is the measure of motion, those things that are subject to generation and corruption are necessarily in time.

85. Gersonides, *Milḥamot* 6.1.3, pp. 298–99.

86. Gersonides, *Milḥamot* 6.1.3, p. 298.

87. See the arguments listed in *Milḥamot* 6.1.3, pp. 298–302.

88. Gersonides, *Milḥamot* 6.1.3, p. 302.

89. The various characteristics of time are elaborated in Gersonides, *Milḥamot* 6.1.10, p. 329ff.

90. For a discussion of this point, see Jacob Staub, *The Creation of the World according to Gersonides* (Chico, Calif.: Scholars Press, 1982), p. 30.

91. Gersonides, *Milḥamot* 6.1.10, pp. 329–30. Aristotle's arguments can be found in *Physics* 4.10–14.

92. Gersonides, *Milḥamot* 6.1.10, pp. 330–31.

93. Gersonides, *Milḥamot* 6.1.10, p. 331.

94. Gersonides' notion of the contingency of the future is elaborated in *Milḥamot* 3, in the context of the issue of divine omniscience. For further discussion of this issue, see T. M. Rudavsky, "Divine Omniscience and Future Contingents in Gersonides," *Journal of the History of Philosophy* 21 (1983), pp. 513–36; Norbert Samuelson, "Gersonides' Account of God's Knowledge of Particulars," *Journal of the History of Philosophy* 10 (1972), pp. 399–416.

95. These arguments are contained in Gersonides, *Milḥamot* 6.1.11, pp. 340–41. For further elucidation of these arguments, see Feldman, "Gersonides' Proofs," pp. 130–31.

96. See, for example, Gersonides, *Milḥamot* 6.1.11, p. 342: "It follows that the number of one infinite is neither larger nor smaller than the number of another infinite; for in order for the one to be larger than the other, it would necessarily have to bc finite." The basis for these arguments has been analyzed by Herbert Davidson, "John Philoponus as a Source of Medieval Islamic and Jewish Proofs for Creation," *Journal of the American Oriental Society* 89 (1969), pp. 357–91.

97. Gersonides, *Milḥamot* 6.1.11, p. 343.

98. This distinction is made in Aristotle, *Categories* 4. See my discussion of this distinction in chapter 1.

99. Gersonides, *Milḥamot* 6.1.11, p. 344.

100. Ibid., 6.1.11, p. 345.

101. Ibid., 6.1.21, pp. 385–86. See Feldman, "Gersonides' Proofs," pp. 134–35, for a discussion of these pages.

102. Gersonides, *Milḥamot* 6.1.21, p. 386. For a description of the Kalām notion of admissibility, see Maimonides, *Guide* 3.15. See the following works for a critical analysis of Maimonides' exposition: Z. Blumberg "Ha-Rambam al Musag al-Tajwiz be-Shittatam shel ha-Mutakallimūn," *Tarbiz* 39 (1970), pp. 268–76; Ivry, "Maimonides on Possibility," pp. 77ff.; and H. A. Wolfson, *The Philosophy of the Kalam* (Cambridge, Mass.: Harvard University Press, 1976), p. 43ff.

103. Gersonides, *Milḥamot* 6.1.21, pp. 389–90.

104. *Milḥamot* 6.1.17, pp. 267–71. For general discussions of Gersonides' theory of creation and matter, see Seymour Feldman, "Gersonides' Proofs"; Idem, "Platonic Themes in Gersonides' Cosmology," in *Salo Whitmayer Baron Jubilee Volume* (Jerusalem, 1975), pp. 383–405.

105. *Milḥamot* 6.1.17, pp. 367–68; 374. For further elaboration of these arguments, see Feldman, "Platonic Themes," pp. 394–95.

106. *Milḥamot* 6.1.18, p. 372.

107. Ibid., 6.1.17, p. 367.

108. Ibid., cf. also ibid., 6.2.7, p. 425: "there was body which does not preserve its shape, from which the upper and lower [waters] came to be."

109. See Ruth Glasner, "The Early Stages in the Evolution of Gersonides' *The Wars of the Lord*," *The Jewish Quarterly Review* 87 (1996), pp. 1–46. Glasner argues, further, that Gersonides' theological and astronomical projects are interrelated: "Gersonides' theology and astronomy are deeply involved with each other through the hypothesis of the body that does not preserve its shape. The existence of this body before the creation of the world is a basic premise of Gersonides' theology; its existence between the celestial spheres is a basic premise of his cosmology and essential for the justification of his eccentric astronomy. Thus, the introduction of the concept of the body that does not preserve its shape was a breakthrough for both projects." Ibid., p. 41.

110. Gersonides, *Milḥamot* 5.2.5, p. 200.

111. Ibid., 5.2.4, p. 197.

112. Goldstein remarks that Gersonides may have had in mind discussions in *Ptolemy's Planetary Hypotheses Book II*, or more certainly Maimonides' remarks in *Guide* 2.24, which intimate the existence of bodies between the spheres to account for the motion of one sphere not affecting that of another. See Bernard R. Goldstein, "Levi ben Gerson's Theory of Planetary Distances," *Centaurus* 29 (1986), p. 277.

113. Gersonides, *Milḥamot* 5.2.2, p. 193.

114. Ibid., 5.1. p. 130, in Goldstein, "Theories of Planetary Motion," p. 288. See also *Milḥamot* 6.2.7 "That is, the body which does not preserve its shape which is in the spheres so that all the motions do not reach each other."

115. Gersonides, *Milḥamot* 5.2.2, p. 194.

116. Detailed discussion of this computational exercise are provided in Goldstein, "Theory of Planetary Distance," pp. 272–313.

117. Gersonides, *Milḥamot* 6.1.21, p. 387. Wolfson points out that this distinction can be traced back to Aristotle's *Physics* 4.11.2I9a22–30: "But we apprehend time only when we have marked motion, marking it by 'before' and 'after'; and it is only when we have perceived 'before' and 'after' in motion that we say that time has elapsed. Now we mark them by judging that A and B are different, and that some third thing is intermediate to them. When we think of the extremes as different from the middle and the mind pronounces that the 'nows' are two, one before and one after, it is then that we say that there is time, and this that we say is time. For what is bounded by the 'now' is thought to be time—we may assume this." See H. A. Wolfson, *Crescas' Critique of Aristotle* (Cambridge, Mass.: Harvard University Press, 1929), p. 653.

118. Gersonides, *Milḥamot* 6.1.21, p. 388.

119. Gersonides, *Milḥamot* VI.1.21, pp. 387–88. This diagram is based on that of C. Touati's discussion in his *Pensée de Gersonide,* p. 236.

120. This discussion occurs in *'Or 'Adonai [The Light of the Lord*], 1.2.11 and 1.1.15 in Wolfson, *Crescas' Critique,* pp. 282–91. Recent discussions of Crescas' theory of time and its relation to Aristotle can be found in the following works: Warren Zev Harvey, "Albo's Discussion of Time," *Jewish Quarterly Review* 71 (1980), pp. 210–38; Warren Zev Harvey, "The Term 'Hitdabbekut' in Crescas' Definition of Time, "*Jewish Quarterly Review* 71 (1981), pp. 44–47; Eliezer Schweid, "*Dibre Mabo,*" in *Maqor* ed. of Hasdai Crescas, *Light of the Lord* (Jerusalem, 1972); and Wolfson, *Crescas' Critique of Aristotle.*

121. Crescas, *'Or 'Adonai,* 1.1.15 (in Wolfson, *Crescas' Critique,* p. 283)

122. Crescas, *'Or 'Adonai,* 1.1.15 (in Wolfson, *Crescas' Critique,* p. 285).

123. The text for this argument is found in Crescas, *'Or 'Adonai,* I.1.15 (in Wolfson, *Crescas' Critique,* p. 285).

124. For Aristotle's version of this argument, see *Physics* 4.10:217b32–218a3.

125. Crescas, *'Or 'Adonai,* 1.2.15 (in Wolfson, *Crescas' Critique,* p. 287).

126. Crescas, *'Or 'Adonai,* 1.2.15 (in Wolfson, *Crescas' Critique,* pp. 287–89).

127. Crescas, *'Or 'Adonai,* 1.2.15 (in Wolfson, *Crescas' Critique,* p. 289). For a discussion of the term *hitdabequt* and whether it means duration or continuity, see Harvey, "The Term *Hitdabbekut,*" p. 47. In this chapter I follow Harvey's suggestion to translate the term *hitdabequt* as "continuity."

128. Wolfson emphasizes the importance of the term *hitdabequt* in Crescas' definition of time and traces it back to Neoplatonic notions of duration or extension, expressed by the term *diastema.* See Wolfson, *Crescas' Critique,* p. 97ff., 655ff., for

discussion of this point. Pines emphasizes the similarities among Saadia Gaon, Abraham bar Hiyyah, Crescas, and Abu-l Barakāt and warns against giving a reading of Crescas that overemphasizes the Plotinian to the exclusion of Peripatetic influences. See Shlomo Pines, "De Quelques Doctrines Relatives au temps chez les Philosophes Arabes Juifs," in *Studies in Abu'l Barakāt al-Baghdādī: Physics and Metaphysics* (Jerusalem: The Magnes Press, 1979), pp. 63–83.

129. Wolfson, *Crescas' Critique,* p. 289.

130. Ibid.

131. Peter Aureol's discussion can be found in his *Commentariorum in secundum librum Sententiarum Pars Secundus,* dist. II, quaest. I, art. 1, quoted in Pierre Duhem, *Le Système du Monde* (Paris: Hermann), p. 300ff.

132. For a sustained discussion of William Ockham's theory of time, see Marilyn McCord Adams, *William Ockham* (Notre Dame, Ind.: University of Notre Dame Press, 1987), p. 853ff; Pierre Duhem, *Le Système du Monde,* p. 305ff. In his influential article "Scholasticism and the Teachings of Hasdai Crescas and His Predecessors," *Proceedings of the Israel Academy of Sciences and Humanities* 1.10, p. 15, Shlomo Pines notes as well the marked similarity between Crescas' theory of time and that of Aureol, Peter Olivi, and William Ockham, but he does not explore this point.

133. Crescas, *'Or 'Adonai,* 1.2.15 (in Wolfson, *Crescas' Critique,* p. 291).

134. Ibid.

135. Ibid.

136. Crescas, *'Or 'Adonai,* 3.1.5, p. 69a.

137. Crescas' argument is contained in *'Or 'Adonai* 3.1.5, p. 69a. For a critical discussion of these arguments, see Feldman, "Eternal Creation," p. 304ff.

138. Crescas, *'Or 'Adonai* 3.1.1, pp. 61b–62a.

139. Ibid., p. 62b. For further discussion of this version of the argument, see Herbert A. Davidson, *Proofs for Eternity: Creation and the Existence of God in Medieval Islamic and Jewish Philosophy* (New York: Oxford University Press, 1987), p. 24.

140. Crescas, *'Or 'Adonai* 3.1.5:70a.

141. Ibid., 3.1.4, 66a–68b. Commentators have tried to make sense of Crescas' apparently contradictory theory. For further discussion of this theory of creation, see Seymour Feldman, "The Theory of Eternal Creation in Hasdai Crecas and Some of His Predecessors," *Viator* 11 (1980), pp. 289–320; Schweid, *"Dibrei Mabo,"* p. 44.

142. Feldman, "Eternal Creation," p. 317.

143. Schweid, *"Dibrei Mabo,"* p. 44.

144. Albo, *Sefer ha-'Ikkarim,* 1. p. 130.

145. Ibid., 2.18 pp. 108–9.

146. For a discussion of Albo's sources, see Harvey, "Albo's Discussion of Time," p. 213. In what follows, I am very much indebted to Harvey's article.

147. Albo, *Sefer ha-'Ikkarim,* 2.18, p. 109.

148. Ibid., 2.18, p. 110.

149. Ibid., 3.27, p. 259.

150. Ibid., 2.18, p. 110–11.

151. Ibid., 2.18, p. 111.

152. Ibid.

153. Ibid., 2.18 p. 111–12.

154. Jacob Neusner, *Genesis Rabbah: The Judaic Commentary to the Book of Genesis,* Vol. 1 (Chico, Calif.: Scholars Press, 1985), p. 33

155. Sara Klein-Braslavy, "The Existence of Time and the First Days of Creation in Medieval Jewish Philosophy" (Heb.) *Tarbiz* 45 (1975), pp. 107–27.

156. Ibid., p. 115.

157. For further elaboration of this point, see Harvey, "A Third Approach," p. 296; Klein-Braslavy, "Maimonides' Interpretation," pp. 81–82, 86–87.

158. See Harvey, "Maimonides on Creation," p. 296. In his *Commentary on the Guide* 2.30 Narboni comments upon Maimonides' distinction between two meanings of "first": For Narboni the term *hatḥala* connotes the idea of efficient cause. Second, Narboni looks at the passage where Maimonides says that to take the term *rei'shit* in the sense of initial instant introduces eternity of the world, since by definition an instant separates anterior from posterior. Thus one cannot take the expression *be-rei'shit* in this temporal sense since the instant would have been preceded by another instant. According to Narboni, Maimonides conceded to the vulgar creation *ex nihilo* whereas in reality he knew that time, which is the number of movement, is eternal. In other words, when Maimonides writes that creation of beings accompanied that of time, he knew that the world had been created in time since he established a connection between time and created beings. See Maurice Ruben Hayoun, *La Philosophie et la théologie de Moise de Narbonne* (Tübingen: J. C. B. Mohr), p. 150.

159. Maimonides, *Guide* 2.30, p. 349.

160. Ibid.

161. Ibid., 2.26, p. 331.

162. Ibid.

163. See Klein-Braslavy, "Maimonides' Interpretation," pp. 235–38, for further discussion of this point.

164. Maimonides, *Guide* 2.30, p. 349.

165. Ibid.

166. Ibid.

167. See Maimonides, *Guide,* 2.30, p. 350: "Accordingly everything was created simultaneously; then gradually all things became differentiated."

168. Gersonides, *Milḥamot* 6.2.2, p. 420. Gersonides returns to this distinction in his *Torah Commentary.* These different ways of understanding 'beginning' reflect Aristotle's discussion in *Categories* 2.1a–4a 25ff.

169. Gersonides, *Milḥamot* 6.2.8, p. 427. See also Gersonides, *Milḥamot* 6.1.18, p. 374: "Prior to the creation of the world there was no time at all." This point is reiterated in his *Torah Commentary,* p. 21.

170. Gersonides, *Milḥamot* 6.2.8, p. 427. Again, the point is repeated in his *Torah Commentary,* p. 21.

171. Gersonides, *Milḥamot* 6.2.8, p. 427. Gersonides gives as examples the

movers of the heavenly bodies that are prior in cause to the heavenly bodies, and the heavenly bodies that are prior in cause to the nature of the elements.

172. Gersonides, *Milḥamot* 6.2.8, p. 428.

173. See Wolfson, *Crescas' Critique,* p. 633.

174. Ibid., p. 291.

175. Albo, *Sefer ha-'Ikkarim* 2.18, p. 113.

176. Ibid., 2. 18, p. 112.

177. Harvey, "Albo's Discussion," p. 223.

178. Arama, *'Aqedat Yitzḥaq* 3.24b.

179. See Sara Heller Wilensky, *R. Yitzhak Arama and His Philosophy* (Heb.) (Jerusalem: Bialik Institute, 1956), p. 118

180. Klein-Braslavy, "The Existence of Time," notes the influence of Averroës upon Arama.

181. Klein-Braslavy, "The Existence of Time," p. 125

182. Arama, *'Aqedat Yitzḥaq* 3.24b–25a.

183. Ibid., p. 24b

184. See ibid., p. 24b. for this example.

185. Klein-Braslavy, "The Existence of Time," p. 123; See Arama's discussion in *'Aqedat Yitzḥaq* 3.24a and 3.26a.

186. Arama, *'Aqedat Yitzḥaq* 3.25a. See Heller-Wilensky, *R. Yitzhak Arama,* p. 119.

Chapter 3. Time, Motion, and the Instant

1. For a survey of the vast literature associated with Zeno's paradoxes, see Jonathan Barnes, *The Presocratic Philosophers* (London: Routledge and Kegan Paul, 1982); Norman Kretzmann, *Infinity and Continuity in Ancient and Medieval Thought* (Ithaca, N.Y.: Cornell University Press, 1982); Wesley Salmon, *Zeno's Paradoxes* (New York: Bobbs-Merrill, 1970); and Richard Sorabji, *Time, Creation and the Continuum* (Ithaca, N.Y.: Cornell University Press, 1983), p. 32ff.

2. For a discussion of this point, as well as an elaboration of the theory of infinity, see J.M.M.H. Thijssen, "The Response to Thomas Aquinas in the Early Fourteenth Century: Eternity and Infinity in the Works of Henry of Harclay, Thomas of Wilton and William of Alnwick O.F.M.," in *The Eternity of the World In the Thought of Thomas Aquinas and His Contemporaries,* ed. J.B.M. Wissink (Leiden: E. J. Brill, 1990). See also Robin Small, "Cantor and the Scholastics," *American Catholic Philosophical Quarterly* 66 (1992), p. 407–28, for an examination of Cantor's own application of his theory of the infinite to traditional philosophical and theological problems.

3. The major source for Zeno's paradoxes is Aristotle, most notably in *Physics* 6.9.

4. Barnes, *The Presocratic Philosophers,* p. 262.

5. F. Miller, "Aristotle Against the Arabs," in Kretzmann, *Infinity and Continuity*, p. 90.

6. For further elaboration of this last point, see Shmuel Sambursky, *The Physical World of Late Antiquity* (London: Routledge and Kegan Paul, 1962), p. 1ff.

7. See Plato's description of Zeno's intentions in his dialogue *Parmenides* 127b–128d: "It is actually a defense of Parmenides' argument against those who make fun of it. . . . Now my treatise opposes the advocates of plurality." Barnes argues convincingly that it is not Zeno's intention to support monism, but rather to show that monism and pluralism lead equally to absurdities.

8. Barnes, *The Presocratic Philosophers*, p. 264. Barnes points out that this question raises the much broader issue of the geometry of space. He argues (p. 245) that in fact this assusmption is unwarranted, and that the moral to be drawn from Zeno's argument precisely is that space and time are not continuous.

9. Aristotle, *De Generatione et Corruptione* 1.2.315b26.

10. D. Furley, "The Greek Commentators' Treatment of Aristotle's Theory of the Continuum," in Kretzmann, *Infinity and Continuity*, p. 34.

11. Aristotle, *De Generatione et Corruptione* 1.2.316a10–317a12. See Sorabji, *Time, Creation and the Continuum*, p. 211.

12. Aristotle, *De Generatione et Corruptione* 1.2.316b15.

13. Ibid., 1.2.317a8–10.

14. Aristotle, *Physics* 5.3. 227a11–14. A similar definition is given in *Categories* 4b25–6.

15. Aristotle, *Physics* 6.1. 231b14–18.

16. "For there are two senses in which length and time and generally anything continuous are called infinite: they are called so either in respect of divisibility or in respect of their extremities. So while a thing in a finite time cannot come in contact with things quantitatively infinite, it can come in contact with things infinite in respect of divisibility; for in this sense the time itself is also infinite." Aristotle, *Physics* 6.2.333a25–30.

17. Sorabji, *Time, Creation and the Continuum*, p. 322.

18. Sorabji critiques this move on the part of Aristotle by arguing that his whole analysis of infinity as merely an extendible finitude is meant to avoid his very conclusion in *Physics* 8.8 that there are infinite collections, or more specifically, that the number of potentially existing divisions is implied to be "more than finite ." What he fears most of all is that "a more than finite number . . . will imply something he considers absurd, that an extended line is composed of nothing but unextended points." Sorabji, *Time, Creation and the Continuum*, p. 323. Sorabji points to Aristotle's *De Generatione et Corruptione* 1.2.316a10–317a12, where this argument is made.

19. Aristotle, *Physics* 8.8.263a18–23.

20. Aristotle, *Physics* 3.6.206a16–18ff: "magnitude is not actually infinite. But by division it is infinite . . . the alternative then remains that the infinite has a potential existence."

21. Aristotle, *Physics* 8.8.263b2–8. For a discussion of these passages, see Furley, "The Greek Commentators' Treatment," pp. 32–33.

22. For an application of infinity to the eternity of time, see Aristotle, *Physics* 4.4–7.

23. Aristotle, *Physics* 3.6.206a27–9.

24. See Furley's discussion of this point in "The Greek Commentators' Treatment," pp. 33–4.

25. Aristotle, *Physics* 3.6.206b33–207a2. see Sorabji, *Time, Creation and the Continuum*, p. 210.

26. Aristotle, *Physics* 3.5.204b7–10. See Sorabji, *Time, Creation and the Continuum*, p. 213.

27. "I believe [1.b] to be false: of the many arguments designed to support [1.b], all are wanting in one or more particulars. But I cannot show that [1.b] is false; indeed, the reason why Zeno's Dichotomy is so fascinating an argument is to be sought in [1.b]: men want to believe [1.b]; they cannot believe that we possess infinite powers; and they keep producing ever more ingenious arguments in favor of Zeno. For all that, until a new batch of arguments comes forward I shall continue to reject Zeno's conclusion by rejecting [1.b]," in Barnes, *The Presocratic Philosophers*, p. 273.

28. For a detailed discussion of the history and transmission of these texts, see Herbert Davidson, *Proofs for Eternity: Creation and the Existence of God in Medieval Islamic and Jewish Philosophy* (Oxford: Oxford University Press, 1987), p. 86ff.; Sorabji, *Time, Creation and the Continuum*, p. 197ff. Some of Philoponus' relevant texts can be found in *Philoponus: Against Aristotle on the Eternity of the World*, trans. Christian Wildberg (Ithaca, N.Y.: Cornell University Press, 1987).

29. For a history of this transmission see Davidson, *Proofs for Eternity*, pp. 86–116. Davidson notes that at least thirteen medieval discussions draw upon Philoponus' position that infinity cannot be traversed.

30. See texts in Sorabji, *Time, Creation and the Continuum*, pp. 214–15.

31. Wildberg, *Philoponus*, p. 144.

32. Ibid., p. 145. See Davidson, *Proofs for Eternity*, p. 88. Philoponus expands this argument, claiming that "if on the one hand the ascent (*anodos*) took place *ad infinitum*, complete things would not precede the incomplete, and the actual not the potential; but if on the other hand ⟨the motions⟩ are limited ⟨in number⟩, then the first ⟨motion⟩ which evidently exists together with the universe, has made a beginning which starts from something actual and complete for the subsequent motions." Wildberg, *Philoponus*, p. 145.

33. Ibid., p. 146.

34. Some scholars have taken the inability to provide demonstrative argument to lie at the heart of Maimonides' discussion of creation in *The Guide of the Perplexed*. See below.

35. Sorabji, *Time, Creation and the Continuum*, p. 177.

36. See Aristotle, *Physics* 3.6.206a10.

37. For a discussion of different types of atomism and their implications, see

Sorabji, *Time, Creation and the Continuum,* p. 350ff.; H. A. Wolfson, *The Philosophy of the Kalam* (Cambridge, Mass.: Harvard University Press, 1976), pp. 472–86.

38. A. Altmann, "Isaac Israeli," *Encyclopedia Judaica,* dates this from 875–904.

39. Ibid. For the impact of Isaac Israeli and the Kairouan School upon medieval science, see Charles Singer, "Science and Judaism," in *The Jews: Their Role in Civilization,* ed. Louis Finkelstein (New York: Schocken, 1978).

40. The entire treatise exists in Hebrew and Latin translations; only a portion survives in the original Arabic. It opens with an account of Aristotle's four types of inquiry (whether, which, what, why) and an elaboration of al-Kindi's definitions of philosophy.

41. Discovered by A. Borisov and edited by S. M. Stern, this work seems to have been written in Arabic characters, though the extant manuscripts are in Hebrew script. See A. Altmann and S. M. Stern, *Isaac Israeli: A Neoplatonic Philosopher of the Early Tenth Century* (Oxford: Clarendon Press, 1958), p. 80.

42. Attributed to Israeli by Altmann and Stern, this text is a commentary on a work by Aristotle. The *explicit* says that the aim of the text is to explain the words of the philosopher by way of arguments and proofs. See Altmann and Stern, *Isaac Israeli,* p. 118.

43. Ibid., p. 119.

44. As in Plotinus, emanation does not imply change; the source remains unaffected. Plotinus used this fact to explain how multiplicity arises from unity; however, because Israeli introduces a notion of creation, he cannot do the same. Instead he tries to harmonize the two motifs. Hence Israeli ignores Plotinus' important distinction of the two moments in emanation: the pure uninformed moment, and the turning back to the source in contemplation.

45. Altmann and Stern, *Isaac Israeli,* p. 159.

46. Other Neoplatonic sources also equate the sphere with nature. There is no warrant for this in Plotinus—he only identifies nature with the vegetative soul. Following his pseudo-Aristotelian source, Israeli transfers Aristotle's divisions of the individual soul (rational, animal, vegetative) to the universal soul, giving us three hypostases of soul, to which he adds as a final quasispiritual substance the "sphere" or heaven, representing the Plotinian hypostasis of nature. Altmann and Stern cite many examples of how this scheme can be seen in his writings; it also occurs in *Ibn Ḥasdai's Neoplatonist* and the *Long Theology.*

47. See Israeli, *Sefer Yesodot* 2.3–8, p. 40: "After we have reached this point in our treatise and have explained all that the philosopher has said concerning the essence of the elements and its limits . . . it is now incumbent that we mention in the second part what Galen says and his beliefs on this; and this is that Galen's view was similar to the philosopher and said that the element is the minimum part of a thing."

48. Israeli, *Sefer Yesodot* 2.3–5, p. 43.

49. Israeli, *Sefer Yesodot* 2.5–10, p. 43.

50. For details on the history of the transmission of these texts from Democritus, see Wolfson, *The Philosophy of the Kalam,* p. 478ff.; H. A. Wolfson, *Reper-*

cussions of the Kalam in Jewish Philosophy (Cambridge, Mass.: Harvard University Press, 1979), pp. 162–65.

51. Israeli, *Sefer Yesodot* 2.12–17, p. 43. The final point of this argument is that we cannot speak of indivisible magnitudes.

52. This argument can be found in Aristotle, *Physics* 6.1.231b2–5; *De Generatione et Corruptione* 1.2.316b14–18; *Metaphysics* 13.8.1083b13–16.

53. This text has had a checkered history. The original work was written in Arabic, and has come down to us in a Latin translation of the twelfth century made by John of Spain, in collaboration with Dominicus Gundissalinus. Hebrew extracts were compiled in the thirteenth century by Shem Tov ben Josef ibn Falaqera, and then subsequently translated into Latin under the author's name of "Avicebrol" or "Avicebron." Latin Scholastics reading the *Fons Vitae,* as it had become known by the thirteenth century, had no idea that this work was written by a Spanish Jew. In 1857, a French scholar named S. Munk edited and translated the Hebrew extracts once again. It was while comparing the various editions that Munk noted that the appellations "Avicebron," "Avencebrol," "Avicebrol" in fact referred to the great Jewish poet Salomon Ibn Gabirol. He did this by comparing passages in the Hebrew translation by Falaqera with certain other quotations in Albert the Great. Before Munk's discovery, it had been assumed that Ibn Gabirol and Avicebron were separate writers. Perhaps one reason for the obscurity of the text lies in its form. As Pines has pointed out, *Meqor Ḥayyim* is unique among Jewish medieval works in that it contains virtually no references to any other Jewish texts, ideas, or sources: it is wholly lacking in Jewish content. Inasmuch as nothing in the work belies the Jewish predilections of its author, later readers had no reason to suspect that the author was in fact Ibn Gabirol, a noted Jewish poet.

54. Ibn Gabirol, *Meqor Ḥayyim,* 5.4.

55. Ibid., 4.2.

56. Ibid., 5.32.

57. Ibid., 5.42.

58. See Ibid., 5.42, and 5.36–38 for examples of these two depictions. The status of divine will is suitably ambiguous. The will is clearly one of the hypostases inserted between God and universal form and matter. But whether will is identical with wisdom is not clear. In *Meqor Ḥayyim* will and wisdom are construed as identical. But as we shall see in *Keter Malkhut* a distinction is made between the two. As Hyman has pointed out, Ibn Gabirol speaks of creation by will in several places, suggesting that creation took place through a volitional act. Yet he also uses models for creation like water flowing from a fountain, reflection of light in a mirror, and the issuing of human speech, suggesting that creation should be understood as necessary emanation. In both cases, however, it is clear that will is posited as an intermediary between God and Intellect in order to explain how multiplicity can come to be from a unitary being. Ultimately, the ultimate metaphysical principle in Gabirol's system is not intellect, but will; this voluntarism becomes an important motif in later Kabbalistic thought.

59. Ibn Gabirol, *Meqor Ḥayyim* 2.16. Schlanger notes that Ibn Gabirol is posing the problem of atomism here, and hopes to prove that there is quantity only where there is substance. He also notes that this entire discussion of atomism in chapters 16–19 was omitted from the Falaqera edition presumably because he was not interested in this discussion. See Salomon Ibn Gabirol, *Livre de la Source de Vie (Fons Vitae)*, trans. and ed. Jacques Schlanger (Paris: Aubier Montaigne, 1970), p. 86, for further discussion of these passages.

60. Ibn Gabirol, *Meqor Ḥayyim* 2.16.

61. Ibid., 2.18.

62. Ibid.

63. Ibid.

64. The text for this argument is found in ibid., 2.18.

65. Ibid.

66. Ibid., 2.19.

67. See the discussion in Israel Efros, *The Problem of Space in Jewish Medieval Philosophy* (New York: Columbia University Press, 1917), pp. 44, 53.

68. Ibn Gabirol, *Meqor Ḥayyim* 4.6.

69. Ibid., 4.20; See also 5.28. In answer to how it is that forms are unities, and duality convenes to matter, the Master offers a number of explanations, among which we find the following: "The second property of matter is multiplicity and divisibility; this because the form divides and multiplies because of the matter; and matter possesses this property because of a second one which is half of two." Ibid., 4.11.

70. Ibid., 3.57.

71. Ibid.

72. Ibid., 4.19.

73. Wolfson, *Philosophy of the Kalam,* p. 468.

74. In other chapters Maimonides deals with the problem of God's attributes (*Guide* 1.50–60) and predestination (*Guide* 1.73 and 2.17).

75. Wolfson discusses the question of the transmission of Kalam thought into Jewish philosophy fairly extensively in his work *Philosophy of the Kalam,* p. 82ff.; see also L.E. Goodman, "Time in Islam," *Asian Philosophy* 2 (1992), pp. 3–19; Wolfson, *Repercussions of the Kalam.*

76. These statements are paraphrased from the text in Maimonides, *Guide* 1.73, p. 195.

77. Maimonides, *Guide* 1.73, pp. 195–96.

78. See Maimonides, *Guide* 2.73, p. 196.

79. Maimonides, *Guide* 1.73, p. 212.

80. Ibid.

81. Ibid.

82. The epistemological grounds for rejecting infinity are developed in premises ten and twelve of Maimonides' account. See *Guide* 1.73, pp. 206–12, 213–14.

83. C.1.a–C.1.e are quoted directly from Maimonides, *Guide* 1.73, p. 197.

84. Maimonides, *Guide* 1.74.2, p. 216.

85. Ibid., 1.74.4, p. 217.

86. For an analysis of other influential Kalām arguments, see Davidson, *Proofs for Eternity,* pp. 117–53.

87. In this paper, unless otherwise noted, Saadia's text will based on the English translation of Alexander Altmann, *The Book of Doctrines and Beliefs,* in *Three Jewish Philosophers* (New York: Atheneum Press, 1972). This edition contains excellent footnotes. A complete English translation of the Arabic text *Amānāt wāl-i'tiqādāt* can be found in *The Book of Beliefs and Opinions,* trans. Samuel Rosenblatt (New Haven, Conn.: Yale University Press, 1948).

88. The actual text of the argument is presented in Saadia Gaon, *The Book of Doctrines and Beliefs,* p. 56: "I know that time is threefold: past, present, and future. Although the present is shorter than any instant, I take the instant as one takes a point, and say: If a man should try in his thought to ascend from that point in time to the uppermost points, it would be impossible for him to do so, inasmuch as time is now assumed to be infinite and it is impossible for thought to penetrate to the furthest point of that which is infinite. . . . The same reason will also make it impossible that the process of generation should traverse an infinite period down to the lowest point so as ultimately to reach us. Yet if the process of generation did not reach us, we would not be generated, from which it necessarily follows that we, the multitude of generated beings, would not be generated and the beings now existent would not be existent. And since I find myself existent, I know that the process of generation has traversed time until it has reached us, and that if time were not finite, the process of generation would not have traversed it."

89. Saadia Gaon, *The Book of Doctrines and Beliefs,* p. 57. "It has come to my notice that a certain heretic in conversation with one of the Believers in the Unity of God, objected to this proof. He said, 'It is possible for a man to traverse by walking that which has an infinite number of parts. For if we consider any distance which a man walks, be it a mile, or an ell, we should find that it can be divided into an infinite number of parts.' "

90. Maimonides describes al-Nazzām's theory of the leap in greater detail in *Guide* 1.73 prop 3. See also the discussions in Goodman, "Time in Islam," p. 8–9; Sorabji, *Time, Creation and the Continuum,* p. 385ff.

91. Sorabji, *Time, Creation and the Continuum,* p. 388.

92. Cf. Aristotle, *Physics* 3.6.206a16–18ff.; 263b2–8.

93. Davidson has pointed to the similarity between Saadia's formulation and al-Kindī's formulation of the same argument. In response to the challenge posed by assuming (3.4), Al-Kindī tries to show that it is not possible for "time to have infinity in actuality, either in the past or future." In Al-Kindī's argument, time is construed as a continuum capable of both infinite addition and division. Assuming that the infinite cannot be traversed, however, Al-Kindī simply rejects the possibility of (3.4) altogether on the grounds that "since there is a definite time (the present) all time must be finite." In this way he rejects the Aristotelian distinction between actual and

potential infinity as applied to time. See Alfred Ivry, *Al-Kindi's Metaphysics* (Albany: State University of York Press, 1974), p. 74: "Before every temporal segment there is (another) segment, until we reach a temporal segment before which there is no segment, i.e. a segmented duration before which there is no segmented duration. It cannot be otherwise—if it were possible, and after every segment of time there was a segment, infinitely, then we would never reach a given time—for the duration from past infinity to this given time would be equal to the duration from this given time regressing in times to infinity; and if (the duration) from infinity to a definite time was known, then (the duration) from this known time to temporal infinity would be known, and then the infinite is finite, and this is an impossible contradiction." Al-Kindī's argument then continues as follows: "Furthermore, if a definite time cannot be reached until a time before it is reached, nor that before it until a time before it is reached, and so to infinity; and the infinite can neither be traversed nor brought to an end; then the temporally infinite can never be traversed so as to reach a definite time. However its termination at a definite time exists, and time is not an infinite segment, but rather is finite necessarily, and therefore the duration of body is not infinite, and it is not possible for body to be without duration. Thus the being of a body does not have infinity; the being of a body is, rather, finite, and it is impossible for body to be eternal."

94. Saadia Gaon, *The Book of Doctrines and Beliefs*, p. 57. Altmann notes that the term imagination (*wahm; maḥshavah*) is sometimes used by the Arabic philosophers in the sense of "potentially."

95. Saadia Gaon, *The Book of Doctrines and Beliefs*, p. 57.

96. "This acceptance of the ontological legitimacy of potential existence enables Aristotle to consider the universe as eternal, though in actuality we perceive only finite time, movements and magnitude," in Ivry, *Al-Kindi's Metaphysics*, p. 151. See Aristotle, *Physics* 3.6.206a16ff.; 7.207b2.

97. In contradistinction, Maimonides would view Saadia's and al-Kindī's refutations of infinity as spurious on the grounds that they are too restrictive, viz. potential infinity. See further discussion of this point in Ivry, *Al-Kindi's Metaphysics*, p. 153.

98. Davidson, *Proofs for Eternity*, p. 96.

99. Ibid., p. 97.

100. Ibid., p. 97; See also Ivry, *Al-Kindi's Metaphysics*.

101. Saadia Gaon, *The Book of Doctrines and Beliefs*, p. 62.

102. Davidson, *Proofs for Eternity*, p. 98.

103. Halevi, *Kuzari* 5.18, pp. 275–76.

104. Ibid., 5.18, p. 276.

105. Davidson notes that with the exception of Ghazāli and Saadia, most philosophers conflated the second and third proofs, as does Halevi. Cf. Davidson, *Proofs for Eternity*, p. 122.

106. Ibn Daud, *'Emunah Ramah* 1.4, p. 16. Note Gersonides' refutation of this argument on the grounds that it makes no sense for line CE to correspond to line AB

in such a way that E corresponds with C. Further that one cannot use the terms *equal, part, whole* of an infinite, since these terms refer to the finite.

107. See the discussion in T. A. M. Fontaine, *In Defense of Judaism: Abraham Ibn Daud* (Assen: Van Gorcum Press, 1990), pp. 34–40.

108. See Fontaine, *In Defense of Judaism,* p. 42.

109. Maimonides, *Guide* 1.73.3, p. 196.

110. Ibid., 1.73.3 p. 198.

111. Maimonides, *Guide* 1.74.7, p. 222. See Wolfson's discussion of these arguments in *Philosophy of the Kalam,* p. 427ff.

112. Ibid., 1.74.7, p. 222.

113. Ibid., 1.74.4, p. 217.

114. Ibid., 1.71, p. 182.

115. Gersonides, *Milḥamot* 3.4, p. 143. Cf. *Wars,* pp. 123–24.

116. Gersonides, *Milḥamot* 3.4, p. 143. Cf. *Wars,* p. 123.

117. Gersonides, *Milḥamot* 3.4, p. 144; *Wars,* p. 124.

118. Gersonides, *Milḥamot* 3.4, p. 144. Cf. *Wars,* p. 125. Gersonides' summary of Aristotle's argument can be restated as follows: Assume that a body is divisible completely. It is logically possible that this body be divided (that is, that the division be completed). If this complete division is assumed, then body must be divided into either divisible or indivisible parts. If these parts are divisible, body would not be divided into every possible part (since the results of the division would be further divisible). But this is impossible in light of (5.2). Hence it follows that body must be divided into what is indivisible. Hence indivisible parts (atoms) must exist. Aristotle's own argument is stated in *Gen Corr.* 316a.15–317a.2. See Feldman's discussion on p. 125 of *Wars.*

119. For Averroës' discussion of the divisibility of body, see Averroës, *On Aristotle's De Generatione et Corruptione: Middle Commentary and Epitome,* ed. and trans. S. Kurland (Cambridge. Mass: Medieval Academy of America, 1958), p. 12ff. Gersonides summarizes Averroës' argument as follows: A body can be divided at all its points simultaneously only if all the points meet each other. But according to *Physics* 6, one point cannot be immediately next to another. So when we divide body at point A, it is not possible for division to occur at A_1 (when A_1 is contiguous to A). Before the division occurred at A_1 division was possible at both A and A_1. But when the division occurred at A division at A_1 was precluded. See Gersonides, *Milḥamot* 3.4, p. 145; *Wars,* pp. 125–26.

120. Gersonides, *Milḥamot* 3.4, p. 145; *Wars,* p. 126.

121. Ibid.

122. Gersonides, *Milḥamot* 3.4, p. 145; *Wars,* pp. 127–28.

123. Feldman points out that this twofold distinction between potentiality *in facto esse,* which at some time or other will actually be realized, and potentiality *in fieri esse,* which can never be realized completely, can be found in the Latin scholastics as well. This distinction is crucial for Gersonides' argument. See Feldman, *Wars,* p. 128, note 28.

124. See Gersonides, *Milḥamot* 6.1.11, p. 345 for an elaboration of the continuous quality of matter. For a discussion of Gersonides' theory of matter, see Touati, *Pensée de Gersonide,* pp. 243ff.; T. M. Rudavsky, "Individuals and the Doctrine of Individuation in Gersonides," *The New Scholasticism* 51 (1982), pp. 30–50.

125. Gersonides, *Milḥamot* 6.1.11, p. 334. See Touati, *La Pensée de Gersonide,* p. 224, for further discussion of this point.

126. Gersonides, *Milḥamot* 6.1.11, p. 345. For a slightly different interpretation of this statement, see Efros, *The Problem of Space,* p. 101.

127. Gersonides, *Milḥamot* 3.4, p. 145: "It could be said that it is possible for there to be a number incapable of further increase and that this number is known by God. For a number is augmentable *ad infinitum,* and if we assume that God knows completely all these increments or that the potential [increments] in them have [all] been actualized, there would then be a number that is no longer augmentable." Aristotle makes a similar comparison between the divisibility and augmentability of number in *Physics* 2.6.206b4ff.

128. Gersonides, *Milḥamot* 3.4, p. 146; *Wars,* p. 128.

129. For a discussion of Averroës' argument, see C. Touati, trans., *Les Guerres du Seigneur, Livres 111 et IV* (Paris, 1968), p. 93.

130. Gersonides, *Milḥamot* 3.4, p. 146; *Wars,* pp. 128–29.

131. Gersonides, *Milḥamot* 3.4, p. 146; *Wars,* p. 129.

132. Ibid.

133. Ibid.

134. Ibid.

135. Ibid.

136. Gersonides, *Milḥamot* 6.1.11, p. 334.

137. As Feldman points out (*Wars,* pp. 130–31, note 34), this notion of indefinitely divisible infinite is picked up by Descartes in his *Principles of Philosophy* 1:26–27. We shall return to these passages in Descartes in chapter 5.

138. Aristotle, *Physics* 4.1 208a31.

139. Aristotle, *Physics* 4.4 212a5–7.

140. See Aristotle, *De Caelo* 1.9.278b35–279a18: "There is neither place, nor void, nor time, outside the heaven;" *De Caelo* 2.10.291b1–5: "It is established that the outermost revolution of the heavens is a simple movement and the swiftest of all."

141. Aristotle, *Physics* 4.5.212b10.

142. Ibid., 4.5.212b11–13.

143. For a study of the reactions to Aristotle, see Max Jammer, *Concepts of Space: The History of Theories of Space in Physics* (Cambridge, Mass.: Harvard University Press, 1969), p. 72ff.; Edward Grant, *Planets, Stars, and Orbs: The Medieval Cosmos, 1200–1687* (Cambridge: Cambridge University Press, 1994), pp. 122–35; Julia Schwartzmann, "Where Is the Heaven? The Story of an Aristotelian Problem and Its Presentation in Jewish Medieval Sources," *Revue des études juives* 63 (1994), pp. 67–85; Wolfson, *Crescas' Critique,* pp. 433–41.

144. Crescas, *'Or 'Adonai* 1.1.2 (in Wolfson, *Crescas' Critique,* p. 195).

145. For a detailed analysis of Crescas' conception of space, see H. A. Wolfson, *Crescas' Critique of Aristotle* (Cambridge, Mass.: Harvard University Press, 1929), pp. 38–69. See also Davidson, *Proofs for Eternity,* p. 253ff.

146. Crescas, *'Or 'Adonai,* 1.1.2 (in Wolfson, *Crescas' Critique,* p. 199).

147. See Wolfson, *Crescas' Critique,* p. 42.

148. See Crescas' statements of these propositions in Wolfson, *Crescas' Critique,* pp. 135, 219, and 221.

149. Crescas, *'Or 'Adonai* 1.1.1 (in Wolfson, *Crescas' Critique,* pp. 143–47).

150. Crescas, *'Or 'Adonai* 1.1.2 (in Wolfson, *Crescas' Critique,* pp. 183–85).

151. Aristotle's argument is summarized by Crescas in Crescas, *'Or 'Adonai* 1.1.1 (in Wolfson, *Crescas' Critique,* p. 147); his critique is on p. 187.

152. Crescas, *'Or 'Adonai* 1.1.2 (in Wolfson, *Crescas' Critique,* p. 179).

153. See Wolfson's discussion of this text on pp. 62–63.

154. Crescas, *'Or 'Adonai* 1.1.1 (in Wolfson, *Crescas' Critique,* p. 149).

155. Crescas, *'Or 'Adonai* 1.1.2 (in Wolfson, *Crescas' Critique,* p. 189).

156. Crescas, *'Or 'Adonai* 3.1.4.67b. See Davidson, *Proofs for Eternity,* pp. 125–27.

157. Crescas, *'Or 'Adonai* 1.1.1 (in Wolfson, *Crescas' Critique,* pp. 171–73). Another version of this argument is expressed as follows: "If an infinite body could have circular motion, it would be possible that any radius moving in a circle would traverse an infinite line from one end to the other, if, e.g. a line drawn perpendicular to the diameter were assumed to be infinite. But that is impossible, for that perpendicular line is assumed to be infinite, and an infinite line cannot be traversed in finite time. Hence an infinite body cannot have circular motion." Crescas, *'Or 'Adonai* 1.1 (in Wolfson, *Crescas' Critique,* p. 175).

158. Crescas, *'Or 'Adonai* 1.2 (in Wolfson, *Crescas' Critique,* p. 205).

159. The text of the argument is found in Crescas, *'Or 'Adonai* 1.2 (in Wolfson, *Crescas' Critique,* p. 211). The argument is presented by Aristotle in *De Caelo* 1.5 272a7–20, and restated by Averroës; cf. Wolfson, *Crescas,* p. 383ff., for the history of the transmission of this argument through Averroës and Altabrizi.

160. Crescas, *'Or 'Adonai* 1.1.2, in Wolfson, *Crescas' Critique,* p. 207.

161. Crescas, *'Or 'Adonai* 1.1.2, in Wolfson, *Crescas' Critique,* p. 213.

162. Crescas, *'Or 'Adonai* 1.3.1, in Wolfson, *Crescas' Critique,* p. 221.

163. See his application of this dictum to intellects and souls in Crescas' *Or 'Adonai,* 1.3.1 (in Wolfson, *Crescas' Critique,* p. 223).

164. Crescas, *'Or'Adonai* 1.3.2 (in Wolfson, *Crescas' Critique,* p. 227). See Davidson, *Proofs for Eternity,* pp. 365–66.

165. Ibid.

166. Max Black's article "Achilles and the Tortoise" is contained in Wesley Salmon, *Zeno's Paradoxes* (New York: Bobbs-Merrill, 1970), p. 67 ff. Black's article has led to wide discussion and controversy among contemporary philosophers. For a recent discussion of these issues, cf. Alba Papa-Grimaldi, "Why Mathematical Solu-

tions of Zeno's Paradoxes Miss the Point: Zeno's One and Many Relation and Parmenides' Prohibition," *Review of Metaphysics* 50 (1996), pp. 299–314.

167. For a discussion of Georg Cantor's groundbreaking theories of infinity and the continuum, cf. Georg Cantor, *Contributions to the Founding of the Theory of Transfinite Numbers,* trans. P. E. B. Jourdain (New York: Dover, 1952); Stephen Korner, *The Philosophy of Mathematics* (London: Hutchinson University Library, 1960). Salmon has a valuable introduction to the main elements of Cantorian views on p. 251ff. For a different reading of Gersonides' views with respect to contemporary accounts, see the discussion by S. Feldman in "Gersonides' Proofs," p. 130, and in "Platonic Themes," p. 404.

Chapter 4. Temporality, Human Freedom, and Divine Omniscience

1. For an introductory survey to the vast primary and secondary literature dealing with issues connected with God's omniscience, the following works should be consulted: L. Baudry, *La Quérelle des Futurs Contingents* (Paris: J. Vrin, 1950); A. Kenny, *The God of the Philosophers* (Oxford: Clarendon Press, 1979); C. Normore, "Future Contingents," in *The Cambridge History of Later Medieval Philosophy,* ed. N. Kretzmann, A. Kenny, and J. Pinborg (Cambridge: Cambridge University Press, 1982); W. Ockham, *Predestination, God's Foreknowledge and Future Contingents,* trans. and ed. M. Adams and N. Kretzmann (New York: Appleton, Century Crofts, 1969); T. M. Rudavsky, ed., *Divine Omniscience and Omnipotence in Medieval Philosophy* (Dordrecht: D. Reidel, 1984); R. Sorabji, *Necessity, Cause and Blame* (Ithaca, N.Y.: Cornell University Press, 1980); L. Stein, *Die Willensfreiheit and ihr Verhältniss zur Göttlichen Präscienz und Providenz Bei den Jüdischen Philosophen des Mittelalters* (Berlin, 1882).

2. Aristotle, *De Interpretatione* 9.8.

3. Ibid., 9.30–32.

4. Ibid., 9.23a.

5. Ibid., 9.26–28.

6. For recent discussions of Aristotle's *De Interpretatione* 9, see Ackrill's discussion and notes in *Aristotle's Categories and De Interpretatione* (Oxford: Clarendon Press, 1964).

7. For a discussion of the relevance of these commentaries upon Jewish philosophers, see Seymour Feldman, "Crescas' Theological Determinism," *Da'at* 9 (1982), pp. 2–28. I shall allude below to Gersonides' commentary upon *De Interpretatione* 9.

8. See C. Normore, "Divine Omniscience, Omnipotence and Future Contingents: An Overview," in *Divine Omniscience and Omnipotence in Medieval Philosophy,* ed. T. M. Rudavsky, p. 4.

9. Normore, "Divine Omniscience, Omnipotence and Future Contingents: An Overview," p. 6.

10. This particular form of fatalist argument is suggested by A. Falk in his paper "The State of the Question about Fate," *Philolosophical Research Archives* (1980), p. 13. Its virtue lies in its identifying elements common to logical, precognitive and causal fatalism. It should be noted that premise 3 is commonly accepted by the majority of medieval philosophers. For other formulations of fatalism, see S. Cahn, *Fate, Logic and Time* (New Haven, Conn.: Yale University Press, 1967).

11. That God's knowledge incorporates not only belief but infallibility as well has been argued by R. Sorabji, *Necessity, Cause and Blame,* p. 112.

12. A similar point is made about seventeenth-century philosophers. Taking determinism to be the view that "whatever happens in the world is brought about by causes other than itself," seventeenth-century philosophers adopted a compatibilist position with respect to human freedom, namely that "being free is logically compatible with being causally determined." See Robert Sleigh Jr., Vere Chappell, and Michael Della Rocca, "Determinism and Human Freedom," in *The Cambridge History of Seventeenth-Century Philosophy,* ed. Daniel Garber and Michael Ayers (Cambridge: Cambridge University Press, 1998), p. 1195.

13. I am very much indebted to Normore's discussion in "Divine Omniscience, Omnipotence and Future Contingents: An Overview," p. 9, for this formulation.

14. Again, compare this account with the analysis found in Sleigh, Chappell, and della Rocca, "Determinism and Freedom," p. 1195, in which a similar distinction is drawn between the two indeterminist camps.

15. Aristotle, *Metaphysics* 12.7.1072b18; cf. also *De Anima* 3.6.431b16, where Aristotle says that the intellect, when it thinks, is the thing it thinks.

16. Maimonides, *Guide* 1.54, p. 125.

17. "[T]he terms 'knowledge,' 'power,' 'will,' and 'life,' as applied to him, may he be exalted, and to all those possessing knowledge, power, will, and life, are purely equivocal, so that their meaning when they are predicated of him is in no way like their meaning in other applications." *Guide* 1.56, p. 131. For recent discussions of Maimonides' doctrine of divine attributes, see Alexander Broadie, "Maimonides on Divine Knowledge," in *Of Scholars, Savants and Their Texts,* ed. R.-L. Salinger (New York: Peter Lang, 1989), pp. 47–55; Kenneth Seeskin, *Maimonides: A Guide for Today's Perplexed* (West Orange, New Jersey: Behrman House, 1991), pp. 19–39; idem, "The Positive Contribution of Negative Theology," in *Jewish Philosophy in a Secular Age* (Albany: State University of New York Press, 1990), pp. 31–69.

18. The secondary literature on Aquinas' theory of analogy is vast. For a comparison of his view with that of Maimonides, see Seymour Feldman, "A Scholastic Misinterpretation of Maimonides' Doctrine of Divine Attributes," *Journal of Jewish Studies* 19 (1968), pp. 23–39; C. L. Miller, "Maimonides and Aquinas on Naming God," *Journal of Jewish Studies* 28 (1977), pp. 65–71; Idit Dobbs-Weinstein, *Maimonides and St. Thomas on the Limits of Reason* (Albany: State University of New York Press, 1995); Avital Wohlman, *Thomas d'Aquin et Maimonide: un dialogue exemplaire* (Paris: Cerf, 1988).

19. Gersonides, *Wars* 3. 3, p. 107.

20. Maimonides, *Guide* 1.58, p. 134. For an exposition of this theory of negative predication, see Maimonides, *Guide,* 1.58–60. The details of this theory, and a comparison with Gersonides, are developed more fully by H. A. Wolfson in his article "Maimonides and Gersonides on Divine Attributes as Amphibolous Terms," in *Mordecai M. Kaplan Jubilee Volume,* English Section ed. Moshe Davis (New York: Jewish Publication Society, 1953), pp. 515–530.

21. Maimonides, *Guide* 1.59, p. 139.

22. Recent attempts to define God's omniscience along such lines can be found in N. Pike's article "Divine Omniscience and Voluntary Actions," *Philosophical Review* 74 (1965), pp. 27–46; cf. also the discussions of A. N. Prior in *Past, Present and Future* (Oxford: Clarendon Press, 1967), pp. 113–36.

23. See Aristotle *De Generatione et Corruptione* 2.10.336b10ff.*; De Generatione Animalia* 4.10.777b24–32. For discussions of natural astrology, see Ptolemy, *Tetrabiblios* 1.2. These issues are discussed by Gad Freudenthal in "Maimonides' Stance on Astrology in Context: Cosmology, Physics, Medicine and Providence," in *Moses Maimonides: Physician, Scientist and Philosopher,* ed. Fred Rosner and Samuel Kottek (Northvale, N.J.: J Aronson, 1993), pp. 78–79.

24. See 2 Kings 17:16; 2 Kings 23:5; Jeremiah 7:18; Jeremiah 8.8; Jeremiah 10:2; Jeremiah 44:17–19; Isaiah 14:12; 47:13; Amos 5:26. In these passages the term *mazalot* is a comprehensive term used to refer to the planets in general, and may extend to the zodiac as well. See Stadelmann, *Hebrew Conception of the World,* pp. 85–86, for a discussion of the etymology of this term. Stadelmann spends considerable time trying to identify the individual planets mentioned in these passages.

25. For an extensive discussion of this issue, see R.C. Kiener, "The Status of Astrology in the Early Kabbalah," *Jerusalem Studies in Jewish Thought* 6 (1987), p. 2; E. E. Urbach, *The Sages—Their Concepts and Beliefs* (Jerusalem: Magnes Press, 1975), p. 277ff.

26. See for example *Genesis Rabbah* 44:12; *Shabbat* 119a . Further discussion of these and other passages can be found in the article "Astrology," *Encyclopedia Judaica,* 2:789.

27. *Shabbat* 53b; *Megillah* 31; *Sanhedrin* 94a: *Nedarim* 39b; *Genesis Rabbah* 10.6.

28. *B. T. Eruvin* 56a.

29. See for example *B. T. Shabbat* 156a: "Not the constellation of the day but that of the hour is the determining influence." Numerous examples are given to connect a person's overall propensities to the position of the celestial bodies at the time of his birth.

30. *B. T. Shabbat* 156a; *B. T. Yebamot* 21b.

31. Kiener, "Status of Astrology," p. 4.

32. Raphael Jospe, "Early Philosophical Commentaries on the *Sefer Yetzirah:* Some Comments," *Révue des études juives* 144 (1990), p. 376.

33. For a succinct study of the *Sefer Yetzirah,* see Gershom Scholem, "Yetzirah, Sefer," in the *Encyclopedia Judaica* 16 (Jerusalem, Keter Publ, 1972), pp. 782–88. A

critical edition can be found in Ithamar Gruenwald, "A Preliminary Critical Edition of *Sefer Yetzirah*," *Israel Oriental Studies* (1971), pp. 132–77.

34. *Sefer Yetzirah* 4.6.

35. *Sefer Yetzirah* 4.11.

36. *Sefer Yetzirah* 5.3.

37. *Sefer Yetzirah* 6.1: "He set them in the Teli, the Cycle, and the Heart."

38. From the mention of the *teli*, Pines suggests that the astronomical views of the group within which the *Sefer Yetzirah* originated may have been influenced by Syriac Christian milieu. Cf. Pines, "Points of Similarity Between the Exposition of the Doctrine of the Sefirot in the *Sefer Yetzirah* and a Text of the Pseudo-Clementine Homilies: The Implications of this Resemblance," *Proceedings of the Israel Academy of Sciences and Humanities* 7 (1989), p. 111.

39. See the following editions of Saadia Gaon's commentary: *Sefer Yetzirah 'im Perush Rabbenu Sa'adia ben Yosef Fayyumi (Ga'on)*, Arabic text edited with Hebrew translation by Yosef Kafih (Jerusalem: M. Atiyeh, 1961/62); *Commentaire sur le Sefer Yesira,* trans. Mayer Lambert (Paris, repr. editions bibliophane 1986).

40. Donnolo's commentary can be found in *Il Commento di Sabbatai Donnolo sul Libro della Creazione,* ed. D. Castelli Firenze 1880 (Heb. Sec). Selected passages and critical discussion can be found in A. Sharf, *The Universe of Shabbetai Donnolo* (New York: Ktav, 1976).

41. *Sefer Yetzirah,* 1.1.

42. Saadia Gaon, *Commentary,* ed. Kafih, pp. 35–36; quoted in Jospe, "Philosophical Commentaries," p. 389.

43. For a history of subsequent commentaries and appropriation of Saadia's conception, see Jospe, "Philosophical Commentaries," p. 390ff.

44. Cf. Jospe, "Commentaries," p. 394, for an elaboration of this identification and its importance for subsequent epistemological and psychological doctrines in Maimonides.

45. Halevi, *Kuzari* 4.25. The apparent uniqueness of Halevi's reading is discussed by Jospe.

46. See Ibn Gabirol, *Meqor Ḥayyim* 2.21.

47. In his article "The Sefirot That Are above the Sefirot," *Tarbiz* (1982) pp. 239–80, Moshe Idel suggests the existence of ten *Sefirot* above the usual set of *Sefirot.* Pines suggests that it is these *Sefirot* that are mentioned by Ibn Gabirol in his poem "*Shokhen 'ad me-'az*"; see Pines, "Points of Similarity," p. 123, for a discussion of this point.

48. Gabirol, "*Shokhen 'ad me-'az,*" in Zangwill, *Selected Religious Poems of Solomon Ibn Gabirol* (Philadelphia: Jewish Publication Society, 1924), p. xxxiii; also in Yarden, *Shirei Ha-Qodesh Le-Rabbi Shelomo ibn Gabirol* (Jerusalem, D. Yarden, 1973), vol. 1, p. 9.

49. See Zangwill, *Selected Religious Poems,* p. xxxiv for discussion of this point.

50. For detailed discussion of Gabirol's sources, see Raphael Lowe, *Ibn Gabirol* (New York: Grove Weidenfeld, 1989), p. 113; also Raphael Lowe, "Ibn Gabirol's

Treatment of Sources," in *Studies in Jewish Religious and Intellectual History Presented to Alexander Altmann,* ed. S. Stein and R. Loewe (1979), pp. 183–94.

51. Ibn Gabirol, *Keter Malkhut,* cento 11.

52. Ibid., cento 16.

53. Ibid., cento 22.

54. See Lowe, "Ibn Gabirol's Treatment of Sources," p. 184.

55. Ibn Gabirol, *Keter Malkhut,* cento 25; See a similar discussion in *Meqor Ḥayyim* 5.1.

56. See Sirat, *History of Jewish Philosophy,* p. 104, for further discussion.

57. For a detailed analysis of Ibn Ezra's astrological tendencies, see Y. Tzvi Langermann, "Some Astrological Themes in the Thought of Abraham Ibn Ezra," in *Rabbi Abraham Ibn Ezra: Studies in the Writings of a Twelfth-Century Jewish Polymath,* ed. I. Twersky and J. M. Harris (Cambridge, Mass.: Harvard University Press, 1993), pp. 28–85.

58. Langermann, "Ibn Ezra," p. 51.

59. Ibn Ezra, *Sefer ha-Mivḥarim,* quoted in Langermann "Ibn Ezra," p. 52.

60. Langermann "Ibn Ezra," p. 68.

61. Halevi, *ha-Kuzari* 1.1.

62. Ibid., 4.9.

63. Ibid., 4.23.

64. Cf. Altmann, "Climatology," for further discussion of this point.

65. For a systematic study of these issues, see T. M. Rudavsky, "Divine Omniscience, Contingency and Prophecy in Gersonides," in *Divine Omniscience and Omnipotence in Medieval Philosophy,* ed. T. M. Rudavsky (Dordrecht: D. Reidel, 1985), pp. 161–84.

66. Gersonides, *Wars* 2.2, p. 33.

67. Ibid.; see also 1.12.

68. Gersonides, *Wars* 2.2, p. 34.

69. Ibid., 2.2, p. 36.

70. Gersonides, *Milḥamot* 5.3.8, p. 272.

71. Gersonides, *Wars* 2.6, p. 53. Gersonides goes on to explain that the information transmitted is of a general nature and does not pertain the individual qua particular. For the theological implications of this point, see T. M. Rudavsky, "Divine Omniscience."

72. The separate intellects which attach to each body are described in more detail in *Milḥamot* 5.3.6–7.

73. Gersonides, *Wars* 2.6, p. 64.

74. Ibid.

75. See Feldman's comments, *Wars,* p. 64, note 33, on the astrological force of the term *yishtatfu,* which connotes the "the joint participation of various heavenly bodies to bring about some event on earth."

76. Gersonides, *Wars* 2.7, pp. 69–70.

77. Ibid., 2.2, p. 34.

78. For further discussion of Gersonides' theory of contingency, see Sarah Klein-Braslavy, "Determinism, Possibility, Choice and Foreknowledge in Ralbag's Works" (Heb.), *Da'at* 22 (1989), pp. 5–53; T. M. Rudavsky, "Divine Omniscience."

79. In the first case Gersonides gives the example of the sun, which, when in the northern and southern regions of the zodiac, will affect sublunar elements differently. For a discussion of this passage, see Gad Freudenthal's review of Feldman's translation in *Révue des études juives,* 148.3–4 (1989), p. 381. In the second case he gives the example of Mars and the Moon, which dominate different elements, and therefore will affect individual temperaments differently depending upon their mutual positions.

80. Gersonides, *Wars* 2.2, p. 35.

81. Gersonides, *Milḥamot* 5.2.3, p. 194.

82. Ibid., p. 196.

83. Ibid., 5.2.6, p. 201. For an excellent examination of action at a distance in Gersonides' works, including the effects of magnetism as well as the sun's heating the earth's atmosphere, see Y. Tzvi Langermann, "Gersonides on the Magnet and the Heat of the Sun," in *Studies on Gersonides,* ed. Gad Freudenthal (Leiden: E. J. Brill, 1992), pp. 267–84. Langermann notes the importance of the "divine force" that functions in this process, and is mentioned by Gersonides in *Milḥamot* 5.2.6, p. 203.

84. Gersonides, *Milḥamot* 5.2.8, pp. 207–8. For a brief discussion of these principles, see Bernard R. Goldstein, *Theory and Observation in Ancient and Medieval Astronomy* (London: Variorum Reprints, 1985), p. 222. See also Langermann, "Gersonides on Astrology."

85. Joseph Albo, *Sefer ha-'Ikkarim; Book of Principles,* ed. and trans. Isaac Husik (Philadelphia: Jewish Publication Society, 1930), 3.8, p. 65.

86. Ibid., 4.4, pp. 24; 27.

87. Ibid., 4.4, pp. 4, 25.

88. Ibid., 3.8, p. 72. Albo gives as examples of natural astrology that the sun warms the air, and the moon cools and moistens the air.

89. Ibid., 4.4, p. 27.

90. Ibid., 4.4, p. 28.

91. Ibid., 4.4, p. 32.

92. Ibid., 4.43, p. 434: The astrologers "are not capable of knowing correctly all the causes which determine the future, and hence they can not judge correctly. Their intellect is too weak to judge correctly even of things past, all of whose causes have been completed and realized. . . . Surely, then, they are unable to know correctly all the causes of the future."

93. Albo, *Sefer ha-'Ikkarim* 4.43, p. 436.

94. For a detailed discussion of the Kalām arguments surrounding free will and divine omniscience, see H. A. Wolfson, *The Philosophy of the Kalam* (Cambridge, Mass.: Harvard University Press, 1976), p. 655ff. Recent discussion of Kalām thought

can be found in Richard Frank, "The Science of Kalām," *Arabic Sciences and Philosophy* 2 (1992), pp. 7–37.

95. For a history of the concept of divine providence in Jewish thought, cf. Alexander Altmann's article "*Hashgaḥah*" in the *Hebrew Encyclopedia,* pp. 476–90.

96. See the discussion by Manekin in Charles H. Manekin and Menachem Kellner, ed., *Freedom and Moral Responsibility: General and Jewish Perspectives* (Baltimore, Md.: University Press of Maryland, 1997), p. 13ff. On p. 168 Manekin points out that the Hebrew phrase for free choice, *beḥirah ḥofshit,* does not appear until the fifteenth century, and can be taken as a sign of the influence of scholastic discussions.

97. *Avot* 3.19. For a sustained discussion of Akiba's dictum and its influence upon the Rabbis, see E. Urbach, *The Sages: Their Concepts and Beliefs* (Jerusalem: The Magnes Press, 1975), pp. 255–85.

98. Saadia Gaon, *'Emunot ve-De'ot* 4.2, p. 125 (in Altmann, p. 119). Unless otherwise noted in this chapter, all translations of Saadia Gaon's work are mine. The Kalām influences upon Saadia's discussion are examined by Altmann in his work "Free Will and Predestination in Saadya; Bahya and Maimonides," in *Religion in a Religious Age,* ed. S. D. Goiten (Cambridge, Mass.: American Jewish Studies, 1973), p. 30ff.

99. For a discussion of the implications of this point, see Altmann, note 1 in his edition of *Book of Doctrines and Beliefs,* p. 119.

100. Saadia Gaon, *'Emunot ve-De'ot* 4.3, p. 127 (in Altmann, p. 118).

101. Ibid. (in Altmann, p. 119). Altmann suggests that Saadia may have been influenced in this distinction by Aristotle's discussions in the *Metaphysics.* See Altmann, "Free Will," p. 31 for further discussion of this point.

102. Saadia Gaon, *'Emunot ve-De'ot* 4.3, p. 128 (in Altmann, p. 119).

103. Halevi, *ha-Kuzari* 5.20, p. 315. Unless otherwise noted in this chapter, all translations, although influenced by Heinemann, are mine.

104. Ibid.

105. Ibid., p. 316.

106. Ibid.

107. Saadia Gaon, *'Emunot ve-De'ot* 4.4, p. 128 (in Altmann, p. 120).

108. Ibid., p. 130 (in Altmann, p. 122).

109. Ibid. Compare to Augustine's statement in *De Libero Arbitrio* that God's knowledge does not bring about the event in question.

110. Saadia Gaon, *'Emunot ve-De'ot* 4.4, p. 130 (in Altmann, p. 122).

111. Boethius, *The Consolation of Philosophy,* ed. James J. Buchanan (New York: Frederick Ungar, 1957), book 5, prose 6, p. 63. We will return below to the doctrine of God's eternal knowledge in the context of Crescas' theory of divine knowledge, which incorporates a notion of timelessness much like that of Boethius and Aquinas.

112. Saadia Gaon, *'Emunot ve-De'ot* 4.4, p. 130 (in Altmann, 122).

113. Halevi, *ha-Kuzari* 5.20, p. 318.

114. Ibid.

115. Ibid., p. 319.

116. Ibid.

117. Ibid., p. 326.

118. Saadia Gaon, *'Emunot ve-De'ot* 4.4, pp. 130–31 (in Altmann, p. 123). Altmann notes in "Free Will," pp. 31, 48, that a similar example can be found in Ghazālī, and suggests historical precedents for this example.

119. Halevi, *ha-Kuzari* 5.4, p. 281.

120. See for example Alexander Altmann, "The Religion of the Thinkers: Free Will and Predestination in Saadia, Bahya, and Maimonides," in *Religion in a Religious Age*, ed. S. D. Goitein (Cambridge, Mass.: American Jewish Studies Press, 1973), pp. 25–52.

121. Maimonides, *Guide* 2.48, p. 410.

122. Altmann, "Free Will," p. 43.

123. See J. Gellman, "Freedom and Determinism in Maimonides' Philosophy," in *Moses Maimonides and His Time,* ed. E. L. Ormsby (Washington, D.C.: Catholic University of America Press, 1989), pp. 139–50. This article is subjected to penetrating critique by Josef Stern in his recent work "Maimonides' Conceptions of Freedom and the Sense of Shame," in *Freedom and Moral Responsibility,* ed. C. Manekin and M. Kellner (Bethseda, Md.: University Press of Maryland, 1997), pp. 217–66. Analyzing the term *reshut* (power to act), Stern suggests that Maimonides incorporates the notion of "self-determining action, free of all external constraints, in which the human agent, through his deliberative rational judgment, acts as and because he so desires or wills" (pp. 241–42). The question, of course, and one that Stern does not address in this paper, is the extent to which Maimonides' notion of *reshut* is compatible with divine omniscience.

124. Arthur Hyman, "Aspects of the Medieval Jewish and Islamic Discussion of 'Free Choice,'" in *Freedom and Moral Responsibility,* ed. Charles Manekin and Menachem Kellner (Bethesda, Md.: University Press of Maryland, 1997), p. 152.

125. Josef Stern, "Maimonides' Conception of Freedom," in *Freedom and Moral Responsibility,* ed. Charles Manekin and Menachem Kellner, p. 231.

126. For a brief discussion of this text, see Altman, "Free Will," pp. 35–36. Maimonides basically warns his readers against taking too strong a stand on such a difficult topic, and promises to discuss the matter further in subsequent works.

127. Maimonides, *Shemoneh Peraqim* 8, p. 95.

128. For a discussion of this passage, and the interpretations of Saadia Gaon, Ibn Ezra, and Ibn Daud, see Wolfson, *Jewish Kalam,* p. 204ff. Wolfson correctly points out that Maimonides recognizes the logical difficulty in his position.

129. See Maimonides, *Shemoneh Peraqim* 8, p. 101.

130. Additional background material on the status of astrology in the time of Maimonides can be found in Alexander Marx, "The Correspondence between the Rabbis of Southern France and Maimonides about Astrology," *Hebrew Union College Annual* 3 (1926), pp. 311–358.

131. Maimonides, "Letter on Astrology," in *A Maimonides Reader,* ed. Isadore

Twersky (New York: Behrman House, 1972), p. 465. The Hebrew text can be found in Marx, "Correspondence."

132. Ibid., p. 466.

133. Ibid.

134. Ibid., p. 470.

135. Ibid., p. 471.

136. Maimonides alludes to the fact that there are such lucid counterarguments to astrology, but does not utilize them in his own attack. See Twersky, *Maimonides Reader,* p. 467. A similar point is made in the *Epistle to Yemen,* in Twersky, *Maimonides Reader,* p. 453: "Its postulates [of astrology] can be refuted by real proofs on rational grounds. But this is not the place to enter into a discussion of them."

137. See Pessin, "Maimonides on Astrology," for more extensive discussion of these issues.

138. Maimonides, *Guide* 2.10, p. 269.

139. Ibid., p. 270. Maimonides quotes *Genesis Rabbah* 10 in support of this contention.

140. Maimonides, *Guide* 1.72, 2.12, 2.6.

141. See Y. Tzvi Langermann, "Maimonides' Repudiation of Astrology," *Maimonidean Studies* 2 (1991), p. 141, for discussion of this crucial point.

142. See for example his discussion in *Guide* 2.10, p. 271.

143. Langermann, "Maimonides' Repudiation," p. 150. The implications of this overflow are examined at length by Josef Stern in his recent article "The Fall and Rise of Myth in Ritual: Maimonides versus Nahmanides on the *Huqqim,* Astrology, and the War against Idolatry," *The Journal of Jewish Thought and Philosophy* 6 (1997), pp. 185–263.

144. Maimonides, *Mishneh Torah* I.5, quoted in Twersky, *Maimonides Reader,* p. 78. A similar point is made in *Shemoneh Peraqim:* Maimonides asserts that humans are not born endowed with either vice or virtue, in an attempt to reject the "absurd idea of astrologers who falsely assert that the constellation at the time of ones birth determines whether one is to be virtuous or vicious, the individual being thus necessarily compelled to follow out a certain line of conduct." In Twersky, *Maimonides Reader,* p. 379. See also 4.4 above.

145. Maimonides, *Guide* 2.11, p. 275.

146. See ibid., 3.15.

147. See Freudenthal, "Maimonides' Stance," p. 89.

148. For a more extensive discussion of these distinctions in Maimonides, and analogous discussion in Averroës, see Rudavsky, "Maimonides and Averroes on God's Knowledge of Possibles," p. 30ff. The notion of possibility is connected to the Kalām theory of admissibility (*al-tājwiz*) and is fairly complex; it can be pursued in a number of discussions: Wolfson, *The Philosophy of the Kalām;* Z. Blumberg, "*Ha Rambam al-Musag al-Tājwiz beshittatam shel ha mutakalimûn,*" *Tarbiz* 39 (1970), pp. 268–76; Alfred Ivry, "Maimonides on Possibility," in *Mystics, Philosophers and Politicians,* ed. J. Reinhartz et al. (Durham, N.C.: Duke University Press, 1982); Shalom Rosenberg,

"Possible and Assertoric in Medieval Logic," *Iyyun* 28 (1978), pp. 56–72; Shalom Rosenberg, "Necessary and Possible in Medieval Logic," *Iyyun* 28 (1978), pp. 103–155.

149. Maimonides, *Guide* 3.16, p. 463.

150. Ibid.

151. Ibid., 3.20, p. 481.

152. Ibid.

153. For a sustained examination of Avicenna's position and its implications with respect to God's knowledge of future events, see Michael Marmura, "Some Aspects of Avicenna's Theory of God's Knowledge of Particulars," *Journal of the American Oriental Society* 83 (1962), pp. 299 - 312; Rudavsky, "Maimonides and Averroës on God's Knowledge of Possibles." Avicenna's position is succinctly summarized by Ghazālī as follows: "and as to those who believe that he knows things besides himself (and this is the theory which Avicenna has chosen) they believe that God knows other things in a universal knowledge which does not fall under the concept of time and which is not differentiated through past, future, and present although, nevertheless, Avicenna affirms that not the weight of a grain escapes God's knowledge either in earth or in the heavens, since he knows individual things in a universal way." Al-Ghazālī, *Tahāfūt* 13, p. 275. Thus according to Avicenna, God's knowledge of particulars is atemporal, unchanging, and occurs through his knowledge of universal causes. The implications of this position will become more apparent when we turn to Gersonides.

154. Maimonides, *Guide* 3.18, p. 474.

155. Ibid., p. 480.

156. Ibid., p. 483.

157. Ibid., p. 480.

158. Ibid., p. 485. For a comparison of Averroës and Maimonides on this point, see T. M. Rudavsky, "Maimonides and Averroës on God's Knowledge of Possibles," p. 23ff.

159. Maimonides, *Guide* 3.20, p. 480.

160. Ibid., p. 481.

161. Ibid.

162. Ibid., p. 482.

163. Ehud Benor, *Worship of the Heart: A Study in Maimonides' Philosophy of Religion* (Albany: State University of New York Press, 1995), p. 149.

164. Benor, *Worship of the Heart*, pp. 150–51.

165. The five theories are: that of Epicurus and his followers who deny God's providence altogether; that of Aristotle who believes that providence extends only to the superlunar sphere but not to the sublunar sphere; that of the Ash'arīah, who claim that every event in the world is predestined by God, thus denying any contingency; that of the Mu'tāzilites, who believe in limited free will for humans; and that of Mosaic Law, according to which humans have perfect freedom, while at the

same time everything that happens to humans is the result of God's justice. See *Guide* 3.17 for the details of these theories as presented by Maimonides.

166. Maimonides, *Guide*, 3.17, p. 472.

167. Ibid., 3.51, p. 625.

168. Ibid. I shall not try, in this context, to attempt to articulate what are the necessary criteria constituting "excellence" for Maimonides. This topic has been addressed in great detail by Menachem Kellner and others.

169. Benor, *Worship of the Heart*, p. 142.

170. See ibid., p. 141.

171. Maimonides, *Guide*, 3.51, p. 626.

172. Ibid., 3.18, p. 476.

173. For a discussion of these arguments, see Sorabji, *Necessity, Cause and Blame;* Cahn, *Fate, Time and Logic;* Rudavsky, *Divine Omniscience and Omnipotence.*

174. Feldman, "Crescas' Theological Determinism," p. 9.

175. See, for example, Maimonides, *Treatise on Logic*, chapter 4; Shalom Rosenberg "Necessity and Contingency in Medieval Philosophy," *Iyyun* 28 (1978), pp. 56–72, 103–55.

176. Ibn Daud, *Sefer ha-'Emunah ha-Ramah* (*The Exalted Faith*). For a recent study of this work, see T. A. M. Fontaine, *In Defense of Judaism: Abraham Ibn Daud* (Assen, Netherlands: van Gorcum, 1990).

177. Ibn Daud, *Sefer ha-'Emunah ha-Ramah*, intro, p. 1.

178. Ibid., 2.6, p. 96.

179. Ibid.

180. Ibid. This example is used by Avicenna but in a different context. For Avicenna, the issue is whether particulars like a lunar eclipse can be known without implying a change in the knower.

181. Ibid.

182. Ibid. See Fontaine, *In Defense*, pp. 13–14, for a discussion of this position.

183. References to specific articles will be made in the present chapter when relevant; however, the following works should be noted for their treatment of Gersonides' theory of divine omniscience: N. Samuelson, "Gersonides' Account of God's Knowledge of Particulars," *Journal of the History of Philosophy* 10 (1972), pp. 399–416; T. M. Rudavsky, "Divine Omniscience and Future Contingents in Gersonides," *Journal of the History of Philosophy* 21 (1983), pp. 513–36; C. Sirat, *Les Théories des Visions Surnaturelles dans la Pensée Juive du' Moyen-Âge* (Leiden: E. J. Brill, 1969); C. Touati, *La Pensée philosophique et théologique de Gersonides* (Paris: Les Éditions de Minuit, 1973); *Gersonides, Les Guerres du Seigneur, Livres III et IV*, trans. C. Touati (Paris: Mouton, 1969).

184. For further discussion of the difference between individuals and particulars, cf. Feldman's discussion in *Wars*, p. 89; cf., also T. M. Rudavsky, "Individuals and the Doctrine of Individuation in Gersonides," *New Scholasticism* 56 (1982), pp. 30–50; T. M. Rudavsky, "Theories of Individuation in Medieval Jewish Philosophy," *Indi-*

viduation in Late Scholasticism and the Counter-Reformation, ed. J. J. Gracia (Albany: State University of New York Press, 1993).

185. "We assert that it is already clear that contingent events (*be-miqreh ve-hizdamen*) have a certain structure, inasmuch as there are already many individuals to whom many good things happen, all of them occurring by chance and it is these who are called fortunate. . . . This necessarily entails that contingent events (*le-mah she-be-miqreh*) will have a mode of determination and structure. As to what that mode is, I wish I knew!" Gersonides, *Milḥamot* 2.2, p. 95.

186. "There is little doubt that according to this assumption (determinism) all voluntary affairs connected to will (*ratzon*) are necessary and there would be no free choice. And this is the lie from which Aristotle fled in the *De Interpretatione.* In truth there will be no choice since when what is chosen is necessary, the choice is not called a choice, since 'choice' refers to (a situation in which) two parts of a contradictory are possible." A. Altmann, "Gersonides' Commentary on Averroës' *Epitome of Parva Naturalia* II.3," *Proceedings of the American Academy for Jewish Research* 66–67 (1979–80), pp. 1–31.

187. Gersonides, *Supercommentary on Averroës' Commentary on Aristotle's De Interpretatione, Chapter 9* (found in the 1562 edition of *Aristotelis omnia quas extant opera . . . Averrois Cordubensis in ea opera omnes qui ad haec usque tempora pervenere commentarii . . . cum Levi Gersonidis in libros logicos annotationibus . . . a Iacob Mantino in Latinum conversi.* (*Venetiis, apud Iunctas,* 1562) (vol. 1, ff. 82^{vb}K–83^{rb}F). An English translation of this portion of Averroës' commentary and Gersonides' supercommentary has been prepared by N. Kretzmann. The quoted passage is found in 83rb D 46ff.

188. Gersonides, *Milḥamot* 2.2, p. 97.

189. See Charles Manekin, "Freedom Within Reason? Gersonides on Human Choice," in *Freedom and Moral Responsibility,* ed. C. Manekin and M. Kellner, pp. 165–204.

190. Clearly, astrological determinism and human freedom are contradictory states of affairs; however, aside from baldly stating that in some cases human freedom is preserved in the face of determinism, Gersonides does not explain how these two contradictory states can be reconciled. Little hint is given in *Milḥamot* to suggest that Gersonides is even aware of the tension between the two. Instead of philosophical explanation, the most we have is Gersonides' repeated phrase "Would that I knew!" in the face of perplexity. Gersonides might be suggesting that human volition is independent of material causation and hence is not subject to the influence of the stars. This point is not developed in detail, however in *Milḥamot* 2.2, p. 96.

191. "Of those misfortunes which arise externally and not from temperament and choice, such as the overturning of a country, earthquakes or lightning, and the like, it is clear that they cause evil only by chance (*be-miqreh*). For example, a fire may accidentally fall upon a man and kill him, or the earth may envelop its inhabitants who found themselves there by chance." *Milḥamot* 4.3, pp. 160–61.

192. Gersonides, *Milḥamot* 2.2, p. 96. Cf. also Gersonides' *Supercommentary on Averroës' Commentary on Aristotle's De Interpretatione, Chapter 9,* 83^{ra}A 11.5–10. "a thing is

said to be by chance because it is the result of an incomplete cause—e.g., when someone digs a ditch and finds a treasure. For this is classified as that which comes about by chance because he does not dig a ditch of that sort for the sake of that end, and it is not necessary that a man who digs a ditch should find a treasure. And so we say that finding a treasure while he is digging that ditch is by chance."

193. The problems associated with contingency are discussed briefly by Touati, *Pensée,* pp. 377–78. Brief allusions to contingency can be found in *Milḥamot* 2.2, p. 94: "Contingency occurs only with respect to human affairs. In all other things, contingency can be said to occur only insofar as it concerns human beings." Cf. also *Milḥamot* 3.4, pp. 142–43, 148–49.

194. It should be noted that Gersonides' summary of Aristotle's position is based on an interpretation of how Aristotle might have treated the problem of divine omniscience. Any number of passages from the Aristotelian commentators, Themistius, Alexander of Aphrodisias and Averroës could be used to support Gersonides' characterization of Aristotle.

195. Gersonides, *Milḥamot* 3.2, pp. 92–93; *Wars,* pp. 92–93.

196. Gersonides, *Milḥamot* 3.2, pp. 125–26; *Wars,* pp. 98–105. It should be noted that although Gersonides bases these on the *Guide,* he has inverted Maimonides' order. For Maimonides' own characterization, cf. *Guide* 3.20, pp. 482–83.

197. Gersonides, *Milḥamot* 3.2, p. 127; *Wars,* p. 100.

198. Gersonides, *Milḥamot* 3.2, p. 127; *Wars,* pp. 100–101.

199. That actual knowledge is an identity between knowledge and object known is stated by Aristotle in *De Anima* 3.5.430a 19: "Actual knowledge is identical with its object." The correspondence theory of truth is stated in a number of places; cf. *Metaphysics* 4.6.1011b26; *De Interpretatione* 9. Gersonides is aware, of course, that not all instances of knowledge involve actually existing particulars, namely, our knowledge of numbers. For a discussion of this point, cf. *Milḥamot,* 3.2, p. 127ff.

200. Gersonides, *Milḥamot* 3.2, p. 130; *Wars,* p. 105.

201. Ibid.

202. Ibid.

203. Gersonides, *Milḥamot* 3.4, p. 143; *Wars,* p. 122.

204. Gersonides, *Milḥamot* 3.2, p. 126; *Wars,* p. 102.

205. Gersonides, *Milḥamot* 3.4, p. 138ff; *Wars,* p. 116.

206. Gersonides, *Milḥamot* 3.4, p. 139; *Wars,* p. 118. See Abraham Ibn Daud, who in *Sefer ha-'Emunah ha-Ramah,* p. 96, develops a similar view according to which God's lack of knowledge of future contingents is not a deficiency.

207. Gersonides, *Milḥamot* 3.4, p. 138; *Wars,* p. 117.

208. See Samuelson, "Gersonides' Account of God's Knowledge of Particulars," p. 173.

209. Gersonides does feel that objects are individuated on the basis of their constitutive matter. For further elaboration of this point, cf. Rudavsky, "Theories of Individuation in Medieval Jewish Philosophy."

210. 107 Leibniz, in his theory of individuals, developed a notion of individual

concepts similar to that suggested here. He too was constrained to explain God's knowledge of individual concepts. Cf. *Leibniz's Philosophical Papers and Letters,* ed. L. Loemker (Dordrecht: D. Reidel, 1970), p. 307ff., 331–35; E. M. Curley, "The Root of Contingency," in *Leibniz: A Collection of Critical Essays,* ed. H. G. Frankfurt (New York, 1972), pp. 69–98.

211. Avicenna's theory of divine omniscience, propounded in a number of works, is summarized by Ghazālī in his *Tahāfūt al-Falāsifah;* both Avicenna's theory and Ghazālī's comments are presented by Averroës in his *Tahāfūt al Tahāfūt* as follows: "and as to those who believe that he knows things besides himself (and this is the theory which Avicenna has chosen) they believe that God knows other things in a universal knowledge which is not differentiated through past, future and present although, nevertheless, Avicenna affirms that not the weight of a grain escapes God's knowledge either in earth or in the heavens, since he knows individual things in a universal way. [T]hey say that God does not know the accidents of Zaid, Amr and Khalīd and that he knows only men in general, through a universal knowledge." Since Gersonides refers to so few contemporary writers by name, the problem of citing influences upon him is difficult. It is likely, however that he was aware of various texts in which Avicenna's views were expounded. For a discussion of this point, cf., C. Touati, *Pensée,* p. 38ff.; G. Vajda, *Isaac Albalag* (Paris: Vrin, 1960), 8, p. 267, 269ff.

212. Gersonides presents a modified version of the problem of prophecy in *Milḥamot* 2.1.

213. "If every affirmation and negation were to divide the true from the false, it would have to be the case that as regards future things one of the propositions would be determinately true before the occurrence of the actual thing. And so the greatest absurdity would follow." Gersonides, *Supercommentary on Averroës' Commentary on Aristotle's De Interpretatione, Chapter 9,* 82^{vb}K 34–36.

214. Ibid., 82^{vb}L 15.

215. Gersonides, *Milḥamot* 2.2, p. 95.

216. Ibid., p. 94.

217. Ibid.

218. Gersonides, *Milḥamot* 2.6, p. 104.

219. Averroës, *Epitome,* p. 43. Averroës claimed that "it is not in [its] nature to comprehend the particular." So how does the prophet receive from the Active Intellect the knowledge of particular events? According to Averroës, the Active Intellect transmits messages of a general nature that are picked up by the imaginative faculty of the prophet as pertaining to particulars: "[The Active Intellect] endows the imaginative soul with the universal nature that the individual that comes into being possesses, that is to say, a comprehension of its causes, and the imaginative soul will receive it as a particular by virtue of the fact that it is in matter" (Averroës, *Epitome,* p. 60). In other words, according to Averroës, the Active Intellect disperses general knowledge and the imaginative faculty receives this knowledge in the mode

of particularity. Averroës does not explain how the information is transformed from the general to the particular mode; nor does he explain why the Active Intellect disperses messages at all.

220. Gersonides, *Milḥamot* 2.6, p. 106.

221. Ibid.

222. Gersonides, *Milḥamot* 2.6, pp. 105–6.

223. The transmission of prophetic messages according to Gersonides can be summarized as follows: The Active Intellect, which functions as the intermediary between God and the sublunar world, grasps certain essences of particulars. These essences, which for the most part are determined by the heavenly bodies, are transmitted in the form of universal statements to the prophet, who then instantiates them with particular details. These messages are not transmitted to any particular prophet, since the Active Intellect has no awareness of individual prophets. Rather, according to Gersonides, prophets are persons predisposed to receive prophetic messages; because they have been thinking about those very messages being transmitted by the Active Intellect, their intellects are attuned to receive the general messages broadcast at random by the Active Intellect. For an interesting modern analogue, see Samuelson, "Gersonides' Account of God's Knowledge of Particulars."

224. Gersonides, *Milḥamot* 3.5, pp. 147–48; *Wars*, p. 133.

225. Gersonides discusses this example in *Milḥamot* 6.2.13, pp. 461–62. An analogous discussion occurs in his *Commentary on the Pentateuch, Exodus* 32.10 (Venice, 1547).

226. Ibid.

227. Touati, *Pensée,* p. 463.

228. Gersonides, *Milḥamot* 2.2, p. 94.

229. R. Gaskin, "Conditionals of Freedom and Middle Knowledge," *The Philosophical Quarterly* 43 (1993), p. 413. For recent discussion of this text, and its importance in late scholastic sources such as Suarez and Molina, see R. M. Adams, "Middle Knowledge and the Problem of Evil," *American Philosophical Quarterly* 14 (1977), pp. 109–17; W. Craig, *The Problem of Divine Foreknowledge and Future Contingents from Aristotle to Suarez* (Leiden: E. J. Brill, 1988); W. Hasker, *God, Time and Knowledge* (Ithaca, N.Y.: Cornell University Press, 1989), p. 22ff.

230. See Ibn Daud, *Sefer ha-'Emunah ha-Ramah,* 2.6, p. 97.

231. For a discussion of "might-be" conditionals, see J. Pollock, *Subjunctive Reasoning* (Dordrecht: D. Reidel, 1976), pp. 25–44.

232. For a more extensive critique and analysis of indeterminism and prophecy, see Rudavsky, "Divine Omniscience, Contingency and Prophecy in Gersonides."

233. Feldman, "Crescas' Theological Determinism" p. 6. For recent studies of Crescas' doctrine of free will, see as well S. Feldman, "A Debate Concerning Determinism in Late Medieval Jewish Philosophy," *Proceedings American Academy for Jewish Research* 51 (1984), pp. 15–40; C. Touati, "Hasday Crescas et Ses Paradoxes sur la Liberté," in *Mélanges d'Histoire des Religions Offerts à Charles-Henry Puech* (Paris: Presses

Universitaires de France, 1974), pp. 573–78, A. Ravitsky, "Hitpatkhut Hashkafuto shel R' Hasdai Crescas Bishe-elat Hofesh Haratzon," *Tarbiz* 51 (1981–82), pp. 445–70.

234. Crescas, *'Or 'Adonai,* 2.1.1, p. 28a.

235. Ibid.

236. Ibid.

237. Crescas, *'Or 'Adonai* 2.1.2, p. 29a.

238. Ibid.

239. Crescas, *'Or 'Adonai* 2.1.3, p. 31a.

240. Crescas, *'Or 'Adonai* 2.1.3, p. 32b (in Harvey, p. 247).

241. Feldman, "Crescas' Theological Determinism," p. 13.

242. Crescas, *'Or 'Adonai,* 32b.

243. Waxman, *Crescas,* p. 109.

244. Crescas, *'Or 'Adonai* 2.1.4, p. 33a.

245. Ibid.

246. Ibid.

247. Ibid.

248. Feldman, "Crescas' Theological Determinism," p. 17; *'Or 'Adonai* 2.1.4, p. 33a.

249. Ibid. (in Harvey, p. 251).

250. Ibid., 2.5, p. 45b.

251. Ibid., 2.5.1, pp. 45b–47b.

252. Ibid., 2.5, p. 48a.

253. Ibid., 2.5.3, p. 48b.

254. Feldman suggests that contingency has vanished because "whatever is eternalized has become past, and thus necessary, at least in the sense of being irrevocable and unpreventable." See Feldman, "Crescas Theological Determinism," p. 25

255. For a discussion of this move on the part of Crescas, see Feldman, "Crescas' Theological Determinism," p. 27.

256. Crescas, *'Or 'Adonai* 2.5.3.

257. Boethius, *Consolation of Philosophy,* 5.6, p. 64.

258. Thomas Aquinas, *Summa Theologica,* trans. Fathers of the English Dominican Province, 2nd ed. (London: Burnes Oates & Washbourne, 1920), 1.14.13, reply Obj 3.

259. Feldman, "Crescas' Theological Determinism," p. 18.

260. See N. Kretzmann's classic article "Omniscience and Immutability," *Journal of Philosophy* 63 (1966), pp. 409–21.

261. Anthony Kenny, *The God of the Philosophers* (Oxford: Clarendon Press, 1979), pp. 38–39.

262. Eleanore Stump and Norman Kretzmann, "Eternity," *Journal of Philosophy* 79 (1981), pp. 429–528. For a survey of the recent philosophical literature, and a

discussion of attempts to reconcile the notion of simultaneity with that of human freedom, see Hasker, *God, Time and Knowledge,* esp. chapters 8–10.

263. Feldman, "Crescas' Theological Determinism," p. 21.

264. The question arises whether, according to Gersonides, prophetic events can be foreknown by anybody? It would appear not. The most anybody can know (God included) is that certain events might occur in light of certain antecedent conditions. but surely we don't need a prophet to tell us that! Ultimately Gersonides' system leaves little room for a theory of prophecy in which future contingent events are determinately prophesied. The most a prophet can tell us is *that* certain events are contingent; he cannot tell us *which* contingent possibility will be actualized. This inability on the part of the prophet to transmit the content of future contingent events is a severe limitation on the content of prophetic knowledge. For further discussion of this point, see Rudavsky, "Divine Omniscience, Contingency and Prophecy in Gersonides," in *Divine Omniscience and Omnipotence in Medieval Philosophy,* ed. T. M. Rudavsky, pp. 161–81.

265. Albo, *Sefer ha-'Ikkarim* 4.1, p. 7.

266. Ibid., p. 8.

Chapter 5. Prelude to Modernity

1. For a recent study of the relation of Spinoza to matters pertaining to Judaism and theology, see Richard Mason, *The God of Spinoza: A Philosophical Study* (Cambridge: Cambridge University Press, 1997).

2. Sarah Hutton, "Some Renaissance Critiques of Aristotle's Theory of Time," *Annals of Science* 34 (1977), p. 348. For an excellent discussion of the impact of the Condemnation of 1277 upon the development of medieval philosophy and science, see John Murdoch, "Pierre Duhem and the History of Late Medieval Science and Philosophy in the Latin West," in *Gli Studi di Filosofia Medievale Fra Otto e Novecento* (Rome: Edizioni di Storia e Letteratura, 1991), pp. 253–302, esp. p. 259ff.

3. It may be instrumental to compare my reading of Aristotle with that of Bas van Frassen, *An Introduction to the Philosophy of Time and Space* (New York: Random House, 1970), p. 16ff. Van Frassen suggests that Aristotle's theory of time is basically a theory of duration, arguing that time measures "how far" and "how long" local motion takes place.

4. M. Ficino, *Opera Omnia* (Basel, 1576; repr. Turin, 1962), esp. p. 1722.

5. Pico della Mirandola, *Examen vanitatis doctrina gentium* in *Opera omnia* (Basel, 1573; repr. Hildesheim, 1969), 2:1198, quoted in Hutton, "Some Renaissance Critiques," p. 353.

6. Van Frassen, *An Introduction to the Philosophy of Time and Space,* p. 21.

7. Piero E. Ariotti, "Toward Absolute Time: The Undermining and Refutation of the Aristotelian Conception of Time in the Sixteenth and Seventeenth Centuries," *Annals of Science* 30 (1973), p. 37. Ariotti argues that Kepler's first two laws

seriously undermine Aristotle's theory of time in emphasizing the nonuniformity of the motions of the planets.

8. Hillel Levine, "Paradise Not Surrendered: Jewish Reactions to Copernicus and the Growth of Modern Science," in *Epistemology, Methodology and the Social Sciences,* ed. R. S. Cohen and M. W. Wartofsky (Dordrecht: D. Reidel, 1983), p. 204.

9. See David Ruderman, *Jewish Thought and Scientific Discovery in Early Modern Europe* (New Haven, Conn.: Yale University Press, 1995), p. 68ff. Ruderman mentions the work of Moses Isserles, who, writing in Cracow, was clearly fascinated by astronomy. "Was it merely a coincidence that Isserles lived in the same city where Copernicus had written his revolutionary work?" Ruderman asks.

10. For discussions of David Gans, see Levine, "Paradise Not Surrendered," p. 207; André Neher, *Jewish Thought and the Scientific Revolution of the Sixteenth Century: David Gans (1541–1613) and His Times,* trans. D. Maisel (Oxford: Oxford University Press, 1986), p. 216ff.; Ruderman, *Jewish Thought and Scientific Discovery,* p. 83; see Gans, *Nehmad,* p. 9a.

11. Neher, *Jewish Thought,* pp. 216–18.

12. See Ruderman, *Jewish Thought and Scientific Discovery,* p. 134. Ruderman himself does not offer a study of *Sefer Elim,* choosing to concentrate instead upon the *Matzref la-Hokhman,* a defense of Kabbalah, and *Mikhtav Ahuz,* a condemnation of kabbalistic thought.

13. Joseph Delmedigo, *Sefer Elim* (Amsterdam, 1629; repr. Odessa, 1864–67). See Levine, "Paradise Not Surrendered," pp. 208–9. For a survey and discussion of Delmedigo's work, see I. Barzilay, *Yoseph Shlomo Delmedigo (Yashar of Candia): His Life, Works, and Times* (Leiden: E. J. Brill, 1974); Ruderman, *Jewish Thought and Scientific Discovery,* pp. 118–52.

14. Delmedigo, *Sefer Gevurot Hashem,* in *Sefer Elim,* p. 292.

15. For details of Ebreo's work, see Hava Tirosh-Rothschild, "Jewish Philosophy on the Eve of Modernity," in *History of Jewish Philosophy,* ed. D. H. Frank and O. Leaman (London: Routledge, 1997), p. 522ff.

16. Leone Ebreo, *The Philosophy of Love (Dialoghi d'Amore),* trans. F. Friedenberg-Seeley and J. H. Barnes (London: The Soncino Press, 1937), p. 280.

17. For an elaboration of the thesis that "the introduction of the concept of absolute independent time means the realization on the part of the seventeenth century scientists of the refutation of the Aristotelian view of time by the discoveries of Copernicus, Kepler, Galileo, Heygens and Hooke," see Ariotti, "Toward Absolute Time," p. 31.

18. See Hutton, "Some Renaissance Critiques," p. 363, for the progression of this change.

19. Isaac Barrow, *The Geometrical Lectures of Isaac Barrow,* trans. J. M. Child (La Salle, Ill.: Open Court, 1916), pp. 35–37, quoted in Van Frassen, *An Introduction to the Philosophy of Time and Space,* p. 23.

20. See the recent study by Matt Goldish, "Newton on Kabbalah," in *The Books of Nature and Scripture: Recent Essays on Natural Philosophy, Theology, and Biblical Crit-*

icism in the Netherlands of Spinoza's Time and the British Isles of Newton's Time, ed. J. E. Force and R. Popkin (Dordrecht: Kluwer Academic, 1994), pp. 89–104. See also idem, *Judaism in the Theology of Sir Isaac Newton* (Boston: Kluwer Academic, 1998).

21. See, for example, Richard Popkin, "Newton and Maimonides," in *A Straight Path: Studies in Medieval Philosophy and Culture—Essays in Honor of Arthur Hyman,* ed. Ruth Link-Salinger (Washington, D.C., 1988), pp. 216–29; idem, "Some Further Comments on Newton and Maimonides," in *Essays on the Context, Nature and Influence of Isaac Newton's Theology,* ed. J. E. Force and R. H. Popkin (Dordrecht: Kluwer Academic, 1990). In these works Popkin traces the possible influences of Maimonides' theological works upon Newton.

22. Newton's unpublished manuscript *Tempus et Locus* is presented in part and discussed by J. E. McGuire in his two essays "Existence, Actuality and Necessity: Newton on Space and Time," *Annals of Science* 35 (1978), reprinted in J. E. McGuire, *Tradition and Innovation: Newton's Metaphysics of Nature* (Dordrecht: Kluwer Academic, 1995), pp. 1–51; and "Newton on Place, Time, and God: An Unpublished Source," *The British Journal for the History of Science* 11 (1978), pp. 115–29.

23. McGuire, "Existence, Actuality and Necessity: Newton on Space and Time," p. 2.

24. Newton, *Tempus et Locus,* p. 117 (in McGuire, "Newton on Place, Time, and God").

25. Sir Isaac Newton, *Philosophia naturalis principia mathematica,* in F. Cajori, ed., *Sir Isaac Newton's Mathematical Principles of Natural Philosophy and His System of the World* (New York: Greenwood Press, 1969), pp. 6, 8; Latin text in *Isaac Newton's Philosophiae Naturalis Principia Mathematica,* ed. Alexandre Koyre and I. Bernard Cohen (Cambridge, Mass.: Harvard University Press, 1972), p. 46.

26. Isaac Newton, *Philosophia naturalis,* p. 6. See also *Tempus et Locus,* p. 117 (in McGuire, "Newton on Place, Time, and God"), where a similar distinction is made between times of the vulgar and time of the astronomers.

27. Isaac Newton, *Philosophia naturalis,* p. 8.

28. Ibid.

29. Ibid. For a discussion of the philosophical implications of this point, see van Frassen, *An Introduction to the Philosophy of Time and Space,* p. 24.

30. McGuire, "Existence, Actuality and Necessity," p. 26.

31. Newton, *General Scholium,* in *Philosophia naturalis* (Latin), p. 761.

32. McGuire, "Existence, Actuality and Necessity: Newton on Space and Time," p. 27.

33. Newton, *General Scholium* in *Philosophia naturalis,* (Latin), p. 761.

34. McGuire, "Existence, Actuality and Necessity," p. 16.

35. See Spinoza, *Cogitata Metaphysica* 2: "the quantity of duration applies to modes but not to God. To ascribe duration to God is to render God's essence mutable."

36. McGuire, "Existence, Actuality and Necessity," p. 27.

37. Brian Copenhaver, "Jewish Theologies of Space in the Scientific Revolu-

tion: Henry More, Joseph Raphson, Isaac Newton and their Predecessors," *Annals of Science* 37 (1980), p. 505. In this important work Copenhaver traces the influence of Lurianic Kabbalah upon the writings of More, Newton, and others.

38. M. Wartofsky, "Nature, Number and Individuals: Motive and Method in Spinoza's Philosophy," *Inquiry* 20 (1977), p. 459.

39. M. Gueroult, "Spinoza's Letter on the Infinite," in *Spinoza: A Collection of Critical Essays,* ed. M. Grene (Notre Dame, Ind.: University of Notre Dame Press, 1979), p. 182.

40. Ibid.

41. Jonathan Bennett, *A Study of Spinoza's Ethics* (Indianapolis: Hackett, 1984), p. 84.

42. Bennett, *Spinoza's Ethics,* p. 83.

43. Ibid., p. 87.

44. Bennett, Ibid., p. 95.

45. Spinoza, *Appendix, Cogitata Metaphysica* 1, p. 300. Unless noted otherwise, English translations of Spinoza's works will be taken from the critical edition of E. M. Curley, *The Collected Works of Spinoza,* vol. 1 (Princeton N.J.: Princeton University Press, 1985). For the Latin text of Spinoza's *Ethics* [*Ethica*], I have used the edition of J. Van Vloten and J. P. N. Land, *Ethica* (Martinus Nijhoff, 1905).

46. Descartes, *Principles* 55, p. 178.

47. Spinoza, *Appendix, Cogitata Metaphysica* 1.4, p. 310.

48. Ibid., p. 316.

49. Ibid., p. 317.

50. Spinoza, *Ethics* 2.d5.

51. Spinoza, *Appendix, Cogitata Metaphysica* 2.10, p. 334.

52. Ibid., p. 335.

53. See Mason's introduction to important influences upon Spinoza, in Richard Mason, *The God of Spinoza* (Cambridge: Cambridge University Press, 1997), pp. 1–17.

54. Descartes, *Meditation* 2, p. 82.

55. Descartes, *Meditation* 3, p. 93

56. Descartes, *Meditation* 3, p. 96. The standard reading of this passage is that Descartes regards time as composed of moments; that these parts are mutually independent and so the moments are separate and distinct from each other; and that this discontinuity thesis is supported by the doctrine of continuous creation.

57. Descartes, *Principles* 21, p. 167.

58. AT 10, 407

59. See Martial Gueroult, *Descartes' Philosophy Interpreted According to the Order of Reasons I: The Soul and God,* trans. Roger Ariew (Minneapolis: University of Minnesota Press, 1984), p. 196ff.

60. Descartes, *Principles,* p. 179.

61. Arthur, "Continuous Creation, Continuous Time," p. 353.

62. Ibid., p. 355.

63. Descartes, *Reply to Objection V,* in Haldane and Ross, 2, p. 219.

64. This point is emphasized in Arthur, "Continuous Creation, Continuous Time," p. 362.

65. J. E. K Secada, "Descartes on Time and Causality," *The Philosophical Review* 94 (1990), p. 47.

66. Daniel Garber, *Descartes' Metaphysical Physics* (Chicago: University of Chicago Press, 1992), p. 270.

67. Ibid., p. 271.

68. Ibid.

69. Ibid., p. 273. See also Arthur's contention that Descartes "does not even believe that there is a problem here, and certainly does not set out to solve the problem of the continuum." Arthur, "Continuous Creation, Continuous Time," p. 350.

70. See H. A. Wolfson, *The Philosophy of Spinoza,* vol. 1 (New York: Schocken Books, 1959), pp. 262–95; Mason, *The God of Spinoza.*

71. Descartes, *Principles* 1.24, p. 168.

72. See Descartes, *Principles* 1.26, p. 169.

73. Ibid. See also "I say . . . that the world is indeterminate or indefinite, because I recognize no boundaries in it; but I would not dare to call it infinite, because I perceive that God is greater than the world, not *ratione extensione,* which, as I have often said, I do not understand as a property in God, but *ratione perfectionis*" (15 April 1649: Adam & Tannery 5, 344)

74. Margaret Wilson, "Can I Be the Cause of the Idea of the World? Descartes on the Infinite and Indefinite," in *Essays on Descartes' Meditations,* ed. A. O. Rorty (Berkeley: University of California Press, 1986), p. 342.

75. Descartes, *Principles* 1.27, p. 169 . See also *Principles* 2.34.

76. Ariew, "The Infinite in Spinoza's Philosophy," p. 18.

77. Ibid., p. 17.

78. Spinoza, *Descartes' Principles of Philosophy,* 5s, p. 269.

79. Ibid.

80. In his article, Ariew correlates Spinoza's discussion of infinity with scholastic discussions of Gregory of Rimini, John Buridan and Albert of Saxony. Of course, there is no direct evidence that Spinoza read and knew of these texts. But Ariew suggests that many of Spinoza's arguments and distinctions reflect these discussions even more than Crescas. Ariew concentrates upon the puzzle in *Ethics* 1p15s, which echoes some of the elements found in Crescas' puzzle about the impossibility of the infinite as incorporeal substance (in Wolfson, *Crescas' Critique,* p. 137).

81. Spinoza, *Descartes' Principles of Philosophy,* 5s, p. 270.

82. Curley states in his notes that this argument is not to be found in any of the ancient sources. It is close in form, however, to Maimonides' example of the millstone, which we discussed in chapter 3.

83. Spinoza, *Descartes' Principles of Philosophy,* 5s, p. 271.

84. Ibid., p. 273.

85. Spinoza, *Ethics* 1p13, p. 420.

86. See E. Curley's note 35 to the text, in Spinoza, *Ethics*, p. 421.

87. Ibid.

88. The text reads as follows: "If corporeal substance is infinite, they, say, let us conceive it to be divided into two parts. Each part will be either finite or infinite. If the former, then an infinite is composed of two finite parts, which is absurd. If the latter, then there is one infinite twice as large as another, which is also absurd." *Ethics* 1p14s, p. 422. (*Ethica*, p. 10: Si substantia corporea, ajunt, est infinita, concipiatur in duas partes dividi; erit unaquaeque pars vel finita, vel infinita. Si illud, componitur ergo infinitum ex duabus partibus finitis, quod est absurdum. Si hoc, datur ergo infinitum duplo majus alio infinito, quod etiam est absurdum.)

89. "Again, if an infinite quantity is measured by parts [each] equal to a foot, it will consist of infinitely many such parts, as it will also, if it is measured by parts [each] equal to an inch. And therefore, one infinite number will be twelve times greater than another [which is no less absurd]." *Ethics* 1p14s, p. 422. (*Ethica*, p. 10: Porro, si quantitas infinita mensuretur partibus pedes aequantibus, infinitis talibus partibus constare debebit; ut et, si partibus mensuretur digitos aequantibus; ac propterea unus numerus infinitus erit duodecies major alio infinito.)

90. See Wolfson, *Spinoza*, p. 286, who discusses medieval versions of this argument that are found in works of Saadia Gaon, Gersonides, Crescas, and Bruno. A second version of the argument is found in Saadia, Al-ghāzāli, Bahya, ibn Daud, and Altabrizi. Modern philosophers would, of course, disagree.

91. "Finally, if we conceive that from one point of a certain infinite quantity two lines, say AB and AC, are extended to infinity, it is certain that, although in the beginning they are a certain, determinate distance apart, the distance between B and C is continuously increased, and at last, from being determinate, it will become indeterminable. Since these absurdities follow—so they think—from the fact that an infinite quantity is supposed, they infer that corporeal substance must be infinite, and consequently cannot pertain to God's essence." *Ethics* 1p14s. (*Ethica*, p. 10: Denique, si ex uno puncto infinitae cujusdam quantitatis concipiatur, duas lineas, ut AB, AC certa ac determinata in initio distantia in infinitum protendi; certum est, distantiam inter B et C continuo augeri, et tandem ex determinata indeterminabilem fore. Cum igitur haec absurda sequantur, ut putant, ex eo, quod quantitas infinita supponitur; inde concludunt, substantiam corpoream debere esse finitam, et consequenter ad Dei essentiam non pertinere.)

92. Wolfson, *Spinoza*, p. 294.

93. Spinoza, *Ethics* 1p14s, p. 422. (*Ethica*, p. 10: "sed quod quantitatem infinitam mensurabilem, et ex partibus finitis conflari supponunt.")

94. Spinoza, *Ethics* 1p14s, p. 423. (*Ethica*, pp. 10–11: "Sic etiam alii, postquam fingunt, lineam ex punctis componi, multa sciunt invenire argumenta, quibus ostendant, lineam non posse in infinitum dividi. Et profecto non minus absurdum est ponere, quod substantia corporea ex corporibus sive partibus componatur, quam

quod corpus ex superficiebus, superficies ex lineis, lineae denique ex punctis componantur.")

95. Ibid., p. 423.

96. Ibid. This argument, incidentally, was often exploited by ancient and medieval philosophers. See for example, discussions in N. Kretzmann, *Infinity and Continuity in Ancient and Medieval Thought* (Ithaca, N.Y.: Cornell University Press, 1982).

97. Spinoza, *Ethics* 1p14s, p. 424. (*Ethica*, p. 11: "nescio, cur divina natura indigna esset, quandoquidem (per Prop. 14) extra Deum nulla substantia dari potest, a qua ipso pateretur. . . . [Q]uare nulla ratione dic potest Deum ab alio pati, aut substantiam extensam divina natura indignam esse, tametsi divisibilis supponatur, dummodo aeterna et infinita concedatur.")

98. Spinoza, *Letter 12*, p. 200.

99. Ibid.

100. Ibid., pp. 202–3.

101. Spinoza, *Ethics* 1p14s, p. 424.

102. We cannot in this study examine the importance of the concept of imagination to Spinoza's overall epistemology. Suffice it to say that for Spinoza imagination represents a less adequate form of knowledge, in that understanding knows essences, whereas imagination knows existences.

103. Spinoza, *Letter 12*, p. 203.

104. Ibid., p. 204.

105. Ibid., pp. 203–4.

106. Ibid.

107. Ibid.

108. Gueroult, "Spinoza's Letter on the Infinite," p. 210.

109. Spinoza, *Letter 12*, p. 205.

110. See ibid., p. 204. This example is further discussed in *Letters* 80 and 81. See also Descartes' discussion in his *Principles* 1.23, pp. 198–9.

111. Spinoza, *Letter 12*, p. 205.

112. For a representative analysis of these issues, see Don Garrett, "Spinoza's Ethical Theory," in *The Cambridge Companion to Spinoza*, ed. D. Garrett (Cambridge: Cambridge University Press, 1996), pp. 267–314. Although Garrett agrees that Spinoza's determinism is connected with his necessitarianism, namely that whatever is true is necessarily true, he tries to argue that Spinoza is not a fatalist and that his system does not preclude moral freedom and responsibility.

113. The argument is found in *Ethics* 1p17s.

114. Spinoza, *Ethics* 1p17s, p. 427. Compare with Maimonides' theory of negative predication discussed in chapter 3.

115. Spinoza, *Ethics* 1p17s, p. 427.

116. Ibid., p. 426.

117. Ibid., 1p33s1, p. 436.

118. Ibid., 1p28, p. 433. See also *Cogitata Metaphysica* 1.3, p. 308.

119. Ibid., *Ethics* 1p33s1, p. 436.

120. Ibid., 1p4c1, p. 480. This is echoed in *Appendix* 1, where Spinoza claims that humans think themselves free because they are ignorant of the causes motivating their wishes, desires, etc. He is not denying that people do in fact deliberate (cf. *Appendix* 2, example of Buridan's ass), but rather that these deliberations are caused.

121. M. Geuroult, "Spinoza's Letter on the Infinite," in *Spinoza: A Collection of Critical Essays*, ed. M. Grene (Notre Dame, Ind.: University of Notre Dame Press, 1979), p. 193.

122. Ibid., p. 210.

123. Bennett, *A Study of Spinoza's Ethics*, p. 105.

Conclusion

1. See Aristotle, *De Caelo.*

2. Saadiah Gaon, *The Book of Beliefs and Opinions*, 1.1.

3. Roslyn Weiss, "Maimonides on the End of the World," *Maimonidean Studies* (New York: Ktav, 1992–93), p. 217.

4. Maimonides, *Guide* 2.29, p. 344.

5. Ibid., 2.27, p. 333.

6. See Seymour Feldman, "The End of the Universe in Medieval Jewish Philosophy," *American Jewish Studies Review* 11 (1986), pp. 53–77.

7. Gersonides, *Milḥamot* 6.1.16.

8. Before turning to the details of this discussion, let me point out some of the differences in scope between Jewish and scholastic discussions of issues of individuation. Part of the difference is due to the fact that until the fourteenth century, Jewish writers had little access to the logical writings of Aristotle, and so the specific logical issues related to individuation that arose out of the *Categories* and *On Interpretation* were of little direct concern to them. Not until the commentaries of Alfārābī and Averroës were translated into Hebrew in the mid-twelfth to early thirteenth centuries were Jewish philosophers exposed to specifically logical concerns. In fact, with the exception of Maimonides's early treatise on logic, which relied heavily upon Alfārābī's books, there are few Jewish writings prior to the fourteenth century devoted to matters of logic. Moreover, inasmuch as Jewish philosophers were obviously not concerned with those ontological issues that arose out of a trinitarian conception of God, they did not feel as obliged as did their scholastic counterparts to construct elaborate theories of identity and individuation to account for the unity within diversity of the Godhead.

9. See Udo Thiel, "Individuation," in *The Cambridge History of Seventeenth-Century Philosophy*, ed. Daniel Garber and Michael Ayers (Cambridge: Cambridge University Press, 1998), pp. 212–62; and idem, "Personal Identity," in *The Cambridge History of Seventeenth-Century Philosophy*, ed. Daniel Garber and Michael Ayers (Cambridge: Cambridge University Press, 1998), pp. 868–912.

10. A helpful discussion of this problem can be found in Wolfson, *Spinoza,* p. 390ff. These ontological concerns carried with them theological ramifications that were manifested in the examination of God's knowledge of these particulars. For if, as we have seen in chapter 5, the majority of Jewish philosophers were wont to argue that God is an immaterial Being and not subject to change, then how can he know the world of changing, material particulars? Concomitantly, if these particulars are in some way individuated on the basis of their matter alone, then in what sense can God individuate them? And if God cannot individuate particulars, how can he bestow providence upon them? For, within the medieval Jewish tradition, divine providence is bestowed only upon individuals who satisfy certain criteria, that is, moral and intellectual qualities.

11. A preliminary summary of this problem in Islamic thought, and its implications for Jewish and scholastic discussions, can be found in D. Callus, "The Origins of the Problem of the Unity of Form," *Thomist* 24 (1961), pp. 257–85; O. Hamelin, *La Théorie de l'Intellect d'après Aristote et Ses Commentateurs* (Paris, 1953).

12. The 'Active' (or 'Agent') Intellect is a term that refers back to Aristotle's actual intellect as described in *De Anima* 3.5. The term was transmitted to Jewish and scholastic writers through Islamic philosophers, and it came to represent not only a part of the human soul but the domain of Divine intellectual cognition as well.

13. Maimonides, *Guide* 2.19, p. 304.

14. Ibid., p. 306.

15. Ibid., 1.74, pp. 218–19.

16. Ibid., 1.72, p. 184. Compare this analogy to similar ones drawn in Aristotle's *De Caelo* 1.7, 8.

17. Maimonides, *Guide* 1.58, p. 136.

18.Ibid., 3.19, p. 305.

19.Ibid., 2.4, p. 256.

20.Ibid., p. 257.

21.Ibid., p. 258.

22.Ibid., 1.70, p. 174.

23.Ibid., 1.74, p. 221. For a discussion of the influence of Ibn Bājja upon Maimonides, see Alexander Altmann, "Ibn Bājja on Man's Ultimate Felicity."

24. Maimonides, *Guide* 2.16, p. 237.

25. Spinoza, *Ethics* 5p23, p. 607. Note that Spinoza does not use the term immortality to express the survival of part of the mind. The only time that the term "immortal" (*immortalis*) appears in the *Ethics* is in 5p41s: "There may be someone who, because he does not believe he can nourish his body with good food to eternity, should prefer to fill himself with poisons and other deadly things, or because he sees that the mind is not eternal or immortal . . . (*vel, quia videt Mentem non esse aeternam seu immortalem*)." Other discussion use the term eternal (*aeternam*) in talking about the continued existence of the soul.

26. Spinoza, *Ethics* 5p23s, pp. 607–8.

27. Stuart Hampshire, *Spinoza* (New York: Penguin Books, 1951), p. 175.

28. Bennett, *Spinoza,* pp. 372, 374.

29. Aristotle, *De Anima* 3.5.430a10–25. Ivry suggests that Gersonides may have started his work with this issue in an attempt to "butter up" his readers by showing them that, in contradistinction to Maimonides who denied personal immortality, he in fact upheld a more orthodox view of immortality. See Alfred Ivry, "Gersonides and Averroes on the Intellect," in *Gersonide en Son Temps,* ed. Gilbert Dahan (Paris: E. Peeters, 1991), p. 235.

30. A discussion of these questions, and the positions of representative commentators, can be found in the introduction by B. Zedler to Thomas Aquinas' *On the Unity of the Intellect Against the Averroists* (Milwaukee: Marquette University Press, 1968).

31. Gersonides, *Wars* 1.1, p. 110; see Feldman's discussion on page 72.

32. Gersonides, *Wars* 1.1, p. 110.

33. For further discussion of this position, see *Wars,* p. 72; also Hamelin, *La Théorie de l'Intellect,* pp. 38–43.

34. Gersonides, *Wars* 1.1, p. 110.

35. For further discussion, see *Wars,* p. 74; also A. Ivry, "Averroës on Intellection and Conjunction," *Journal of the American Oriental Society* 86 (1966), pp. 76–85.

36. Gersonides, *Wars* 1.10, p. 110. This view has been identified by some with Christian thinkers of Gersonides' day. See *Wars* p. 75, note 10.

37. Gersonides, *Wars* 1.2, pp. 114–15.

38. Ibid., 1.4, p. 79.

39. Ibid., 1.8, p. 170.

40. Ibid., 1.4, p. 136; see also p. 170.

41. Ibid., 1.6, pp. 162–63.

42. In my interpretation of *Ethics* 5, I reflect the suggestion of Wolfson who already pointed to a similarity between Spinoza and Gersonides. In his zeal to postulate personal immortality, Wolfson failed, however, to capture the important similarities between Spinoza and Gersonides. See Wolfson, *Spinoza,* p. 318ff.

43. See Hampshire, *Spinoza,* p. 137:

44. For a detailed examination of these propositions in the context of Spinoza's theory of moral psychology, see Michael Della Rocca, "Spinoza's Metaphysical Psychology," in *The Cambridge Companion to Spinoza,* ed. Donald Garrett (Cambridge: Cambridge University Press, 1996), pp. 192–266.

45. In fact, although I cannot argue the point here, I would maintain that this very theory is shared by Ibn Gabirol as well: what unites these three thinkers is their recognition and acceptance of the implications of a hylomorphic ontology.

46. Rice, for one, has argued that for Spinoza, to be an individual is to be "a center of action connected in various ways with a network of other individuals." See Lee Rice, "Spinoza on Individuation," in *Spinoza: Essays in Interpretation,* ed. E. Freeman and M. Mandelbaum (La Salle, Ill.: Open Court, 1975), p. 205.

47. Mason, *The God of Spinoza,* p. 241.

48. For a depiction of the three levels of knowledge, see Spinoza, *Ethics* 2p40s2 ff.

49. See Margaret Wilson, "Spinoza's Theory of Knowledge," in *The Cambridge Companion to Spinoza,* ed. Donald Garrett (Cambridge: Cambridge University Press, 1996), pp. 89–141.

50. I owe this objection to M. della Rocca, who raised it in private conversation. The response of Spinoza, as well as that of Gersonides, would have to be that *x*'s "idea of *b*" and *y*'s "idea of *b*" can never be identical inasmuch as the content of the knower enters into the relation between knower and object known. But this leads us far afield from the matter at hand.

51. Wolfson, *Spinoza,* pp. 318, 322.

52. For an engaging and comprehensive discussion of Spinoza's life and subsequent *ḥerem,* see Steven Nadler, *Spinoza: A Life* (Cambridge: Cambridge University Press, 1998).

53. E.E. Harris, "Spinoza's Theory of Human Immortality," in *Spinoza: Essays in Interpretation,* ed. E. Freeman and M. Mandelbaum (La Salle, Ill.: Open Court, 1975), pp. 114–45.

54. Mason, *The God of Spinoza,* p. 225.

55. Ibid., p. 241.

56. Spinoza, *Ethics* 5p23.

Bibliography

Abbreviations

AJS Review	*Association for Jewish Studies Review*
HUCA	*Hebrew Union College Annual*
JAOS	*Journal of the American Oriental Society*
JHP	*Journal of the History of Philosophy*
JQR	*Jewish Quarterly Review*
PAAJR	*Proceedings of the American Academy for Jewish Research*
PPR	*Philosophy and Phenomenological Research*
REJ	*Revue des études juives*

Primary Sources

Abraham Ibn Ezra. *Commentary on the Pentateuch (Genesis),* trans. and annotated H. Norman Strickman and Arthur M. Silver (New York: Menorah, 1988); *Perushei ha-Torah,* ed. Asher Weiser (Jerusalem: Mossad Ha-Rav Kook, 1976); also in standard rabbinic Bibles.

Abravanel, Isaac. *Perush 'al ha-Torah.* 5 pts. in 3 vols. (New York, Saphrograph Publ. 1959).

Abravanel, Judah (Leone Ebreo). *Dialoghi D'amore* (Jerusalem: Bialik Institute, 1983); *Siḥot 'al ha-'ahavah* (Jerusalem: Bialik Institute, 1983); Engl. trans. F. Friedenberg-Seeley and J. H. Barnes (London: The Soncino Press, 1937).

Albalag, Isaac. *Sefer Tiqqun ha-De'ot* (Jerusalem: The Israel Academy of Sciences and Humanities, 1973).

Albo, Joseph. *Sefer ha-'Ikkarim (Book of Principles),* ed. and trans. Isaac Husik, 4 vols. (Philadelphia: Jewish Publication Society, 1929–30).

Aquinas, Thomas. *On the Unity of the Intellect Against the Averroists,* trans. B. Zedler (Milwaukee, Marquette University Press, 1968).

———. *Opera Omnia* (Rome, 1882–present).

———. *Quaestiones Quodlibetales,* ed. Raymundus Spiazzi (Turin Rome: Marietti, 1956).

———. *Summa Theologiae,* 5 vols. (Ottawa, 1941–45); Engl. trans. Fathers of the English Dominican Province, 2nd ed. (London: Burnes Oates and Washbourne, 1920).

Arama, Isaac ben Moses, *'Aqedat Yitzḥak* (Pressburg [Bratislava, Slovakia]: V. Kittseer, 1849).

Aristotle. *Categories and De Interpretatione,* ed. and trans. Ackrill (Oxford: Clarendon Press, 1964).

———. *De Caelo,* trans. W.K.C. Guthrie, Loeb Classical Series (Cambridge, Mass.: Harvard University Press, 1971).

———. *Metaphysics,* 2 vols., trans. T. Tredennick, Loeb Classical Series (Cambridge, Mass.: Harvard University Press, 1936).

———. *Physics,* 2 vols., trans. P. H. Wicksteed and F. M. Cornford, Loeb Classical Series (Cambridge, Mass.: Harvard University Press, 1957).

———. *The Complete Works of Aristotle, 2 vols.,* ed. Jonathan Barnes (Princeton, N.J.: Princeton University Press, 1984).

Augustine. *Confessions,* comm. by James J. O'Donnell (Oxford: Clarendon Press, 1992).

Aureol, Peter. *Commentariorum in Secundum Librum Sententiarum Pars Secundus,* (Rome: Ex typographia Vaticana, 1596–1605).

Averroës. *Epitome of* Parva Naturalia, trans. Harry Blumberg (Cambridge, Mass.: Medieval Academy of America, 1961).

———. *Incoherence of the Incoherence (Tahāfut al-Tahāfut),* ed. and trans. Simon van den Bergh, 2 vols. (London: E. J. Gibb Memorial, 1954).

———. *Middle Commentary and Epitome of* De Generatione et Corruptione, ed. and trans. S. Kurland (Cambridge, Mass.: Medieval Academy of America, 1958).

Baḥya ibn Paquda. *Ḥovot ha-Levavot,* ed. and Hebrew trans. Yoseph Kafiḥ (Jerusalem: Feldheim Publ., 1973); edition of Judah ibn Tibbon trans. and English trans. Moses Hyamson (Jerusalem: Feldheim Publ., 1970).

Bar Ḥiyyah, Abraham. *Hegyon ha-Nefesh,* ed. Isaac Freiman (Leipzig: Vollrath, 1860).

Bedersi, Yedaiah. *A Treatise upon Personal or Individual Forms.* Bibliothèque Nationale, Paris, MS 984 Heb. fol. 66a–93b.

———*Sefer Beḥinat 'Olam* (first printed in Mantua, before 1480).

Boethius. *The Consolation of Philosophy,* ed. James J. Buchanan (New York: Frederick Ungar, 1957).

Cassuto, A. *Commentary on the Book of Genesis I,* trans. Israel Abrahams (Jerusalem: Magnes Press, 1961–64).

Crescas, Ḥasdai, *'Or 'Adonai* (*The Light of the Lord*) (Ferrara, 1555; repr. Jerusalem: Makhor, 1970).

Delmedigo, Joseph. *Sefer 'Elim* (Amsterdam, 1629; repr. Odessa, 1864–67).

Descartes, René. *Meditations,* trans. John Cottingham (Cambridge: Cambridge University Press, 1986).

Donnolo. *Il Commento di Sabbatai Donnolo sul Libro della Creazione,* ed. D. Castelli (Firenze 1880, Heb. Sec).

Duns Scotus. *Opera Omnia,* 26 vols. (Paris, 1891–95).

Feldman, Leon A. *Rabbi Nissim ben Reuben Gerondi-Commentary on the Bible* (Jerusalem: Makhon Shalem, 1968).

Ficino, M. *Opera Omnia* (Basel, 1576; repr. Turin, Bottega d'Erasmo, 1962).

Gersonides. *Commentary on the Pentateuch, Exodus* 32:10 (Venice, 1547), also in standard rabbinic Bibles.

———. *Perushei ha-Torah le-Rabbeinu Levi ben Gershom,* ed. by Rabbi Yaacov Levi (Jerusalem: Mossad Harav Kook, 1992).

———. *Sefer Maʿaseh Ḥoshev,* ed. and trans. Gerson Lange (Frankfurt: Louis Golde, 1909).

———. *Sefer Milḥamot ha-Shem* (Leipzig, 1866; repr. Berlin, 1923); German trans. *Die Kämpfe Gottes von Lewi ben Gerson,* trans. Benzion Kellerman (Berlin: Mayer and Muller, 1914–16); French trans. *Les Guerres du Seigneur: Livres III et IV,* ed. and trans. Charles Touati (Paris: Mouton and Co., 1968); English trans. S. Feldman, trans. and ed. *The Wars of the Lord (Books 1–4),* trans. and ed. S. Feldman (Philadelphia: Jewish Publication Society of America, 1984, 1987).

———. *Supercommentary on Averroës' Commentary on Aristotle's De Interpretatione, Chapter 9* (found in the 1562 edition of *Aristotelis omnia quas extant opera . . . Averroës Cordubensis in ea opera omnes qui ad haec usque tempora pervenere commentarii . . . cum Levi Gersonidis in libros logicos annotationibus . . . a Iacob Mantino in Latinum conversi.* (Venetiis, apud Iunctas, 1562) (vol. I, ff. 82^{vb}K–83^{rb}F). An English translation of this portion of Averroës' commentary and Gersonides' supercommentary has been prepared by N. Kretzmann.

Heraclitus. "Fragments," in *Philosophy Before Socrates,* ed. Richard D. McKirahan Jr. (Indianapolis: Hackett, 1994).

Hillel of Verona. *Sefer Tagmule ha-Nefesh,* ed. and trans. Joseph Sermoneta (Jerusalem: Israel Academy of Sciences and Humanities, 1981).

Ibn Daud, Abraham. *Sefer ha-Emunah ha-Ramah* (*The Exalted Faith*), trans. Solomon B. Labi, and ed. S. Weil (Frankfurt: Sifriyah le-Maḥshevet Yisra'el, 1852).

Ibn Gabirol, Solomon. *Selected Religious Poems of Solomon Ibn Gabirol,* ed. I. Davidson and trans. I. Zangwill (Philadelphia: Jewish Publication Society, 1924).

———. *Keter Malkhut,* in Raphael Loewe, *Ibn Gabirol* (New York: Grove Weidenfeld, 1989).

———. *Meqor Ḥayyim* (*Fountain of Life*) in A. Munk, "La Source de Vie," in *Mélanges de philosophie juive et arabe* (Paris, 1857; repr. 1955); Latin trans. *Avencebrolis Fons Vitae,* trans. Clemens Baeumker, in *Beitrage zur Geschichte der Philosophie des Mittelalters* (vol. 1, fascicules 2–4) (Munster, 1892–95); Hebrew trans. E. Sifroni (Tel Aviv; Mossad Harav Kook, book 3, 1926); Eng. trans. *The Fountain of Life,* trans. H. E. Wedeck (New York: Philosophical Library, 1962) French trans. *Livre de la Source de Vie (Fons Vitae),* trans. and ed. Jacques Schlanger (Paris: Aubier Montaigne, 1970).

Ibn Rushd. See Averroës.

Judah Halevi. *Sefer ha-Kuzari* (*The Kuzari*), 5 pts. in 1 vol. (1880; repr. Tel Aviv, 1959); Arabic and Hebrew, ed. H. Hirschfeld (Leipzig, 1887); English trans. Hirschfeld (London, 1905; New York: Schocken Books, 1964).

Kasher, Menahem M. *Encyclopedia of Biblical Interpretation, A Millennial Anthology* vol. 1: *Genesis* (New York: American Biblical Encyclopedia Society, 1953).

Lassen, Abraham. *The Commentary of Levi ben Gerson on the Book of Job* (New York: Bloch, 1946).

Leone Ebreo. See Abravanel, Judah.

Levi ben Gershom. See Gersonides.

Leibniz. *Leibniz' Philosophical Papers and Letters,* ed. L. Loemker (Dordrecht: D. Reidel, 1970).

Maimonides, Moses. "Letter on Astrology." In *A Maimonides Reader,* ed. Isadore Twersky (New York: Behrman House, 1972).

———. "Maimonides' Treatise on Logic," ed. Israel Efros. *PAAJR* 10 (1938).

———. *Commentary on the Mishnah,* ed. Yoseph Kafih (Jerusalem: Mossad Harav Kook, 1965).

———. *Letters of Maimonides,* trans. Yoseph Kafiḥ (Jerusalem: Mossad ha-Rav Kook, 1972).

———. *Mishneh Torah: The Book of Knowledge,* trans. Moses Hyamson (Jerusalem, 1962); Hebrew ed. *Mishneh Torah,* trans. and commentary E. Touger (Jerusalem: Moznaim, 1989).

———. *Moreh ha-Nevukhim* (*The Guide of the Perplexed, Dalālat al-Ha'irīn*), trans. Samuel Ibn Tibbon with commentaries by Abravanel, Crescas, Profiat Duran, and Shem Tov ibn Shem Tov (Jerusalem, 1959–60); Arabic ed. Munk-Joel (Jerusalem, 1931); French trans. S. Munk. *Le Guide des Égarés* (Paris, 1856; repr. 1970); English trans. Shlomo Pines (Chicago: University of Chicago Press, 1963).

Moses of Narbonne. See Naḥmanides.

Midrash Bereshit Rabba: Critical Edition with Notes and Commentary (Heb.), ed. J. Theodor and Ch. Albeck, 3 vols. (1903–36; repr. Jerusalem Makhon Hamidrash Hamevoar, 1965).

Naḥmanides. *Perushei ha-Torah* ed. Charles B. Chavel 2 vols. (Jerusalem, 1959–60; repr. 1977); Engl. trans. *The Commentary of Naḥmanides on Genesis Chapters 1–6*[8], ed. Jacob Newman (Pretoria Oriental Series 4: Leiden: E.J. Brill, 1960).

———. *Kitve RaMbaN,* ed. Charles B. Chavel, 2 vols. (Jerusalem, 1963).

Neusner, Jacob. *Genesis Rabbah: The Judaic Commentary to the Book of Genesis—A New American Translation,* vol. 1 (Atlanta: Scholars Press, 1985).

Newton, Sir Isaac. *Philosophia Naturalis Principia Mathematica,* in F. Cajori, ed. *Sir Isaac Newton's Mathematical Principles of Natural Philosophy and His System of the World* (New York: Greenwood Press, 1969); Latin text in *Isaac Newton's Philosophiae Naturalis Principia Mathematica,* ed. Alexandre Koyre and I. Bernard Cohen (Cambridge, Mass.: Harvard University Press, 1972).

Ockham, William. *Predestination, God's Foreknowledge and Future Contingents,* trans. and ed. M. M. Adams and N. Kretzmann (New York: Appleton, Century Crofts, 1969).

Philo of Alexandria. *Complete Works,* ed. F. H. Colson et al., Loeb Classical Library, 10 vols. and 2 suppl. vols. (Cambridge, Mass.: Harvard University Press, 1929–53).

Philoponus. *Philoponus: Against Aristotle on the Eternity of the World,* trans. Christian Wildberg (Ithaca, N.Y.: Cornell University Press, 1987).

Pico della Mirandola. *Examen Vanitatis Doctrina Gentium*, in *Opera Omnia* (Basel, 1557; repr. Hildesheim: G. Olms, 1969).

Plato. *The Collected Dialogues of Plato*, ed. E. Hamilton and H. Cairns (Princeton, N.J.: Princeton University Press, 1971).

———. *Timaeus*, ed. and trans. R.G. Bury, Loeb Classical Series (Cambridge, Mass.: Harvard University Press, 1966).

Plotinus. *Enneads*, English trans. Stephen McKenna (London: Faber, 1969).

Ptolemy. *Almagest*, trans. and ed. G. J. Toomer (London: Duckworth, 1984)

Rashi. *Rash'i 'al ha-Torah*, in standard editions of the Hebrew Bible.

Saadia Gaon. *Commentary on Genesis* (Arabic with Hebrew trans.), ed. Moshe Zucker (New York: Bet hamidrash le-rabanim, 1984).

———. *Kitāb al-Amānāt wa' l-i' tiqādāt*, trans. Samuel Rosenblatt as *The Book of Beliefs and Opinions* (New Haven: Yale University Press, 1948); *Sefer Emunot ve-De' ot*, ed. and trans. Yosef Kafiḥ (Jerusalem: Sura Press, 1970–71).

———. Commentary: *Sefer Yetzirah 'im Perush Rabbenu Sa'adia ben Yosef Fayyumi (Ga'on)*, Arabic text edited with Hebrew translation by Yosef Kafiḥ (Jerusalem, M. Atiyeh, 1961/1962); *Commentaire sur le Sefer Yesira*, trans. Mayer Lambert (repr. Paris Editions bibliophane, 1986).

Sefer ha-Bahir, ed. Reuben Margaliot (1951; repr. Jerusalem: Mossad Harav Kook, 1978).

Sefer Yetzirah (Jerusalem: Yeshivat kol Yehudah, 1990).

Sforno, 'Obadia. *Be'ur 'al ha-Torah*, in *Miqra'ot Gedolot: Sefer Berei'shit* (New York: Shulsinger Bros, 1950–78); English trans. *Commentary on the Torah*, trans. Raphael Pelcovitz (New York: Mesorah, 1987).

Spinoza, B. *Ethica*, ed. J. Van Vloten and J. P. N. Land (Martinus Nijhoff, 1905); English trans. *The Collected Works of Spinoza* vol. 1, trans. and ed. E. M. Curley (Princeton, N.J.: Princeton University Press, 1985).

Von Arnim, Hans, ed. *Stoicorum Veterum Fragmenta*, 4 vols. (Leipzig: Lipsiae, in aedibus B. G. Teubneri, 1903–1924).

Secondary Sources

"Astronomy." *Encyclopedia Judaica* (Jerusalem: Keter Publ., 1972), vol. 2, p. 795.

"Astrology." *Encyclopedia Judaica* (Jerusalem: Keter Publ., 1972), vol. 2, p. 789.

Adams, Marilyn McCord. *William Ockham* (Notre Dame, Ind.: University of Notre Dame Press, 1987).

Adams, Robert M. "Middle Knowledge and the Problem of Evil," *American Philosophical Quarterly* 14 (1977): 109–17.

Agassi, Joseph. "Leibniz's Place in the History of Physics," *Journal of the History of Ideas* 30.3 (1969): 331–44.

Allison, Henry E. *Benedict de Spinoza* (Boston: Twayne, 1975).

Altmann Alexander and S. M. Stern. *Isaac Israeli: A Neoplatonic Philosopher of the Early Tenth Century* (Oxford: Clarendon Press, 1958).

Altmann, Alexander. "R. Yehudah ha-Levi's Climatology [Hebrew]," *Melilah* 1 (1944): 1–17.

———. "A Note on the Rabbinic Doctrine of Creation," *Journal of Jewish Studies* 7 (1956): 195–206.

———. "Ibn-Bajja on Ultimate Felicity," in *H.A. Wolfson Jubilee Volume,* ed. Saul Lieberman (Jerusalem: American Academy for Jewish Research, 1965), pp. 47–87.

———. "Isaac Israeli," *Encyclopedia Judaica* (Jerusalem: Keter Publ., 1972).

———. "The Religion of the Thinkers: Free Will and Predestination in Saadia, Bahya, and Maimonides," in *Religion in a Religious Age,* ed. S. D. Goitein (Cambridge, Mass.: AJS Press, 1974), pp. 25–52.

———. "Gersonides' Commentary on Averroës' *Epitome of Parva Naturalia* II.3,'" *PAAJR Jubilee Volume* 66–67 (1979–80): 1–31.

———. "*Hashgaḥah,*" in the *Hebrew Encyclopedia,* pp. 476–90.

———. "Maimonides on the Intellect and the Scope of Metaphysics," in A. Altmann, *Von der mittelalterlichen zur modernen Aufklärung* (Tübingen: Mohr, 1986), pp. 60–129.

Anderson, Bernhard W., ed. *Creation in the Old Testament,* Issues in Religion and Theology 6 (Philadelphia: Fortress Press, 1984).

Angel, R. B. "Some Central Problems in the Philosophy of Space and Time," in *Space Time, and Creation,* ed. R. Swinburne (Boston: Martinus Nijhoff, 1982).

Annas, Julia. "Aristotle, Number and Time," *The Philosophical Quarterly* 99 (1975): 97–113.

Ariew, Roger. "The Infinite in Spinoza's Philosophy," in *Spinoza: Issues and Directions* (Leiden: Brill, 1990), pp. 16–31.

Ariotti, Piero E. "Toward Absolute Time: The Undermining and Refutation of the Aristotelian Conception of Time in the Sixteenth and Seventeenth Centuries," *Annals of Science* 30 (1973): 31–50.

Arthur, Richard T. "Continuous Creation, Continuous Time: A Refutation of the Alleged Discontinuity of Cartesian Time," *JHP* 26 (1988): 349–75.

Baldner, Stephen. "St. Bonaventure on the Beginning of the Temporal World," *The New Scholasticism* 63 (1989): 206–228.

Barbour, Ian. *Religion and Science: Historical and Contemporary Issues* (San Francisco: Harper Books, 1997).

Barkai, Ron. "Theoretical and Practical Aspects of Jewish Astrology in the Middle Ages," in R. Barkai, *Science, Magic and Mythology in the Middle Ages* (Heb) (Jerusalem: Makhon Van Leer, 1987): 7–35.

Barker, Peter, and Bernard R. Goldstein. "The Role of Comets in the Copernican Revolution," *Studies in the History of Philosophy of Science* 19 (1988): 299–319.

Barnes, Jonathan. *The Presocratic Philosophers* (London: Routledge and Kegan Paul, 1982).

Barr, James. *Biblical Words for Time* (Naperville, Ill.: Alec R. Allenson, 1962).

———*The Semantics of Biblical Language* (Oxford: Oxford University Press, 1961).

Barrow, Isaac. *The Geometrical Lectures of Isaac Barrow,* trans. J. M. Child (La Salle, Ill.: Open Court, 1916).

Barzilay, I. *Yoseph Shlomo Delmedigo (Yashar of Candia): His Life, Works, and Times* (Leiden: E.J. Brill, 1974).

Bassler, Bradley O. "Leibniz on the Indefinite as Infinite," *The Review of Metaphysics* 51 (1998): 849–74.

Baudry, L. *La Quérelle des futurs contingents* (Paris: J. Vrin, 1950).

Benjamin, Cornelius. "Ideas of Time in the History of Philosophy," in *The Voices of Time: A Cultural Survey of Man's Views of Time as Expressed by the Sciences and the Humanities,* ed. J. T. Fraser (New York: George Braziller, 1966): 3–30.

Bennett, Jonathan. *A Study of Spinoza's Ethics* (Indianapolis: Hackett, 1984).

Benor, Ehud. *Worship of the Heart: A Study in Maimonides' Philosophy of Religion* (Albany: State University of New York Press, 1995).

Benthem, J.F.A.K. *The Logic of Time: A Model-Theoretic Investigation into the Varieties of Temporal Ontology and Temporal Discourse* (Dordrecht: D. Reidel, 1983).

Blau, Joseph L. "On the Supposedly Aristotelian Character of Gabirol's *Keter Malkut,*" in *Salo Wittmayer Baron Jubilee Volume on the Occasion of His Eightieth Birthday,* ed. S. Lieberman (Jerusalem: American Academy for Jewish Research, 1974): 219–28.

Blumberg Z. Ha-Rambam al Musag al-Tajwīz be-Shitatam shel ha-Mutakalimūn," *Tarbiz* 39 (1970): 268–76.

Blumenfeld, David. "Leibniz on Contingency and Infinite Analysis," *PPR* 45 (1985): 483–514.

Blumenthal, H.J. "Plotinus in Later Platonism," in *Neoplatonism and Early Christian Thought: Essays in Honour of A. H. Armstrong,* ed. H. J. Blumenthal and R. A. Markus (London: Variorum Publications, 1981): 212–22.

Boman, Thorlief. *Hebrew Thought Compared with Greek* (London: Westminster Press, 1960).

Bostock, David. "Aristotle's Account of Time," *Phronesis* 25 (1980): 148–169.

Broadie, Alexander. "Maimonides on Divine Knowledge," in *Of Scholars, Savants and Their Texts,* ed. R.-L. Salinger (New York: Peter Lang, 1989): 47–55.

Brunner, Fernand. "La doctrine de la matière chez Avicebron," *Revue Thélogique et Philosophique* 6 (1956): 261–79.

———. "Sur la philosophie d'Ibn Gabirol," *REJ* 128 (1969): 317–37.

———. *Platonisme et aristotelisme: La Critique d'Ibn Gabirol par St Thomas d'Aquin* (Louvin: Publications Universitaires de Louvin, 1965).

Burnett, Charles. "Al-Kindi on Judicial Astrology: *The Forty Chapters,*" *Arabic Sciences and Philosophy* 3 (1993): 77–117.

Cahn, Steven. *Fate, Logic and Time* (New Haven: Yale University Press, 1967).

Callus, David. "The Origins of the Problem of the Unity of Form," *Thomist* 24 (1961): 257–85.

Cantor, Georg. *Contributions to the Founding of the Theory of Transfinite Numbers,* trans. P. E. B. Jourdain (New York: Dover, 1952).

Capek, Milic. *The New Aspects of Time: Its Continuities and Novelties. Selected Papers in the Philosophy of Science* (Dordrecht, Boston, London: Kluwer Academic, 1991).

Carlebach, Salomon. *Levi ben Gerson als Mathematiker* (Berlin, 1910).

Cassuto, Umberto. *A Commentary on the Book of Genesis,* trans. Israel Abrahams, 2 vols. (Jerusalem: Magnes Press, 1961–64).

Chapman, T. *Time: A Philosophical Analysis* (Dordrecht: D. Reidel, 1982).

Chisholm, Roderick M. "Beginnings and Endings," in *Time and Cause: Essays Presented to Richard Taylor,* ed. Peter van Inwagen (Dordrecht: D. Reidel, 1980): 17–26.

Clagett, Marshall. *Greek Science in Antiquity* (New York: Abelard-Schuman, 1955).

Clark, Gordon H. "The Theory of Time in Plotinus," *The Philosophical Review* 53 (1944): 337–58.

Coats, G.W. and Long Burke, ed. *Canon and Authority: Essays in Old Testament Religion and Theology* (Philadelphia: Fortress Press, 1977).

Code, Murry J. "Zeno's Paradoxes I: The Standard Mathematical Response," *Nature and System* 4 (1982): 45–48.

Copenhaver, Brian. "Jewish Theologies of Space in the Scientific Revolution: Henry More, Joseph Raphson, Isaac Newton and their Predecessors," *Annals of Science* 37 (1980): 489–548.

Cornish, Dennis. "Aristotle's Attempted Derivation of Temporal Order from that of Movement and Space," *Phronesis* 21 (1976): 241–51.

Cottingham, John G. "The Intellect, the Will, and the Passions: Spinoza's Critique of Descartes," *JHP* 26 (1988): 239–57.

Craig, William Lane. *The Problem of Divine Foreknowledge and Future Contingents from Aristotle to Suarez* (Leiden: Brill, 1988).

Curley, E. M. "The Root of Contingency," in *Leibniz: A Collection of Critical Essays,* ed. H.G. Frankfurt (New York, 1972): 69–98.

Dahan, Gilbert, ed. *Gersonide en son temps* (Louvain-Paris: E. Peeters, 1991).

Dales, Richard C. "Discussions of the Eternity of the World during the First Half of the Twelfth Century," *Speculum* 57 (1982): 495–508.

——— "Early Latin Discussions of the Eternity of the World in the Thirteenth Century," *Traditio* 43 (1987): 171–97.

———*Medieval Discussions of the Eternity of the World* (Leiden: Brill, 1990).

Daniel, Stephen H. "Seventeenth-Century Scholastic Treatments of Time," *The Journal of the History of Ideas* 42 (1981): 587–606.

Davidson, Herbert. "John Philoponus as a Source of Medieval Islamic and Jewish Proofs for Creation," *JAOS* 89 (1969): 357–91.

———. "Maimonides' Secret Position on Creation," *Studies in Medieval Jewish History and Literature,* ed. I. Twersky (Cambridge, Mass.: Harvard University Press, 1979): 16–40.

———. *Proofs for Eternity, Creation and the Existence of God in Medieval Islamic and Jewish Philosophy* (Oxford: Oxford University Press, 1987).

———. *Alfarabi, Avicenna, and Averroës on Intellect* (New York: Oxford University Press, 1992).

———. "Gersonides on the Material and Active Intellects," in *Studies on Gersonides: A Fourteenth-Century Jewish Philosopher-Scientist,* ed. Gad Freudenthal (Leiden: Brill, 1992): 195–265.

———. "Maimonides on Metaphysical Knowledge," *Maimonidean Studies* 3 (1992–93): 49–103.

Della Rocca, Michael. "Spinoza's Metaphysical Psychology," in *The Cambridge Companion to Spinoza,* ed. Donald Garrett (Cambridge: Cambridge University Press, 1996): 192–266.

———. *Representation and the Mind-Body Problem in Spinoza* (Oxford and New York: Oxford University Press, 1996).

Dellig. "Kairos" in *Theological Dictionary of the New Testament,* vol. 3, ed. Gerhard Kittel (Grand Rapids, Mich. Eerdmans, 1965): 455–64.

Diesendruck, Zvi. "Saadya's Formulation of the Time-Argument for Creation," in *Jewish Studies in Memory of George A. Kohut,* ed. Salo W. Baron and Alexander Marx (New York: The Alexander Kohut Memorial Foundation, 1935): 145–58.

———. "Samuel and Moses ibn Tibbon on Maimonides' Theory of Providence," *HUCA* 11 (1936): 341–66.

Dillon, John. "Salomon Ibn Gabirol's Doctrine of Intelligible Matter. *Irish Philosophical Journal* 6 (1989): 59–81.

Dobbs-Weinstein, Idit. *Maimonides and St. Thomas on the Limits of Reason* (Albany: State University of New York Press, 1995).

Dodd, C. H. *The Bible and the Greeks* (1935; repr. London: Hodder & Stoughton, 1954).

Duhem, Pierre. *Le Système du monde* (Paris: A. Hermann Press, 1913–59).

———. *To Save the Phenomena: An Essay on the Idea of Physical Theory from Plato to Galileo,* trans. E. Dolan and Ch. Maschler (Chicago: University of Chicago Press, 1969).

———. *Medieval Cosmology,* ed. and trans. Roger Ariew (Chicago: University of Chicago Press, 1985).

Durr, Walter. "Rhythm in Music: A Formal Scaffolding of Time," in *The Voices of Time: A Cultural Survey of Man's Views of Time as Expressed by the Sciences and the Humanities,* ed. J. T. Fraser (New York: George Braziller, 1966): 180–200.

Efros, Israel. *The Problem of Space in Jewish Medieval Philosophy* (New York: Columbia University Press, 1917).

———. "Sa'adyah's Second Theory of Creation in Its Relation to Pythagoreanism and Platonism," in *Louis Ginzberg: Jubilee Volume on the Occasion of His Seventieth Birthday,* ed. Alexander Marx et al. (New York: The American Academy for Jewish Research, 1945): 133–42.

———. *Studies in Medieval Jewish Philosophy* (New York and London: Columbia University Press, 1974).

Eichrodt, W. *Theology of the Old Testament,* trans. J. A. Baker, 2 vols. (Philadelphia: Fortress Press, 1961–67).

———. "In the Beginning: A Contribution to the Interpretation of the First Word of the Bible," in B. Anderson, *Creation in the Old Testament* (Philadelphia and London: Fortress Press, 1984).

Eisen, Robert. *Gersonides on Providence, Covenant, and the Jewish People* (Albany: State University of New York Press, 1995).

Eliade, Mircea. *The Sacred and the Profane* (New York: Harcourt, Brace and Co., 1959).

Fakhry, Majid. "The Antinomy of the eternity of the world in Averroës, Maimonides and Aquinas," *Le Muséon* 66 (1953): 1–17.

———. *A History of Islamic Philosophy* (New York: Columbia University Press, 1983).

Falk, Arthur. "The State of the Question about Fate," *Philosophical Research Archives* (1980).

Falkenstein, Lorne. "Kant, Mendelssohn, Lambert, and the Subjectivity of Time," *JHP* 29 (1991): 227–51.

Feldman, Seymour. "Gersonides' Proofs for the Creation of the Universe," *PAAJR* 35(1967): 113–37; reprinted in *Essays in Medieval Jewish and Islamic Philosophy*, ed. Arthur Hyman (New York: Ktav, 1977): 219–54.

———. "A Scholastic Misinterpretation of Maimonides' Doctrine of Divine Attributes," *Journal of Jewish Studies* 19 (1968): 23–39.

———. "Cosmology," *Encyclopedia Judaica* (Jerusalem: Keter Publ., 1972): 982–83.

———. "Platonic Themes in Gersonides' Cosmology," in *Salo Whittmayer Baron Jubilee Volume* (Jerusalem: American Academy for Jewish Research, 1975): 383–405.

———. "Gersonides on the Possibility of Conjunction with the Agent Intellect," *AJS Review* 3 (1978): 99–120.

———. "The Theory of Eternal Creation in Hasdai Crecas and Some of His Predecessors," *Viator* 11 (1980): 289–320.

———. "Crescas' Theological Determinism," *Da'at* 9 (1982): 2–28.

———. "A Debate Concerning Determinism in Late Medieval Jewish Philosophy," *PAAJR* 51 (1984): 15–54.

———. "The End of the Universe in Medieval Jewish Philosophy," *AJS Review* 11 (1986): 53–78.

———. "In the Beginning God Created: A Philosophical Midrash," in *God and Creation: An Ecumenical Symposium*, ed. David M. Burrell and Bernard McGinn (Notre Dame, Ind.: University Notre Dame Press, 1990): 3–26.

———. "Platonic Themes in Gersonides' Doctrine of the Active Intellect," in *Neoplatonism and Jewish Thought*, ed. Lenn Goodman (Albany: State University of New York Press, 1992): 255–77.

Feldman, W. M. *Rabbinical Mathematics and Astronomy* (New York: Hermon Press, 1975).

Fenton, Paul. "Gleanings from Moseh Ibn Ezra's *Maqalat al-Hadiqa*," *Sefarad* 36 (1976): 285–98.

———. "The Arabic and Hebrew Versions of the *Theology of Aristotle*," *Pseudo-Aristotle in the Middle Ages: The 'Theology of Aristotle' and Other Texts*, ed. J. Kraye (London: Warburg Institute, 1986): 241–64.

———. "Shem Tov Ibn Falaquera and the *Theology of Aristotle*," *Da'at* 29 (1992): 27–39.

Fontaine, T. A. M. *In Defense of Judaism: Abraham Ibn Daud* (Assen, Netherlands: Van Gorcum Press, 1990).

Force, James E. "Spinoza's *Tractatus Theologico Politicus:* A New Way of Looking at the World," *The Southern Journal of Philosophy* 12 (1974): 343–55.

Frank, Richard. "The Science of Kalām," *Arabic Sciences and Philosophy* 2 (1992): 7–37.

Frankel, Lois. "Leibniz on the Foundations of Space and Time," *Nature and System* 3 (1981): 91–98.

Fraser, Julius Thomas. "The Study of Time," in *The Voices of Time: A Cultural Survey of Man's Views of Time as Expressed by the Sciences and the Humanities,* ed. Julius Thomas Fraser (New York: George Braziller, 1966): 582–91.

Freudenthal, Gad. "Épistemologie, astronomie et astrologie chez Gersonide," REJ (1987): 357–65.

———. "Human Felicity and Astronomy: Gersonides' Revolt against Ptolemy" (Heb.) *Da'at* 22 (1989): 55–72.

———. Review of Feldman's translation in *REJ* 148 (1989), p. 381.

———. ed. *Studies on Gersonides: A Fourteenth-Century Jewish Philosopher-Scientist* (Leiden: Brill, 1992).

———. "Les sciences dans les communautés juives médiévales de Provence: Leur appropriation, leur rôle," *REJ* 152 (1993): 29–136.

———. "Maimonides' Stance on Astrology in Context: Cosmology, Physics, Medicine and Providence," in *Moses Maimonides: Physician, Scientist and Philosopher,* ed. Fred Rosner and Samuel Kottek (Northvale, N.J.: J. Aronson, 1993): 78–79.

Funkenstein, Amos. *Theology and the Scientific Imagination from the Middle Ages to the Seventeenth Century* (Princeton, N.J.: Princeton University Press, 1986).

———. *Perceptions of Jewish History* (Berkeley: University of California Press, 1993).

Furley, David J. "Treatment of Aristotle's Continuous Theory," in *Infinity and Continuity in Ancient and Medieval Thought,* ed. Norman Kretzmann (Ithaca, N.Y.: Cornell University Press, 1982): 17–36.

Garber, Daniel. *Descartes' Metaphysical Physics* (Chicago: University of Chicago Press, 1992).

Garber, Daniel, John Henry Lynn Joy, and Alan Gabbey. "New Doctrines of Body and Its Powers, Place and Space," in *The Cambridge History of Seventeenth-Century Philosophy,* ed. Daniel Garber and Michael Ayers (Cambridge: Cambridge University Press, 1998): 553–623.

Garrett, Donald. "Spinoza's Ethical Theory," in *The Cambridge Companion to Spinoza,* ed. D. Garrett (Cambridge: Cambridge University Press, 1996): 267–314.

Gaskin, R. "Conditionals of Freedom and Middle Knowledge," *The Philosophical Quarterly* 43 (1993): 412–30.

Gellman, J. "Freedom and Determinism in Maimonides' Philosophy," in *Moses Maim-*

onides and His Time, ed. E. L. Ormsby (Washington, D.C., Catholic University of America Press, 1989): 139–50.

Gerson, Lloyd. *Plotinus* (London and New York: Routledge, 1994).

Gilead, Amihud. "The Order and Connection of Things (Are They Constructed Mathematically-Deductively according to Spinoza?)," *Kantstudien* 76 (1985): 72–78.

Glasner, Ruth. "The Early Stages in the Evolution of Gersonides' *The Wars of the Lord,*" *The Jewish Quarterly Review* 87 (1996): 1–46.

———. "The Hebrew Version of *De Celo et Mundo* Attributed to Ibn Sina," *Arabic Sciences and Philosophy* 6 (1996): 89–112.

Gockerell, Nina. "Telling Time without a Clock," in *The Clockwork Universe: German Clocks and Automata 1550–1650,* ed. K. Maurice and O. Mayr (New York: Neale Watson, 1980): 131–43.

Goldfeld, L.N. *Moses Maimonides' Treatise on Resurrection: An Inquiry into Its Authenticity* (New York: Ktav, 1986).

Goldish, Matt. "Newton on Kabbalah," in *The Books of Nature and Scripture: Recent Essays on Natural Philosophy, Theology, and Biblical Criticism in the Netherlands of Spinoza's Time and the British Isles of Newton's Time,* ed. J. E. Force and R. Popkin (Dordrecht: Kluwer Academic, 1994): 89–104.

———. *Judaism in the Theology of Sir Isaac Newton* (Boston: Kluwer Academic, 1998).

Goldstein, Bernard R. *The Astronomical Tables of Levi ben Gerson* (New Haven, Conn.: Connecticut Academy of Arts and Sciences, 1974).

———. "The Physical Astronomy of Levi ben Gerson," *Perspectives on Science* (1977): 1–30.

———. "The Status of Models in Ancient and Medieval Astronomy," *Centaurus* 24 (1980): 132–47.

———. *The Astronomy of Levi ben Gerson (1288–1344): A Critical Edition of Chapters 1–20* (New York: Springer Verlag, 1985).

———. *Theory and Observation in Ancient and Medieval Astronomy* (London: Variorum Reprints, 1985).

———. "Levi ben Gerson's Theory of Planetary Distances," *Centaurus* 29 (1986): 272–313.

———. "A New Set of Fourteenth Century Planetary Observations," *Proceedings of the American Philosophical Society* 132 (1988): 371–99.

———. "Galileo's Account of Miracles in the Bible: A Confusion of Sources," *Nuncius* 5 (1990): 3–16.

———. "Levi ben Gerson's Astrology in Historical Perspective," in *Gersonide en son temps,* ed. Gilbert Dahan (Louvain-Paris: E. Peeters, 1991): 287–300.

———. "Levi ben Gerson's Contributions to Astronomy," in *Studies on Gersonides,* ed. Gad Freudenthal (Leiden: E. J. Brill, 1992): 3–20.

———. "Astronomy and Astrology in the Works of Abraham Ibn Ezra," *Arabic Sciences and Philosophy* 6 (1996): 9–21.

Goldstein, Bernard R., and David Pingree. "Prognostication for the Conjunction of 1345," *Transactions of the American Philosophical Society* 80 (1990).

Goldstein, Jonathan A. "The Origins of the Doctrine of Creation *Ex Nihilo*," *Journal of Jewish Studies* 35 (1984): 127–135.

Goodman, Lenn E. "Ghazali's Argument From Creation II," *International Journal of Middle Eastern Studies* 2 (1971): 168–88.

———. "Ghazali's Argument From Creation," *International Journal of Middle East Studies* 2 (1971): 67–85.

———. "Time in Islam," *Asian Philosophy* 2 (1992): 3–19.

———. "Time, Creation, and the Mirror of Narcissus," *Philosophy East and West* 42 (1992).

———. *Neoplatonism and Jewish Thought: Studies in Neoplatonism; Ancient and Modern*, ed. L. E. Goodman (Albany: State University of New York Press, 1992).

Gould, Stephen Jay. *Time's Arrow, Time's Cycle: Myth and Metaphor in the Discovery of Geological Time* (Cambridge, Mass.: Harvard University Press, 1987).

Govrin, Myrtal. Tefisat ha-Zeman be-Maḥshevet R. 'Azriel mi-Gerona," Meḥkere Yerushalayim be-Maḥshevet Yisra'el 6 (1987): 309–36.

Gracia, Jorge J. E. *Introduction to the Problem of Individuation in the Early Middle Ages* (Washington, D.C.: Catholic University of America Press, 1984).

Grant, Edward. "Cosmology," in *Science in the Middle Ages*, ed. David Lindberg (Chicago: University of Chicago Press, 1978): 265–302.

———. "Were There Significant Differences between Medieval and Early Modern Scholastic Philosophy? The Case for Cosmology," *Nous* 18 (1984): 5–16.

———. "Eccentrics and Epicycles in Medieval Cosmology," in *Mathematics and Its Applications to Science and Natural Philosophy in the Middle Ages*, ed. Edward Grant and John E. Murdoch (Cambridge: Cambridge University Press, 1987): 189–213.

———. *Planets, Stars and Orbs: The Medieval Cosmos, 1200–1687* (Cambridge: Cambridge University Press, 1994).

Groarke, Leo. "Zeno's Dichotomy: Undermining the Modern Response," *Auslegung* 9 (1982): 67–75.

Gruenwald, Ithamar. "A Preliminary Critical Edition of *Sefer Yetzirah*," *Israel Oriental Studies* (1971): 132–77.

Grunbaum, Adolf. "The Pseudo-Problem of Creation in Physical Cosmology," *Philosophy of Science* 56 (1989): 373–94.

Gueroult, Martial. "Spinoza's Letter on the Infinite," in *Spinoza: A Collection of Critical Essays*, ed. Marjorie Grene (Notre Dame, Ind.: University of Notre Dame Press, 1979).

———. *Descartes' Philosophy Interpreted According to the Order of Reasons I: The Soul and God*, trans. Roger Ariew (Minneapolis: University of Minnesota Press, 1984).

Gurevich, A. J. "Time as a Problem of Cultural History," in *Cultures and Time*, ed. L. Gardet et al. (Paris: The Unesco Press, 1976): 229–45.

Hahn, Robert. "Continuity, Discontinuity, and Some Paradoxes of Motion: Zeno's Arguments in the Light of Quantum Mechanics," *Southwest Philosophical Studies* (1982): 115–23.

Hamelin, O. *La Theorie de l'intellect d'après Aristote et ses commentateurs* (Paris: J. Vrin: 1953).

Hampshire, Stuart. *Spinoza* (New York: Penguin Books, 1951).

Hardin, C. L. "Spinoza on Immortality and Time," *Southwestern Journal of Philosophy* 15 (1977): 129–38.

Harris, E. E. "Spinoza's Theory of Human Immortality," in *Spinoza: Essays in Interpretation,* ed. E. Freeman and M. Mandelbaum (La Salle, Ill.: Open Court, 1975): 245–62.

———. *The Reality of Time* (Albany: State University of New York Press, 1988).

Harris, R. B. "Preface," in *Neoplatonism and Jewish Thought,* ed. L. E. Goodman (Albany: State University of New York Press, 1992).

Harvey, Warren Z. *Ḥasdai Crescas' Critique of the Theory of the Acquired Intellect* (Ann Arbor, Mich.: University Microfilms, 1973).

———. "Albo's Discussion of Time," *JQR* 71 (1980): 210–38.

———. "A Third Approach to Maimonides' Cosmogony-Prophetology Puzzle," *Harvard Theological Review* 74 (1981): 287–301.

———L'zihui meḥabran shel ha-histaiguyot min ha-determinism be-Sefer 'Or 'Adonai l'rabbi Ḥasdai Crescas'edut ktav yad Firenze," *Kiryat Sefer* 55 (1980): 794–801.

———. "The Term '*Hitdabbekut*' in Crescas' Definition of Time," *JQR* (1981): 44–47.

Hasker, W. *God, Time and Knowledge* (Ithaca, N.Y.: Cornell University Press, 1989).

Hayoun, Maurice Ruben. *La Philosophie et la théologie de Moise de Narbonne* (Tubingen: J.C.B. Mohr, 1989).

Higgins, Anne. "Medieval Notions of Time," *Journal of Medieval and Renaissance Studies* 19 (1989): 227–50.

Horwich, Paul. "On the Existence of Time, Space, And Space-Time," *Nous* 12 (1978): 397–419.

———. *Asymmetries in Time: Problems in the Philosophy of Science* (Cambridge, Mass.: MIT Press, 1987).

Hourani, George F. "The Dialogue between Al Ghazālī and the Philosophers on the Origin of the World," *Muslim World* 48 (1958).

Husik, Isaac. "Studies in Gersonides," *JQR* 7 (1916–17): 553–94; 8 (1917–18): 113–56; pp. 231–68.

Hutton, Sarah. "Some Renaissance Critiques of Aristotle's Theory of Time," *Annals of Science* 34 (1977): 345–63.

Hyman, Arthur. "Maimonides on Creation and Emanation," in *Studies in Medieval Philosophy,* ed. J. F. Wippel (Washington, D.C.: Catholic University of America Press, 1987): 45–61.

———. "From What Is One and Simple Only What is One and Simple Can Come to

Be," in *Neoplatonism and Jewish Thought: Studies in Neoplatonism; Ancient and Modern,* ed. L. E. Goodman (Albany: State University of New York Press, 1992).

———. "Aspects of the Medieval Jewish and Islamic Discussion of 'Free Choice,'" in *Freedom and Moral Responsibility,* ed. Charles Manekin and Menachem Kellner (Bethesda, Md.: University Press of Maryland, 1997): 133–52.

Idel, Moshe. "The Sefirot That Are above the Sefirot," *Tarbiz* (1982): 239–80.

———. "Kabbalah and Ancient Theology in R. Isaac and Judah Abravanel," in *The Philosophy of Love of Leone Ebreo,* ed. M. Dorman and Z. Levy (Haifa: Hakatedra le-moreshet ha-hagut ha-Yehudit be-Universitat Haifa, 1985): 73–112.

———. "The Anthropology of Yohanan Alemanno: Sources and Influences," *Annali di Storia dell'esegesi* 7 (1990): 93–112.

———. "Jewish Kabbalah and Platonism in the Middle Ages and Renaissance," in *Neoplatonism and Jewish Thought: Studies in Neoplatonism; Ancient and Modern,* ed. L. E. Goodman (Albany: State University of New York Press, 1992): 319–52.

Ivry, Alfred L. "Averroës on Intellection and Conjunction," *JAOS* 86 (1966): 76–85.

———. *Al-Kindi's Metaphysics* (Albany: State University of New York Press, 1974).

———. "Maimonides on Possibility," in *Mystics, Philosophers, and Politicians: Essays in Jewish Intellectual History in Honor of Alexander Altmann,* ed. J. Reinhartz et al. (Durham, N.C.: Duke University Press, 1982): 67–84.

———. "Maimonides on Creation," in *Creation and the End of Days,* ed. David Novak and Norbert Samuelson (Lanham, Md.: University Press of America, 1986): 185–214.

———. "Gersonides and Averroës on the Intellect: The Evidence of the *Supercommentary on the De Anima,*" in *Gersonide en son temps,* ed. Gilbert Dahan (Louvain, Belgium: Peeters, 1991): 235–51.

Jacob, B. *The First Book of the Bible: Genesis,* ed. Ernest I. Jacob and Walter Jacob (New York: Ktav Publ., 1974).

James, E. O. *Creation and Cosmology: A Historical and Comparative Inquiry,* Studies in the History of Religions (Supplement to *Numen*) (Leiden: E.J. Brill, 1969).

Jammer, Max. *Concepts of Space: The History of Theories of Space in Physics* (Cambridge, Mass.: Harvard University Press, 1969).

Jenni, E. "Time," in *The Interpreter's Dictionary of the Bible,* vol. 4 (Nashville, Tenn.: Abingdon, 1962): 642–49.

Jospe, Raphael. "Early Philosophical Commentaries on the *Sefer Yezirah:* Some Comments," *REJ* 144 (1990): 369–415.

Kadushin, Max. *The Rabbinic Mind* (New York: The Jewish Theological Seminary, 1952).

Kafiḥ, R. J. "The Order of the Stars and Constellations and Events," in Kafiḥ, ed., *Ketavim,* vol. 1 (Jerusalem: Agudat halikhot am Yisrael, 1988/1989).

Kaplan, Aryeh. *Sefer Yetzirah, The Book of Creation,* trans. and comm. Aryeh Kaplan (York Beach, Maine: Samuel Weiser, 1990).

Kaplan, Lawrence. "Maimonides on the Miraculous Element in Prophecy," *Harvard Theological Review* 70 (1977): 233–56.

Kasher, Menachem. Humash *Torah Shelemah: Talmudic-Midrashic Encyclopedia of the Pentateuch* (Heb.), 44 vols. (Jerusalem and New York, 1927–85).

———. "The Concept of Time in Scripture and the Early Rabbis" (Heb.) *Talpiot* 5 (1952): 799–827.

Kaufman, Yehiel. *Toldot ha-Emunah ha-Yisra'elit, me-Yemei Qedem 'ad Sof Bayit Sheni,* 4 vols. (1937–56; repr. Tel Aviv.: Devir, 1938); trans. in idem. *The Religion of Israel, from Its Beginnings to the Babylonian Exile,* abr. and trans. Moshe Greenberg (Chicago: Univ. of Chicago Press, 1960).

Kaufmann, David. "The Pseudo-Empedocles as a Source of Salomon ibn Gabirol," in *Meḥqarim ba-Sifrut ha-'Ivrit shel Yemei ha-Beinayim,* ed. David Kaufmann (Jerusalem: Mossad Harav Kook, 1962): 78–165.

Kellner, Menachem. "On the Status of the Astronomy and Physics in Maimonides' *Mishneh Torah* and *Guide of the Perplexed:* A Chapter in the History of Science," *British Journal for the History of Science* 24 (1991): 453–63.

———. *Bibliographia Gersonideana: An Annotated List of Writings by and about R. Levi ben Gershom,* in *Studies on Gersonides,* ed. Gad Freudenthal (Leiden: E. J. Brill, 1992): 368–410.

Kenny, Anthony. *The God of the Philosophers* (Oxford: Clarendon Press, 1979).

Kiener, R. C. "The Status of Astrology in the Early Kabbalah," *Jerusalem Studies in Jewish Thought* 6 (1987): 1–42.

Klein-Braslavy, Sara. "The Existence of Time and the First Days of Creation in Medieval Jewish Philosophy" (Heb.), *Tarbiz* 45 (1975): 107–27.

———. "Interpretation of Maimonides of the Term 'Create' and the Question of the Creation of the Universe" (Heb.), *Da'at* 16 (1986): 39–55.

———. *Maimonides' Interpretation of the Story of Creation* (Heb.) (Jerusalem: Israel Society for Biblical Research, 1987).

———. "Gersonides on Determinism, Possibility, Choice and Foreknowledge" (Heb.), *Da'at* 22 (1989): 5–53.

Knorr, Wilbur R. "Infinity and Continuity," in *Infinity and Continuity in Ancient and Medieval Thought,* ed. Norman Kretzmann (Ithaca, N.Y.: Cornell University Press, 1982): 112–45.

Kogan, Barry. *Averroës and the Metaphysics of Causation* (Albany: State University of New York Press, 1985).

———. "Creation and Time in Maimonides and Gersonides: A Response to Tamar Rudavsky," in *God and Creation: An Ecumenical Symposium,* ed. David M. Burrell and Bernard McGinn (Indianapolis: University of Notre Dame Press, 1990): 147–53.

Korner, Stephen. *The Philosophy of Mathematics* (London: Hutchinson University Library, 1960).

Kraemer, Joel. "Maimonides on Aristotle and Scientific Method," in *Moses Maimonides and His Time,* ed. Eric Ormsby (Washington, D.C.: The Catholic University Press of America, 1989): 53–88.

Kretzmann, Norman. "Omniscence and Immutability," *Journal of Philosophy* 63 (1966): 409–21.

———. "Aristotle on the Instant of Change," *Aristotelian Society* 50 (Supp. 76) (1976): 91–114.

———, ed. *Infinity and the Continuum in Ancient and Medieval Thought* (Ithaca, N.Y.: Cornell University Press, 1982).

———. "Continuity, Contrariety and Change," in *Infinity and Continuity in Ancient and Medieval Thought,* ed. Norman Kretzmann (Ithaca, N.Y.: Cornell University Press, 1982): 270–96.

———. "Ockham and the Creation of the Beginningless World," *Franciscan Studies* 45 (1985): 1–33.

Norman Kretzmann, A. Kenny, and J. Pinborg, eds. *The Cambridge History of Later Medieval Philosophy* (Cambridge: Cambridge University Press, 1982).

Kümmel, Friedrich. "Time as Succession and the Problem of Duration," in *The Voices of Time: A Cultural Survey of Man's Views of Time as Expressed by the Sciences and the Humanities,* ed. J. T. Fraser (New York: George Braziller, 1966): 31–55.

Landes, David. *Revolution in Time: Clocks and the Making of the Modern World* (Cambridge, Mass.: Harvard University Press, 1983).

Landes, David. "Clocks: Revolution in Time," *History Today,* 34 (1984): 19–26.

Langermann, Y.Tzvi. "Science, Jewish," in *Dictionary of the Middle Ages* (New York: Scribner, 1989): 89–94.

———. *Ibn al-Haytham's* On the Configuration of the World (New York and London: Garland, 1990).

———. "The True Perplexity: *The Guide of the Perplexed* Part II, ch 24," in *Perspectives on Maimonides,* ed. Joel L. Kraemer (Oxford: Oxford University Press, 1991): 159–74.

———. "Maimonides' Repudiation of Astrology," *Maimonidean Studies* 2 (1991): 123–59.

———. "Gersonides on the Magnet and the Heat of the Sun," in *Studies on Gersonides,* ed. Gad Freudenthal (Leiden: E. J. Brill, 1992): 267–84.

———. "Gersonides on Astrology," in Gersonides. *The Wars of the Lord (Books V–VI)*, ed. and trans. Seymour Feldman, (Philadelphia: Jewish Publication Society, forthcoming).

———. "Some Astrological Themes in the Thought of Abraham Ibn Ezra," in *Rabbi Abraham Ibn Ezra: Studies in the Writings of a Twelfth-Century Jewish Polymath,* ed. Isadore Twersky and Jay M. Harris (Cambridge, Mass.: Harvard University Press, 1993): 28–85.

Lauer, R. H. *Temporal Man: The Meaning and Uses of Social Time* (New York: Praeger, 1981).

Leftow, Brian. "Boethius on Eternity," *History of Philosophy Quarterly* 7 (1990): 23–42.

LeGoff, Jacques. *Time, Work and Culture in the Middle Ages,* trans. Arthur Goldhamer (Chicago: University of Chicago Press, 1980).

Leibowitz, Nehama. *Studies in the Book of Genesis in the Context of Ancient and Modern Jewish Bible Commentary,* trans. Aryeh Newman (Jerusalem: World Zionist Organization, 1972).

Lerner, Ralph. "Maimonides' Letter on Astrology," *History of Religions* 8.2 (1968): 143–58.

Levi, Leo. *Jewish Chrononomy: The Calendar and Times-of-Day in Jewish Law* (Jerusalem: R. Mass, 1992)

Levine, Hillel. "Paradise not Surrendered: Jewish Reactions to Copernicus and the Growth of Modern Science," in *Epistemology, Methodology and the Social Sciences,* ed. R. S. Cohen and M.W. Wartofsky (Dordrecht: D. Reidel, 1983): 203–25.

Lindberg, David, ed. *Science in the Middle Ages* (Chicago and London: University of Chicago Press, 1978).

———. *The Beginnings of Western Science* (Chicago and London: University of Chicago Press, 1992).

Lloyd, G. E. R. *Early Greek Science: Thales to Aristotle* (New York: W.W. Norton, 1970).

———. *Greek Science After Aristotle* (New York: W.W. Norton, 1973).

———. "Views on Time in Greek Thought," in *Cultures and Time,* ed. L. Gardet et al. (Paris: The Unesco Press, 1976): 117–49.

Loeb, Laurence. "Time, Myth and History in Judaism," *Conservative Judaism* 42 (1990): 54–66.

Loewe, Raphael. "Ibn Gabirol's Treatment of Sources in the *Keter Malkhuth,*" in *Studies in Jewish Religious and Intellectual History Presented to Alexander Altmann,* ed. S. Stein and R. Loewe (Cambridge, Mass.: Harvard University Press, 1979): 183–94.

———. *Ibn Gabirol* (New York: Grove Weidenfeld, 1989).

Lucash, Frank. "On the Finite and Infinite in Spinoza," *The Southern Journal of Philosophy* 20 (1982): 61–73.

MacDonald, D. B. "Continuous Re-Creation and Atomic Time in Muslim Scholastic Theology," *Isis* 9 (1927): 326–44.

Madkour, Ibrahim. *L'Organon d'Aristote dans le monde arabe* (Paris: Librairie Philosophique, J Vrin, 1934).

Malina, B. J. "Christ and Time: Swiss or Mediterranean?" *Catholic Biblical Quarterly* 51 (1989): 1–31.

Malino, Jonathan. "Maimonides' Guide to the Perplexities of Creation." (HUC, 1979), unpublished rabbinic dissertation.

———. "Aristotle on Eternity: Does Maimonides Have a Reply?" in *Maimonides and Philosophy,* ed. S. Pines and Y. Yovel (Dordrecht: Martinus Nijhoff, 1986): 52–64.

Manekin, Charles. *The Logic of Gersonides: An Analysis of Selected Doctrines* (Dordrecht: Kluwer Academic, 1992).

———. "Freedom Within Reason? Gersonides on Human Choice," in *Freedom and Moral Responsibility,* ed. Charles Manekin and Menachem Kellner (Bethesda, Md.: University Press of Maryland, 1997): 165–204.

Marmorstein, A. *The Old Rabbinic Doctrine of God,* vol. 1: *The Names and Attributes of God,* Jews College Publications, No. 10 (Oxford: Oxford University Press, 1927).

Marmura, Michael E. "The Logical Role of the Argument from Time in the *Tahāfut*'s Second Proof for the World's Pre-Eternity," *Muslim World* 49 (1959): 306–14.

———. "Some Aspects of Avicenna's Theory of God's Knowledge of Particulars," *Journal of the American Oriental Society* 83 (1962): 299–312.

Marx, Alexander. "The Correspondence between the Rabbis of Southern France and Maimonides about Astrology," *HUCA* 3 (1926): 311–58.

Mason, Richard. "Spinoza on the Causality of Individuals," *JHP* 24 (1986): 197–210.

———. *The God of Spinoza* (Cambridge: Cambridge University Press, 1997).

Matson, Wallace. "Death and Destruction in Spinoza's Ethics," *Inquiry* 20 (1977): 403–17.

McGuire J. E. "Existence, Actuality and Necessity: Newton on Space and Time," *Annals of Science* 35 (1978), reprinted in J. E. McGuire, *Tradition and Innovation: Newton's Metaphysics of Nature* (Dordrecht: Kluwer Academic, 1995): 1–51.

———. "Newton on Place, Time, and God: An Unpublished Source," *The British Journal for the History of Science* 11 (1978): 115–29.

———. "Space, Infinity and Indivisibility: Newton on the Creation of Matter," in *Contemporary Newtonian Research,* ed. Z. Bechler (Dordrecht: Reidel, 1982): 145–90.

McGuire, J. E., and Tamny Martin. *Certain Philosophical Questions: Newton's Trinity Notebook* (Cambridge: Cambridge University Press, 1983).

Merlan, Paul. *Monopsychism, Mysticism, Metaconsciousness* (The Hague: Martinus Nijhoff, 1963).

Miller, C. L. "Maimonides and Aquinas on Naming God," *Journal of Jewish Studies* 28 (1977): 65–71.

Miller, Fred D. "Aristotle against the Atomists," in *Infinity and Continuity in Ancient and Medieval Thought,* ed. Norman Kretzmann (Ithaca, N.Y.: Cornell University Press, 1982): 87–111.

Moesgaard, K. P. "Astronomy," in I. Grattan-Guinness. *Companion Encyclopedia of the History and Philosophy of the Mathematical Sciences* (London and New York: Routledge, 1994): 241–50.

Momigliano, Arnaldo. "Time in Ancient Historiography," in *History and the Concept of Time* (Middletown, Conn: Wesleyan University Press, 1966).

Mumford, Lewis. *Technics and Civilization* (New York: Harcourt Brace and Co., 1934).

Munitz, Milton K. *Space, Time, and Creation: Philosophical Aspects of Scientific Cosmology* (New York: Dover, 1981).

Murdoch, John. "Philosophy and the Enterprise of Science in the Later Middle Ages," in *The Interaction between Science and Philosophy,* ed. Y. Elkana (Atlantic Highlands, N.J.: Humanities Press, 1974): 51–74.

———. "Pierre Duhem and the History of Late Medieval Science and Philosophy in the Latin West," in *Gli Studi di Filosofia Medievale Fra Otto e Novecento* (Rome: Edizioni di Storia e Letteratura, 1991): 253–302.

Nadler, Steven. *Spinoza: A Life* (Cambridge: Cambridge University Press, 1998).

Nasr, Seyyed Hossein. *An Introduction to Islamic Cosmological Doctrines* (Albany: State University of New York Press, 1993).

Neher, André. "The View of Time and History in Jewish Culture," in *Cultures and Time,* ed. Louis Gardet et al. (Paris: The Unesco Press, 1976): 149–68.

———. *Jewish Thought and the Scientific Revolution of the Sixteenth Century: David Gans and His Times,* trans. David Maisel (Oxford: Oxford University Press, 1986).

Neusner, Jacob. *History of the Mishnaic Law of Appointed Times, vol. 5* (Leiden: E. J. Brill, 1981–83).

———. *Genesis Rabbah: The Judaic Commentary to the Book of Genesis,* vol. 1 (Chico, Calif.: Scholars Press, 1985).

Normore, Calvin. "Future Contingents," in *The Cambridge History of Later Medieval Philosophy* ed. N. Kretzmann, A. Kenny, and J. Pinborg (Cambridge: Cambridge University Press, 1982).

———. "Divine Omniscience, Omnipotence and Future Contingents: An Overview," in *Divine Omniscience and Omnipotence in Medieval Philosophy,* ed. T. M. Rudavsky (Dordrecht: D. Reidel, 1984).

North, John. *Horoscopes and History* (London: The Warburg Institute, 1986).

Onians, Richard Broxton. *The Origins of European Thought about the Body, the Mind, the Soul, the World, Time, and Fate* (Cambridge: Cambridge University Press, 1954).

Overbye, D. *Lonely Hearts of the Cosmos: The Scientific Quest for the Secret of the Universe* (New York: HarperCollins, 1991).

Papa-Grimaldi, Alba. "Why Mathematical Solutions of Zeno's Paradoxes Miss the Point: Zeno's One and Many Relation and Parmenides' Prohibition," *Review of Metaphysics* 50 (1996): 299–314.

Pearl, Chaim. *The Medieval Jewish Mind: The Religious Philosophy of Isaac Arama* (London: Vallentine Mitchell, 1971).

Pederson, Olaf. "Astronomy," in *Science in the Middle Ages,* ed. David Lindberg (Chicago and London: University of Chicago Press, 1978): 303–37.

Pessin, Sarah. "Maimonides' View on Astrology: Inconclusive Evidence, Unconvincing Arguments," unpublished paper.

Pike, Nelson. "Divine Omniscience and Voluntary Actions," *Philosophical Review* 74 (1965): 27–46.

Pines, Shlomo. "La Longue Recension de la *Théologie d'Aristotle* dans ses rapports avec la doctrine ismaelienne," *Revue des études islamiques* (1954).

———. "*Sefer ʿArugat ha-Bosem:* Qetaʿim, mi-Tokh *Sefer Meqor Ḥayyim,*" *Tarbiz* 27 (1958): 218–33.

———. "Scholasticism and the Teachings of Hasdai Crescas and his Predecessors," *Proceedings of the Israel Academy of Sciences and Humanities* 1.10 (1967): 1–101.

———. "De quelques doctrines relatives au temps chez les philosophes arabes juifs," in *Studies in Abu'l Barākāt al-Baghdādī: Physics and Metaphysics* (Jerusalem: The Magnes Press, 1979): 63–83.

———. *Studies in Abu'l Barākāt al-Baghdādī: Physics and Metaphysics* (Jerusalem: The Magnes Press, 1979).

———. "The Limits of Human Knowledge according to Al-Farābi, Ibn Bājja and Maimonides," in *Studies in Medieval Jewish History and Literature,* vol 1, ed. Isadore Twersky (Cambridge, Mass.: Harvard University Press, 1979): 82–109.

———. "And He Called Out to Nothingness and It Was Split: A Note on a Passage in Ibn Gabirol's *Keter Malkhut,*" *Tarbiz Anniversary Volume II* (1980–81): 339–97.

———. "The Semantic Distinction between the Terms Astronomy and Astrology according to Al-Birūni," in *The Collected Works of Shlomo Pines,* vol. 2 (Jerusalem: The Magnes Press, 1986): 341–46.

———. *Studies in Arabic Versions of Greek Texts and in Mediaeval Science* (Jerusalem: The Magnes Press, 1986).

———. "Points of Similarity between the Exposition of the Doctrine of the *Sefirot* in the *Sefer Yetzirah* and a Text of the Pseudo-Clementine Homilies: The Implications of This Resemblance," *Proceedings of the Israel Academy of Sciences and Humanities* 7(1989).

Pollock, John. *Subjunctive Reasoning* (Dordrecht: D. Reidel, 1976).

Popkin, Richard. "Some Further Comments on Newton and Maimonides," in *Essays on the Context, Nature and Influence of Isaac Newton's Theology,* ed. J. E. Force and R. H. Popkin (Dordrecht: Kluwer Academic, 1990): 1–8.

———. "Newton and Maimonides," in *A Straight Path: Studies in Medieval Philosophy and Culture—Essays in Honor of Arthur Hyman,* ed. Ruth Link-Salinger (Washington, D.C., 1988): 216–29.

Prior, A. N. *Past, Present and Future* (Oxford: Clarendon Press, 1967).

Rahman, Fazlur. "The Eternity of the World and the Heavenly Bodies in Post-Avicennan Philosophy," in *Essays On Islamic Philosophy and Science,* ed. George F. Hourani (Albany: State University of New York Press, 1975): 222–37.

Ravitsky, Aviezer. "*Hitpatkhut Hashkafato shel R' Ḥasdai Crescas bi-She'elat Ḥofesh ha-Ratzon,*" *Tarbiz* 51 (1981–2): 445–70.

———. *History and Faith: Studies in Jewish Philosophy* (Amsterdam: J.C. Gieben, 1996)

Ray, Christopher. *Time, Space and Philosophy* (London and New York: Routledge, 1991).

Rice, Lee. "Spinoza on Individuation," in *Spinoza: Essays in Interpretation,* ed. E. Freeman and M. Mandelbaum (La Salle, Ill.: Open Court, 1975): 195–214.

Robbins, Ellen. "Time Telling in Ritual and Myth," *The Journal of Jewish Thought and Philosophy* 6 (1997): 71–88.

Rosenberg, Shalom. "Logic and Ontology in Jewish Philosophy in the Fourteenth Century," Ph.D. dissertation, Hebrew University, Jerusalem, 1974.

———. "Possible and Assertoric in Medieval Logic [Heb.]," *Iyyun* 28 (1978): 56–72.

———,. "Necessary and Possible in Medieval Logic [Heb]," *Iyyun* 28 (1978): 103–55.

———. "From Anaximandros to Levinas—The Idea of the Infinite," *A Journal of Jewish Philosophy and Kabbalah* 30 (1993).

Rubenstein, Jeffrey. "Mythic Time and the Festival Cycle," *The Journal of Jewish Thought and Philosophy* 6 (1997): 157–83.

Rudavsky, T. M. "Individuals and the Doctrine of Individuation in Gersonides," *The New Scholasticism* 56 (1982): 30–50.

———. "Divine Omniscience and Future Contingents in Gersonides," *JHP* 21 (1983): 513–36.

———. "Maimonides and Averroës on God's Knowledge of Possibles," *Da'at* (1984), pp. 27–44.

———, ed. *Divine Omniscience and Omnipotence in Medieval Philosophy* (Dordrecht: D. Reidel, 1984).

———. "Divine Omniscience, Contingency and Prophecy in Gersonides," in *Divine Omniscience and Omnipotence in Medieval Philosophy,* ed. T. M. Rudavsky (Dordrecht: D. Reidel, 1984): 161–81.

———. "Creation, Time and Infinity in Gersonides," *JHP* 26 (1988): 25–45.

———. "Creation and Time in Maimonides and Gersonides," in *God and Creation,* ed. B.McGinn and D. Burrell (Notre Dame, Ind.: University of Notre Dame Press, 1990): 122–47.

———. "Time in Maimonides and Crescas," *Maimonidean Studies* 1 (1990):143–162.

———. "The Jewish Tradition: Maimonides, Gersonides and Bedersi," in *Individuation in Late Scholasticism and the Counter-Reformation,* ed. J. J. Gracia (Albany: State University of New York Press, 1993): 69–96.

———. "Jewish Neoplatonism," *Routledge History of World Philosophies,* vol. 2 ed. Daniel H. Frank and Oliver Leaman (London: Routledge, 1997).

———. "Philosophical Cosmology in Judaism," *Early Science and Medicine* (1997), 149–184.

Ruderman, David. *Jewish Thought and Scientific Discovery in Early Modern Europe* (New Haven, Conn.: Yale University Press, 1995).

Saliba, George. "The Astronomical Tradition of Maragha: A Historical Survey and Prospects for Future Research," *Arabic Sciences and Philosophy* 1 (1991): 67–99.

Salmon, Wesley C. *Space, Time and Motion: A Philosophical Introduction* (Minn: University of Minnesota Press, 1980).

———. *Zeno's Paradoxes* (New York: Bobbs-Merrill, 1970).

Sambursky, Shmuel. *The Physical World of Late Antiquity* (London: Routledge and Kegan Paul, 1962).

———. "Place and Space in Late Neoplatonism," *Studies in the History and Philosophy of Science* 8 (1977): 173–87.

Samuelson, Norbert. "Gersonides' Account of God's Knowledge of Particulars," *Journal of the History of Philosophy* 10 (1972): 399–416.

———. "Elements and Matter in Gersonides' Cosmogony," in *Gersonide en son temps,* ed. Gilbert Dahan (Louvain and Paris: E. Peeters, 1991): 199–233.

———. *The First Seven Days: A Philosophical Commentary on the Creation of Genesis* (Atlanta: Scholars Press, 1992).

———. *Judaism and the Doctrine of Creation* (Cambridge: Cambridge University Press, 1994).

Sasse. "*aion*" in *The Theological Dictionary of the New Testament* Vol. I, ed. Gerhard Kittel (Grand Rapids, Mich.: Eerdmans, 1964): 197–209.

Schlanger, Jacques. "Sur le Rôle Du "Tout" dans la Création selon Ibn Gabirol," *REJ* 124 (1965): 125–35.

———. *La Philosophie de Salomon Ibn-Gabirol* (Leiden: E. J. Brill, 1968).

Schlesinger, George N. "Comparing Space and Time Once More," *Mind* 87 (1978): 264–66.

Scholem, Gershom. "Traces of Gabirol in the Kabbalah," *Me'asef Sofrei Erez Yisrael* (1960): 160–178.

———. "*Yezirah, Sefer,*" in *Encyclopedia Judaica* (Jerusalem, 1972): vol. 16, 782–88.

Schroeder, G. L. *Genesis and the Big Bang: The Discovery of Harmony between Modern Science and the Bible* (New York: Bantam, 1990).

Schwartz, Dov. La Magie astrale dans la pensée juive rationaliste en Provence au XIVe siècle," *Archives d'histoire doctrinale et littéraire du moyen-âge* 16 (1994): 31–55.

Schwartzmann, Julia. "Where Is the Heaven? The Story of an Aristotelian Problem and Its Presentation in Jewish Medieval Sources," *REJ* 63 (1994): 67–85.

Schweid, Eliezer. "*Dibre Mabo,*" in *Maqor Ha Pilosophia Ha-Datit Shel R' Hasdai Crescas. Light of the Lord* (Jerusalem: Makor, 1972).

Secada, J. E. K. "Descartes on Time and Causality," *The Philosophical Review* 99 (1990): 45–72.

Seeskin, Kenneth. *Jewish Philosophy in a Secular Age* (Albany: State University of New York Press, 1990).

———, *Maimonides: A Guide for Today's Perplexed* (West Orange, New Jersey: Behrman House, 1991).

Sharf, A. *The Universe of Shabbetai Donnolo* (New York: Ktav, 1976).

Simons, John. "Matter and Time in Plotinus," in *Dionysius* 9 (1985): 53–74,

Singer, Charles. "Science and Judaism," in *The Jews: Their Role in Civilization,* ed. Louis Finkelstein (New York: Schocken, 1978): 216–65.

Sirat, Colette. *Les théories des visions surnaturelles dans la pensée juive du moyen-âge* (Leiden: E. J. Brill, 1969).

———. *A History of Jewish Philosophy in the Middle Ages* (Cambridge: Cambridge University Press, 1985).

Sleigh, Robert, Jr., Vere Chappell, and Michael Della Rocca. "Determinism and Human Freedom," in *The Cambridge History of Seventeenth-Century Philosophy,* ed. Daniel Garber and Michael Ayers (Cambridge: Cambridge University Press, 1998).

Small, Robin. "Cantor and the Scholastics," *American Catholic Philosophical Quarterly* 66 (1992): 407–28.

Sorabji, Richard. "Aristotle on the Instant of Change," *Aristotelian Society* 50 (supp. 1976): 91–114.

———. *Necessity, Cause and Blame* (Ithaca, N.Y.: Cornell University Press, 1980).

———. "Atoms and Time Atoms," in *Infinity and Continuity in Ancient and Medieval Thought,* ed. Norman Kretzmann (Ithaca, N.Y.: Cornell University Press, 1982): 37–86.

———. *Time, Creation and the Continuum* (Ithaca, N.Y.: Cornell University Press, 1983).

Speiser, E. A. *The Anchor Bible: Genesis* (Garden City, N.Y.: Anchor/Doubleday, 1967).

Stadelman, L. *The Hebrew Conception of the World: A Philological and Literary Study* (Rome: Pontifical Biblical Institute, 1970).

Staub, Jacob. *The Creation of the World According to Gersonides* (Chico, Calif.: Scholars Press, 1982).

Steensgaard, P. "Time in Judaism," in *Religion and Time,* ed. A. N. Balslav and J. N. Mohanty (Leiden: E. J. Brill, 1993): 63–108.

Stein, L. *Die Willensfreiheit and ihr Verhältniss zur Göttlichen Präscienz und Providenz Bei den Jüdischen Philosophen des Mittelalters* (Berlin, 1882).

Steinschneider, Moritz. *Die hebräischen Übersetzungen des Mittelalters und die Juden als Dolmetscher* (Berlin, 1893).

Stern, Josef. "Remarks on a Skeptical Theme in Maimonides' *Guide of the Perplexed,*" (unpublished paper).

———. "The Fall and Rise of Myth in Ritual: Maimonides versus Nahmanides on the *Ḥuqqim,* Astrology, and the War against Idolatry," *The Journal of Jewish Thought and Philosophy* 6 (1997): 185–263.

———. "Maimonides' Conceptions of Freedom and the Sense of Shame," in *Freedom and Moral Responsibility,* ed. Charles Manekin and Menachem Kellner (Bethesda, Md.: University Press of Maryland, 1997): 217–66.

———. *Problems and Parables of Law* (Albany: State University of New York Press, 1998).

Stern, Sacha. "Fictitious Calendars: Early Rabbinic Notions of Time, Astronomy, and Reality," *The Jewish Quarterly Review* 88 (1996): 103–29.

Stitzkin, Leo. *Judaism as a Philosophy: The Philosophy of Abraham bar Ḥiyya* (New York: Bloch, 1961).

Strang, Colin, and K. W. Mills, "Plato and the Instant," *Proc Aristotelian Society* 48 (1974): 63–96.

Stump, Eleanore, and Norman Kretzmann. "Eternity," *Journal of Philosophy* 79 (1981): 429–528.

Sweeney, Leo, S.J. "Are Plotinus and Albertus Magnus Neoplatonists?" in *Graceful Reason: Essays in Ancient and Medieval Philosophy Presented to Joseph Owens, CSSR,* ed. L. P. Gerson, Papers in Mediaeval Studies 4 (Toronto: Pontifical Institute of Mediaeval Studies, 1983): 177–202.

———. *Divine Infinity in Greek and Medieval Thought* (New York: Peter Lang, 1992).

Taylor R. C. "A Critical Analysis of the Structure of the *Kalām fī māhd al-Khaīr* (*Liber de Causis*)" in *Neoplatonism and Islamic Thought,* ed. Parviz Morewedge, Studies in Neoplatonism: Ancient and Modern 5 (Albany: State University of New York Press, 1992).

Tester, S.J. *A History of Western Astrology* (Suffolk, England: The Boydell Press, 1987).

Thiel, Udo. "Individuation," in *The Cambridge History of Seventeenth-Century Philosophy,* ed. Daniel Garber and Michael Ayers (Cambridge: Cambridge University Press, 1998): 212–62.

———. "Personal Identity," in *The Cambridge History of Seventeenth-Century Philosophy,* ed. Daniel Garber and Michael Ayers (Cambridge: Cambridge University Press, 1998): 868–912.

Thijssen, J.M.M.H. "The Response to Thomas Aquinas in the Early Fourteenth Century: Eternity and Infinity in the Works of Henry of Harclay, Thomas of Wilton and William of Alnwick O.F.M.," in *The Eternity of the World in the Thought of Thomas Aquinas and His Contemporaries,* ed. J.B.M. Wissink (Leiden: E. J. Brill, 1990).

Tirosh-Rothschild, Hava. "Jewish Philosophy on the Eve of Modernity," in *History of Jewish Philosophy,* ed. Daniel H. Frank and Oliver Leaman (London: Routledge Publ, 1997): 499–573.

Touati, Charles. *La Pensée philosophique et théologique de Gersonides* (Paris: Les Éditions de Minuit, 1973).

———. "Hasday Crescas et ses paradoxes sur la liberté," in *Mélanges d'histoire des religions offerts à Charles-Henry Puech* (Paris: Presses Universitaires de France, 1974): 573–78.

Turnbull, Robert. "Naural Science in the Later Dialogues of Plato" unpublished paper.

Twersky, Isadore, ed. *A Maimonides Reader* (New York: Behrman House, 1972).

———. *Studies in Maimonides, Harvard Judaic Texts and Studies,* vol 7 (Cambridge, Mass.: Harvard University Press, 1992).

Unguru, Sabetai, ed. *Physics, Cosmology and Astronomy, 1300–1700: Tension and Accommodation* (Dordrecht: Kluwer Academic, 1991).

Urbach, Ephraim. *The Sages: Their Concepts and Beliefs* (Jerusalem: The Magnes Press, 1975).

van Frassen, Bas. *An Introduction to the Philosophy of Time and Space* (New York: Random House, 1970).

Vajda, Georges. "Les idées théologiques et philosophiques d'Abraham bar Ḥiyyah," *Archives d'histoire doctrinale et litteraire du moyen âge* 15 (1946): 191–223.

———. *L'Amour de Dieu dans la théologie juive du moyen âge* (Paris: Vrin, 1957).

———. *Isaac Albalag* (Paris: Vrin, 1960).

van Benthem, J.F.A.K. *The Logic of Time: A Model-Theoretic Investigation into the Varieties of Temporal Ontology and Temporal Discourse* (Dordrecht: D. Reidel, 1983).

Verbeke, Gerard. "Some Later Neoplatonic Views on Divine Creation and the Eternity of the World," in *Neoplatonism and Christian Thought,* ed. D. O'Meara (Albany: State University of New York Press, 1982): 45–63.

Vlastos, Gregory. "Creation in the Timaeus: Is It a Fiction?" in *Studies in Plato's Metaphysics,* ed. R. E. Allen (Atlantic Highlands, N.J.: Humanities Press, 1965): 401–19.

———. "The Disorderly Motion in the 'Timaeus,'" in *Studies in Plato's Metaphysics,* ed. R. E. Allen (Atlantic Highlands, N.J.: Humanities Press, 1965): 379–399.

———. "Plato's Testimony Concerning Zeno of Elea," *The Journal of Hellenic Studies* 95 (1975): 136–63.

———. *Plato's Universe* (Seattle: University of Washington Press, 1975).

Wartofsky, Max W. "Nature, Number and Individuals: Motive and Method in Spinoza's Philosophy," *Inquiry* 20 (1977): 457–79.

Weiss, Roslyn. "Maimonides on the End of the World," *Maimonidean Studies* 3 (1992–93): 195–218.

Whitrow, G. J. "Time and the Universe," in *The Voices of Time: A Cultural Survey of Man's Views of Time as Expressed by the Sciences and the Humanities,* ed. J. T. Fraser (New York: George Braziller, 1966): 564–581.

———. *Time in History* (New York: Oxford University Press, 1988).

Wilensky, Sara Heller. *R. Yitzhak Arama and his Philosophy* (Heb.) (Jerusalem: Bialik Institute, 1956).

Wilson, Margaret D. "Can I Be the Cause of My Idea of the World? Descartes on the Infinte and Indefinite," in *Essays on Descartes' Meditations,* ed. A. O. Rorty (Berkeley: University of California Press, 1986): 339–58.

———. "Spinoza's Theory of Knowledge," in *The Cambridge Companion to Spinoza,* ed. Donald Garrett (Cambridge: Cambridge University Press, 1996): 89–141.

Winston, David. "The *Book of Wisdom*'s Theory of Cosmogony," *History of Religions* 11 (1971): 185–202.

Wohlman, Avital. *Thomas d'Aquin et Maimonide: Un dialogue éxemplaire* (Paris: Cerf, 1988).

Wolfson, H. A. *Crescas' Critique of Aristotle* (Cambridge, Mass.: Harvard University Press, 1929).

———. *Philo: Foundations of Religious Philosophy in Judaism, Christianity and Islam,* 2 vols. (Cambridge, Mass.: Harvard University Press, 1947).

———. "Maimonides and Gersonides on Divine Attributes as Amphibolous Terms," in *Mordecai M. Kaplan Jubilee Volume,* English Section, ed. Moshe Davis (New York: Jewish Publication Society Publ, 1953): 515–30.

———. *The Philosophy of Spinoza,* vols. 1 and 2 (New York: Schocken Books, 1959).

———. "The Meaning of *Ex Nihilo* in the Church Fathers, Arabic and Hebrew Philosophy, and St. Thomas," in *Studies in the History of Philosophy and Religion,* vol. 1, ed. I. Twersky and G. Williams (Cambridge, Mass.: Harvard University Press, 1973): 207–11.

———. *The Philosophy of the Kalam* (Cambridge, Mass.: Harvard University Press, 1976).

———. *Repercussions of the Kalam in Jewish Philosophy* (Cambridge, Mass.: Harvard University Press, 1979).

Wood, David. *The Deconstruction of Time* (Atlantic Highlands, N.J.: Humanities Press International, 1989).

Wright, M. R. *Cosmology in Antiquity* (Routledge: London and New York, 1995).

Yarden, Dov. *Shirei ha-Qodesh le-Rabbi Shelomo ibn Gabirol*, vols. 1 and 2 (Jerusalem: D. Yarden, 1973).

Yates, Frances. *Giorduno Bruno and the Hermetic Tradition* (London: Routledge and K. Paul, 1964).

Yerushalmi, Yosef. *Zakhor: Jewish History and Jewish Memory* (Seattle: University of Washington Press, 1982).

Zangwill, Israel. *Selected Religious Poems of Solomon Ibn Gabirol* (Philadelphia: Jewish Publication Society Press, 1923).

Zimmerli, W. *The Old Testament and the World*, trans. John J. Scullion (Atlanta: J Knox Press, 1976).

Zimmerman, Albert. *"Some Aspects of the Reception of Aristotle's Physics and Metaphysics in the Thirteenth Century*, in *Ad Litteram: Authoritative Texts and Their Medieval Readers*, ed. Kent Jordan, Jr. (Notre Dame, Ind.: University of Notre Dame Press, 1992): 217–27.

INDEX